Lecture Notes in Computer Science 4447

Commenced Publication in 1973
Founding and Former Series Editors:
Gerhard Goos, Juris Hartmanis, and Jan van Leeuwen

Editorial Board

Elena Marchiori   Jason H. Moore
Jagath C. Rajapakse (Eds.)

Evolutionary Computation, Machine Learning and Data Mining in Bioinformatics

5th European Conference, EvoBIO 2007
Valencia, Spain, April 11-13, 2007
Proceedings

 Springer

Volume Editors

Elena Marchiori
VU University of Amsterdam, IBIVU
Department of Computer Science
de Boelelaan 1081a, 1081 HV Amsterdam, The Netherlands
E-mail: elena@cs.vu.nl

Jason H. Moore
Dartmouth-Hitchcock Medical Center
Computational Genetics Laboratory
706 Rubin Building, HB 7937, One Medical Center Dr., Lebanon, NH 03756, USA
E-mail: jason.h.moore@dartmouth.edu

Jagath C. Rajapakse
Nanyang Technological University
School of Computer Engineering
Blk N4-2a05, 50 Nanyang Avenue, Singapore 639798
E-mail: asjagath@ntu.edu.sg

Cover illustration: Morphogenesis series #12 by Jon McCormack, 2006

Library of Congress Control Number: 2007923724

CR Subject Classification (1998): D.1, F.1-2, J.3, I.5, I.2

LNCS Sublibrary: SL 1 – Theoretical Computer Science and General Issues

ISSN   0302-9743
ISBN-10   3-540-71782-X Springer Berlin Heidelberg New York
ISBN-13   978-3-540-71782-9 Springer Berlin Heidelberg New York

Springer is a part of Springer Science+Business Media

springer.com

© Springer-Verlag Berlin Heidelberg 2007

Typesetting: Camera-ready by author, data conversion by Scientific Publishing Services, Chennai, India
Printed on acid-free paper   SPIN: 12044597   06/3180   5 4 3 2 1 0

Preface

The field of bioinformatics has two main objectives: the creation and maintenance of biological databases, and the discovery of knowledge from life sciences data in order to unravel the mysteries of biological function, leading to new drugs and therapies for human disease. Life sciences data come in the form of biological sequences, structures, pathways, or literature. One major aspect of discovering biological knowledge is to search, predict, or model specific patterns present in a given dataset and then to interpret those patterns. Computer science methods such as evolutionary computation, machine learning, and data mining all have a great deal to offer the field of bioinformatics. The goal of the Fifth European Conference on Evolutionary Computation, Machine Learning, and Data Mining in Bioinformatics (EvoBIO 2007) was to bring experts in computer science together with experts in bioinformatics and the biological sciences to explore new and novel methods for solving complex biological problems.

The fifth EvoBIO conference was held in Valencia, Spain during April 11-13, 2007 at the Universidad Politecnica de Valencia. EvioBIO 2007 was held jointly with the Tenth European Conference on Genetic Programming (EuroGP 2007), the Seventh European Conference on Evolutionary Computation in Combinatorial Optimisation (EvoCOP 2007), and the Evo Workshops. Collectively, the conferences and workshops are organized under the name Evo* (www.evostar.org).

EvoBIO, held annually as a workshop since 2003, became a conference in 2007 and it is now the premiere European event for those interested in the interface between evolutionary computation, machine learning, data mining, bioinformatics, and computational biology. All papers in this book were presented at EvoBIO 2007 and responded to a call for papers that included topics of interest such as biomarker discovery, cell simulation and modeling, ecological modeling, fluxomics, gene networks, biotechnology, metabolomics, microarray analysis, phylogenetics, protein interactions, proteomics, sequence analysis and alignment, and systems biology. A total of 60 papers were submitted to the conference for double-blind peer-review. Of those, 28 (46.7%) were accepted.

We would first and foremost like to thank all authors who spent time and effort to make important contributions to this book. We would like to thank the members of the Program Committee for their expert evaluation of the submitted papers. Moreover, we would like to thank Jennifer Willies for her tremendous administrative help and coordination, Anna Isabel Esparcia-Alcázar for serving as the Local Chair, Leonardo Vanneschi for serving as Evo* Publicity Chair, Marc Schoenauer and the MyReview team (http://myreview.lri.fr/) for the conference management system.

We would also like to acknowledge the following organizations. The Universidad Politécnica de Valencia, Spain for their institutional and financial support, and for providing premises and administrative assistance; the Instituto

Tecnológico de Informática in Valencia, for cooperation and help with local arrangements; the Spanish Ministerio de Educación y Ciencia, for their financial support; and the Centre for Emergent Computing at Napier University in Edinburgh, Scotland for administrative support and event coordination.

Finally, we hope that you will consider contributing to EvoBIO 2008.

February 2007

Elena Marchiori
Jason H. Moore
Jagath C. Rajapakse

Organization

EvoBIO 2007 was organized by Evo* (`www.evostar.org`).

Program Chairs

Elena Marchiori   (IBIVU, VU University Amsterdam, The Netherlands)
Jason H. Moore   (Dartmouth Medical School in Lebanon, NH,USA)
Jagath C. Rajapakse (Nanyang Technological University, Singapore)

General Chairs

David W. Corne (Heriot-Watt University, Edinburgh, UK)
Elena Marchiori (IBIVU, VU University Amsterdam, The Netherlands)

Steering Committee

David W. Corne   (Heriot-Watt University, Edinburgh, UK)
Elena Marchiori   (IBIVU, VU University Amsterdam, The Netherlands)
Carlos Cotta   (University of Malaga, Spain)
Jason H. Moore   (Dartmouth Medical School in Lebanon, NH,USA)
Jagath C. Rajapakse (Nanyang Technological University, Singapore)

Program Committee

Jesus S. Aguilar-Ruiz (Spain)
Francisco J. Azuaje (UK)
Wolfgang Banzhaf (Canada)
Jacek Blazewicz (Poland)
Marius Codrea (The Netherlands)
Dave Corne (UK)
Carlos Cotta (Spain)
Alex Freitas (UK)
Gary Fogel (USA)
James Foster (USA)
Rosalba Giugno (Italy)
Raul Giraldez (Spain)
Jin-Kao Hao (France)
Antoine van Kampen
  (The Netherlands)
Natalio Krasnogor (UK)
Ying Liu (USA)

Elena Marchiori (The Netherlands)
Andrew Martin (UK)
Jason Moore (USA)
Pablo Moscato (Australia)
Jagath Rajapakse (Singapore)
Menaka Rajapakse (Singapore)
Michael Raymer (USA)
Vic J. Rayward-Smith (UK)
Jem Rowland (UK)
Marylyn Ritchie (USA)
Ugur Sezerman (Turkey)
El-Ghazali Talbi (France)
Andrea Tettamanzi (Italy)
Janet Wiles (Australia)
Andreas Zell (Germany)
Eckart Zitzler (Switzerland)

Table of Contents

Identifying Regulatory Sites Using Neighborhood Species

Claudia Angelini[1], Luisa Cutillo[1], Italia De Feis[1], Richard van der Wath[2], and Pietro Lio'[2,*]

[1] Istituto per le Applicazioni del Calcolo "Mauro Picone" CNR, Napoly Italy
c.angelini@iac.cnr.it, cutillo@na.iac.cnr.it, i.defeis@iac.cnr.it
[2] Computer Laboratory, University of Cambridge, Cambridge UK
rcv23@cam.ac.uk, pl219@cam.ac.uk

Abstract. The annotation of transcription binding sites in new sequenced genomes is an important and challenging problem. We have previously shown how a regression model that linearly relates gene expression levels to the matching scores of nucleotide patterns allows us to identify DNA-binding sites from a collection of co-regulated genes and their nearby non-coding DNA sequences. Our methodology uses Bayesian models and stochastic search techniques to select transcription factor binding site candidates. Here we show that this methodology allows us to identify binding sites in nearby species. We present examples of annotation crossing from *Schizosaccharomyces pombe* to *Schizosaccharomyces japonicus*. We found that the eng1 motif is also regulating a set of 9 genes in *S. japonicus*. Our framework may have an effective interest in conveying information in the annotation process of a new species. Finally we discuss a number of statistical and biological issues related to the identification of binding sites through covariates of genes expression and sequences.

1 Introduction

The identification of the repertoire of regulatory elements in a genome is one of the major challenges in modern biology. Gene transcription is determined by the interaction between transcription factors and their binding sites, called motifs or cis-regulatory elements. In eukaryotes the regulation of gene expression is highly complex and often occurs through the coordinated action of multiple transcription factors. This combinatorial regulation has several advantages; it controls gene expression in response to a variety of signals from the environment and allows the use of a limited number of transcription factors to create many combinations of regulators. Identification of the regulatory elements is necessary for understanding mechanisms of cellular processes. In eukaryotes these sites comprise short DNA stretches often found within non-coding upstream regions. DNA microarrays provide a simple and natural vehicle for exploring the regulation of thousands of genes and their interactions. Genes with similar expression

---

* Corresponding author.

E. Marchiori, J.H. Moore, and J.C. Rajapakse (Eds.): EvoBIO 2007, LNCS 4447, pp. 1–10, 2007.

profiles are likely to have similar regulatory mechanisms. A close inspection of their promoter sequences may therefore reveal nucleotide patterns that are relevant to their regulation.

In order to identify regulative sites several authors have used the following strategy: 1) candidate motifs can be obtained from the upstream regions of the most induced or most repressed genes; 2) a score can be assigned to reflect the matching of each motif to a particular upstream sequence; 3) regression analysis and variable selection methods can be used to detect sets of motifs acting together to affect the expression of genes [4,5,8].

Most of the current focus on microarray analysis is on integrating results from repeated experiments using the same species or using different species. This paper is extending this focus to transcription factor binding site identification. Following [8], we propose the use of Bayesian variable selection models to use the gene expression of an organism to find transcription binding sites of a closely related species or of a different strain. Variable selection methods use a latent binary vector to index all possible sets of variables (patterns). Stochastic search techniques are then used to explore the high-dimensional variable space and identify sets that best predict the response variable (expression). The method provides joint posterior probabilities of sets of patterns, as well as marginal posterior probabilities for the inclusion of single nucleotide patterns. We have chosen to exemplify our methodology using *S. japonicus* and *S. pombe* genomes and microarray data from cell cycle-regulated gene experiments [6].

Similar to a better known Schizosaccharomyces *S. pombe*, which has been a major model organism for cell cycle and cell biology research for thirty years, *S. japonicus* is a simple, unicellular yeast. Unlike the cousin, it readily adopts a invasive, hyphal growth form. Such growth is an important virulence trait in pathogenic fungi, making *S. japonicus* a potentially important model for fungal disease. The comparison of the *S. pombe* genome, which was sequenced several years ago, with those of its close relatives will greatly improve our understanding of the genomes and the proteins they encode. In addition, the three fission yeasts form an early-branching clade among the Ascomycete (ascus-forming) fungi, which includes yeast, hyphal fungi, and truffles [2]. Although a great deal of molecular information is available from *S. pombe*, a model eukaryote, very little is known about the *S. japonicus* cell-cycle regulative network.

Here we show that our methodology allows us to identify binding sites in *S. japonicus* using *S. pombe* gene expression data. As an example of annotation crossing from *S. pombe* to *S. japonicus* we focus on the Eng1 cluster, a set of very strongly cell cycle-regulated genes, which in *S. pombe* contains nine genes, involved in cell separation [6]. The genes are adg1 and adg2 (cell surface glycoproteins), adg3 (glucosidase), agn1 and eng1 (glycosyl hydrolases), cfh4 (chitin synthase regulatory factor), mid2 (an anillin needed for cell division and septin organization), ace2 (a cell cycle transcription factor), and SPCC306.11, a sequence orphan of unknown function. Motif searches showed that each gene of the cluster has at least one binding site for the Ace2 transcription factor (consensus CCAGCC). The Eng1 cluster has a recognizably similar functional cluster in

S. cerevisiae, the SIC1 cluster, which also contains the glycosyl hydrolase eng1 in *S. pombe*, and its ortholog DSE4 in *S. cerevisiae*.

In the next section we briefly describe the data and provide details on the statistical procedures used. Then we describe the analysis and related findings. Finally we discuss statistical issues related to the procedure we have used and the potentialities, which are currently addressed.

2   Methodology

2.1   Motif Selection Procedure

We propose a method for finding DNA binding sites which is an extension of that from [8]. While these authors have shown that variable selection is more effective than the linear regression used by [4], we have extended their procedure to time series analysis and to the use of gene expression from different species/strains. We briefly describe our methodology, pointing to the differences with respect to [8].We consider microarray experiments that explore the transcriptional responses of the fission yeast *S. pombe* [6] to cell cycle. This allows us to compare two organisms with similar biological complexity. We focus on two stress conditions in wild type cultures: oxidative stress caused by hydrogen peroxide and heat shock caused by temperature increase. Our other data consists of the organisms' genome sequences. *S. Pombe* DNA sequence data were obtained from the NCBI's FTP site (`ftp://ftp.ncbi.nih.gov/genomes/`); *S. japonicus* sequence data from (`http://www.broad.mit.edu`). The motif finding algorithms are sensitive to noise, which increases with the size of upstream sequences examined. As reported by [9], the vast majority of the yeast regulatory sites from the TRANSFAC database are located within 800 bp from the translation start site. We therefore extracted sequences up to 800 bp upstream, shortening them, if necessary, to avoid any overlap with adjacent ORF's. For genes with negative orientation, this was done taking the reverse complement of the sequences. Then we used MDScan [5] to search for nucleotide patterns. The algorithm starts by enumerating each segment of width w (seed) in the top t sequences. For each seed, it looks for w-mers with at least n base pair matches in the t sequences. These are used to form a motif matrix and the highest scoring seeds are retained. The updating step is done iteratively by scanning all w-mers in the remaining sequences and adding in or removing from the weight matrix segments that increase the score. This is repeated until the alignment stabilizes. The score of each motif is computed as in [4]. For each organism, the entire genome regions were extracted and used as background models. We searched for nucleotide patterns of length 5 to 15 bp and considered up to 30 distinct candidates for each width.

Our methodology is summarized in Figure 1 and proceed as follows: we first select candidate motifs which are generated from the over-represented nucleotide patterns, then we derive pattern scores for each motif following [4]. We continue by fitting a linear regression model relating gene expression levels (Y) to pattern scores (X), and using a Bayesian variable selection method we select motifs that best explain and predict the changes in expression level.

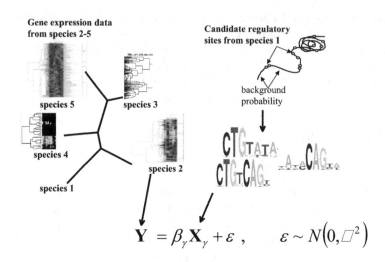

Fig. 1. Graphical representation of methodology

The variable selection method proceeds as follows. A latent vector, $\boldsymbol{\gamma}$, with binary entries is introduced to identify variables included in the model; γ_j takes on value 1 if the jth variable (motif) is included and 0 otherwise. The regression model is then given by:

$$Y = X_\gamma \beta_\gamma + \varepsilon, \qquad \varepsilon \sim N(0, \sigma^2), \tag{1}$$

where the columns of X and Y are mean-centered and (γ) indexes variables included in the model [1].

We specify Bernoulli priors for the elements of $\boldsymbol{\gamma}$:

$$p(\boldsymbol{\gamma}) = \prod_{j=1}^{p} \theta^{\gamma_j} (1 - \theta)^{1 - \gamma_j}, \tag{2}$$

where $\theta = p_{\text{prior}}/p$ and p_{prior} is the number of covariates expected *a priori* to be included in the model. For the other model parameters, we take

$$\beta_\gamma \sim N(0, c\{X'_\gamma X_\gamma\}^{-1})$$
$$\sigma^2 \sim \text{Inv}-\chi^2(a, b), \tag{3}$$

where a, b and c need to be assessed through a sensitivity analysis (see [8]). The scaling value b was taken to be comparable in size to the expected error variance of the standardized data.

Note that, with respect to [8], the choice of the prior is not completely random: it draws suggestions from the motifs already discovered in analysis of the genome data of close species.

2.2   Stochastic Search

Having set the prior distributions, a Bayesian analysis proceeds by updating the prior beliefs with information that comes from the data. Our interest is in the posterior distribution of the vector $\boldsymbol{\gamma}$ given the data, $f(\boldsymbol{\gamma}|X,Y)$. Vector values with high probability identify the most promising sets of candidate motifs. Given the large number of possible vector values (2^p possibilities with p covariates), we use a stochastic search Markov chain Monte Carlo (MCMC) technique to search for sets with high posterior probabilities.

Our method visits a sequence of models that differ successively in one or two variables. At each iteration, a candidate model, $\boldsymbol{\gamma}^{\mathrm{new}}$, is generated by randomly choosing one of these two transition moves:

   (i) Add or delete one variable from $\boldsymbol{\gamma}^{\mathrm{old}}$.

   (ii) Swap the inclusion status of two variables in $\boldsymbol{\gamma}^{\mathrm{old}}$.

The proposed $\boldsymbol{\gamma}^{\mathrm{new}}$ is accepted with a probability that depends on the ratio of the relative posterior probabilities of the new versus the previously visited models:

$$\min\left\{\frac{f(\boldsymbol{\gamma}^{\mathrm{new}}|X,Y)}{f(\boldsymbol{\gamma}^{\mathrm{old}}|X,Y)},1\right\}, \tag{4}$$

which leads to the retention of the more probable set of patterns [8,1].

Our stochastic search results in a list of visited sets and corresponding relative posterior probabilities. The marginal posterior probability of inclusion for a single motif j, $P(\gamma_j = 1|X,Y)$, can be computed from the posterior probabilities of the visited models:

$$p(\gamma_j = 1|X,Y) = \int p\left(\gamma_j = 1, \boldsymbol{\gamma}_{(-j)}|X,Y\right)d\boldsymbol{\gamma}_{(-j)} \tag{5}$$

$$\propto \int p\left(Y|X,\gamma_j = 1, \boldsymbol{\gamma}_{(-j)}\right) \cdot p(\boldsymbol{\gamma})d\boldsymbol{\gamma}_{(-j)}$$

$$\approx \sum_{t=1}^{M} p\left(Y|X,\gamma_j = 1, \boldsymbol{\gamma}_{(-j)}^{(t)}\right) \cdot p\left(\gamma_j = 1, \boldsymbol{\gamma}_{(-j)}^{(t)}\right),$$

where $\boldsymbol{\gamma}_{(-j)}^{(t)}$ is the vector $\boldsymbol{\gamma}$ at the tth iteration without the jth motif.

For each organism and stress condition, we regressed the expression levels on the pattern scores using separate models. In all cases, the analyses were started with a set of around 200 patterns. We chose $p_{\mathrm{prior}} = 10$ for the prior of $\boldsymbol{\gamma}$. This means that we expect models with relatively few motifs to perform well.

For every regression model, we ran 8 parallel MCMC chains. The searches were started with a randomly selected γ_j's set to one. We pooled together the sets of patterns visited by the 8 MCMC chains and computed the normalized posterior probabilities of each distinct visited set. We also computed the marginal posterior probabilities, $P(\gamma_j = 1|X,Y)$, for the inclusion of single nucleotide patterns.

For comparison, we repeated the analysis with MotifRegressor [4], which uses stepwise regression to select motifs.

3   Results

We first assessed the distance between *S. pombe* and *S. japonicus* by means of a phylogenetic tree. A maximum likelihood tree, based on the eng1 protein showing the position of S. pombe with respect other fungi (*S. cerevisiae, S. pombe, Eremothecium gossypii, Kluyveromyces lactis, Debaryomyces hansenii, Candida albicans, Yarrowia lipolytica, Aspergillus oryzae, Phaeosphaeria nodorum, Neurospora crassa*) is shown in Figure 2. We used sequences from the nearest neighborhood fungi. Since Eng1 protein family are globular cytoplasmic proteins we used the JTT [10] model of evolution. The phylogeny shows the short distance between *S. pombe* and *S. japonicus* (branch length is in 100 amino acid replacement units) with respect to the other fungi genomes. The future availability of other genome sequences (Neurospora and Aspergillus sequencing projects are under way) will provide ground for testing the integration of molecular information at larger phylogenetic distances.

Then we have run the variable selection software, which we have implemented in Matlab; we found that, even using a poorly refined genomic background (Markov chain of order 3), we were able to identify the exact pattern of ace2 in *S. pombe* and a nearly exact pattern of ace2 in *S. japonicus*.

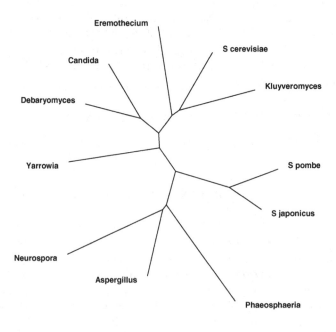

Fig. 2. Maximum Likelihood tree inferred using Eng1 protein sequences

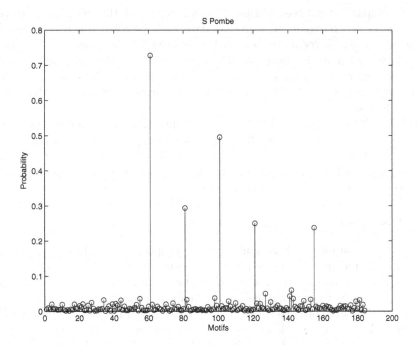

Fig. 3. Marginal posterior probabilities for *S. pombe*

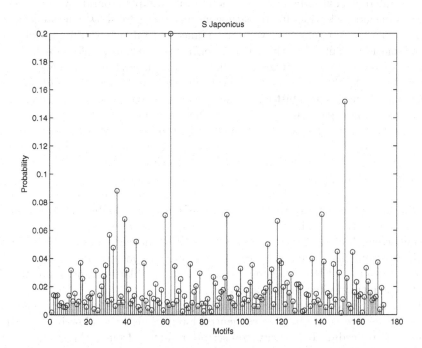

Fig. 4. Marginal posterior probabilities for *S. japonicus*

The comparison between Figure 3, which represent the marginal posterior probabilities computed from *S. pombe* data and Figure 4 which represent *S. japonicus* suggest a regulatory similarity between *S. japonicus* and *S. pombe*: very likely eng1 motif is also regulating a set of 9 genes in *S. japonicus*.

In both figures the x-axes correspond to the pattern indices and the y-axes correspond to $p(\gamma|X, Y)$. The spikes indicate patterns included in the model with high probability.

This result provides example of the possibility of using *S. pombe Eng1* promoters to annotate or simply offer suggestions useful in the annotation of *S. japonicus*. Moreover algorithms such as Gibbs sampling allows one to identify similar patterns in other genes and, therefore, determine other genes putatively regulated by the same transcription factor.

3.1   Discussion

The reasoning of our work is double: 1) we propose a method for annotating the transcription sites in one genome using gene expression data from another species; 2) our statistical framework allows us to estimate the degree of consistency between the motifs identified in the analysis of one set of experiments and those identified in another set. Using the mathematical notation introduced before, given Y_b and X_{γ_a} we determine $\hat{Y}_{b(\gamma_a)}$ and the error in the fitting; 2) given Y_b e X we repeat the variable selection and compute γ_b and the fitting error.

Different groups may work on the same species, which may evolve rapidly due to different environmental conditions (for example laboratory cultures) or interactions and selection due to pathogens and therefore show consistent differences at population genetics level, for instance several classes of mutations and recombination events. Clearly when experiments consider a reference sequence from a model genome, changes may have occurred, affecting the generality of the findings. At the same time, slightly different experimental conditions, microclimate or a different history in media cultures may result in some changes in the expression patterns, at least for the genes of interest. Comparisons of gene expression profiles in yeast's mutants may prove the effective involvement of changes in a DNA binding site in giving a phenotype. Particular focus may be put to unravel the function of orphan genes through the presence of motifs in their upstream region.

It is nowadays not uncommon to replicate gene expression experiments. Finding the best way of using these replicates is however not obvious - one option is to partition the data into training and test sets in order to asses the generalization strength of the fitted model. Clearly if we have a consistent number of replicates from the same strain we can compute the mean and variance which gives an estimate of the stringency of the genetic regulatory control involved in the transcription factor site recognition.

There is a growing trend of using microarrays in re-sequencing, although entire genomes are very rarely re-sequenced lately. This may allow in the future to use our procedure to identify transcription factors from a single microarray experiment [11]. It is known that genes coding for DNA-binding proteins are usually less conserved than genes coding for structural proteins. Rajewsky and collaborators [7] have demonstrated an inverse correlation between the rate of evolution of transcription factors and the number of genes that they regulate. The analysis of variances and covariances of sequences and microarray may allow insights into the evolution and selection pressure on transcription binding sites and gene expression. For example, an interesting possibility is to start with sequences belonging to a family of motifs and let them evolve accordingly to a model of evolution [10]; then the regression coefficients may provide an estimate of the negative selection pressure acting on the motif sequences.

In summary the use of gene expression and sequences from close species together with a better information extraction from replicates [3], may lead to a significant improvement of the quality and speed of genome annotation. Moreover it allows the recognition of different cell conditions, strains, species, similar conditions in different species, based on sequence-gene expression variances and covariances. A change in gene expression due to genome sequence modification (for instance the insertion of a new copy of motif upstream one gene) can be estimated in terms of the regression coefficients. Work in progress consists in the elucidation of the entire cell cycle regulatory control and stress induced genes in *S. japonicus* using all the wealth of information from annotated or semi-annotated genomes and more refined background models (5th order Markov chain or higher).

4   Figures

Figure 1. Graphical representation of methodology - We fit a linear regression model that relates gene expression data to pattern scores. A Bayesian variable selection method, which introduces a latent binary vector γ, is used to identify variables included in the model. A stochastic search MCMC technique is used to update γ. Motifs with high posterior probability are selected and indicate promising sets for further investigation.

Figure 2. Maximum Likelihood tree inferred using Eng1 protein sequences from *S. cerevisiae, S. pombe, Eremothecium gossypii, Kluyveromyces lactis, Debaryomyces hansenii, Candida albicans, Yarrowia lipolytica, Aspergillus oryzae, Phaeosphaeria nodorum, Neurospora crassa*. Branch length is in 100 amino acid replacement units.

Figure 3. Marginal posterior probabilities of all nucleotide patterns found in *S. pombe*. The x-axes correspond to the pattern indices and the y-axes correspond to $p(\gamma|X,Y)$. The spikes indicate patterns included in the model with high probability.

Figure 4. Marginal posterior probabilities of all nucleotide patterns found in *S. japonicus*. See details of Figure 3.

Acknowledgements

We acknowledge the Schizosaccharomyces japonicus Sequencing Project at Broad Institute of Harvard and MIT (http://www.broad.mit.edu). We also acknowledge the European Molecular Biology Organization for awarding Luisa Cutillo of a short term fellowship, during which part of this work was done at the Computer Laboratory of the University of Cambridge.

References

1. P.J. Brown, M. Vannucci, and T. Fearn. Multivariate bayesian variable selection and prediction. *J. R. Stat. Soc. Ser. B*, 60:627641, 1998.
2. C. E. Bullerwell, J. Leigh, L. Forget, and B. F. Lang. A comparison of three fission yeast mitochondrial genomes. *Nucleic Acids Res*, 31(2):759–768, Jan 2003.
3. E M Conlon, Joon J Song, and Jun S Liu. Bayesian models for pooling microarray studies with multiple sources of replications. *BMC Bioinformatics*, 7:247, 2006.
4. Erin M Conlon, X. Shirley Liu, Jason D Lieb, and Jun S Liu. Integrating regulatory motif discovery and genome-wide expression analysis. *Proc Natl Acad Sci U S A*, 100(6):3339–3344, Mar 2003.
5. X. Shirley Liu, Douglas L. Brutlag, and Jun S. Liu. An algorithm for finding protein-dna binding sites with applications to chromatin-immunoprecipitation microarray experiments. *Nat Biotechnol*, 20(8):835–839, Aug 2002.
6. Anna Oliva, Adam Rosebrock, Francisco Ferrezuelo, Saumyadipta Pyne, Haiying Chen, Steve Skiena, Bruce Futcher, and Janet Leatherwood. The cell cycle-regulated genes of schizosaccharomyces pombe. *PLoS Biol*, 3(7):e225, Jul 2005.
7. Nikolaus Rajewsky, Nicholas D Socci, Martin Zapotocky, and Eric D Siggia. The evolution of dna regulatory regions for proteo-gamma bacteria by interspecies comparisons. *Genome Res*, 12(2):298–308, Feb 2002.
8. Mahlet G Tadesse, Marina Vannucci, and Pietro Liò. Identification of dna regulatory motifs using bayesian variable selection. *Bioinformatics*, 20(16):2553–2561, Nov 2004.
9. J. van Helden, B. Andr, and J. Collado-Vides. Extracting regulatory sites from the upstream region of yeast genes by computational analysis of oligonucleotide frequencies. *J Mol Biol*, 281(5):827–842, Sep 1998.
10. S. Whelan, P. Li, and N. Goldman. Molecular phylogenetics: state-of-the-art methods for looking into the past. *Trends Genet*, 17(5):262–272, May 2001.
11. Michael E Zwick, Farrell Mcafee, David J Cutler, Timothy D Read, Jacques Ravel, Gregory R Bowman, Darrell R Galloway, and Alfred Mateczun. Microarray-based resequencing of multiple bacillus anthracis isolates. *Genome Biol*, 6(1):R10, 2005.

Genetic Programming and Other Machine Learning Approaches to Predict Median Oral Lethal Dose (LD$_{50}$) and Plasma Protein Binding Levels (%PPB) of Drugs

Francesco Archetti[1,2], Stefano Lanzeni[1], Enza Messina[1], and Leonardo Vanneschi[1]

[1] D.I.S.Co., Department of Computer Science and Communication
University of Milan-Bicocca, p.zza Ateneo Nuovo 1, 20126,Milan, Italy
{archetti,lanzeni,messina,vanneschi}@disco.unimib.it
[2] Consorzio Milano Ricerche
via Leopoldo Cicognara 7, 20100, Milan, Italy
archetti@milanoricerche.it

Abstract. Computational methods allowing reliable pharmacokinetics predictions for newly synthesized compounds are critically relevant for drug discovery and development. Here we present an empirical study focusing on various versions of Genetic Programming and other well known Machine Learning techniques to predict Median Oral Lethal Dose (LD$_{50}$) and Plasma Protein Binding (%PPB) levels. Since these two parameters respectively characterize the harmful effects and the distribution into human body of a drug, their accurate prediction is essential for the selection of effective molecules. The obtained results confirm that Genetic Programming is a promising technique for predicting pharmacokinetics parameters, both from the point of view of the accurateness and of the generalization ability.

1 Introduction

Because of recent advances in high throughput screening (HTS), pharmaceutical research is currently changing. In the traditional drug discovery process, when a target protein is identified and validated, the search of lead compounds begins with the design of a structural molecular fragment with therapeutic potency. Libraries of millions of chemical compounds similar to the identified effective fragment are then tested and ranked according to their specific biological activity. After these tests some candidate drugs are selected from the library for more specific development (see figure 1.a). In order to have a real pharmacological value, compounds have not only to show a good target binding, but also have to reach the target in vivo. In other words, it is necessary that compounds follow a proper route into the human body without causing toxic behaviors. It is interesting to remark that, both in 1991 [22] and in 2000 [7], a considerable fraction of attritions in pharmacological development were generated at the level of pharmacokinetics and toxicology, producing an unacceptable burden on the budget of drug companies (see figure 1.b). The necessity of deeply characterizing the behaviors of the pharmacological molecules in terms of adsorption, distribution, metabolism, excretion and toxicity processes (collectively referred to as ADMET [4]) makes the

E. Marchiori, J.H. Moore, and J.C. Rajapakse (Eds.): EvoBIO 2007, LNCS 4447, pp. 11–23, 2007.
© Springer-Verlag Berlin Heidelberg 2007

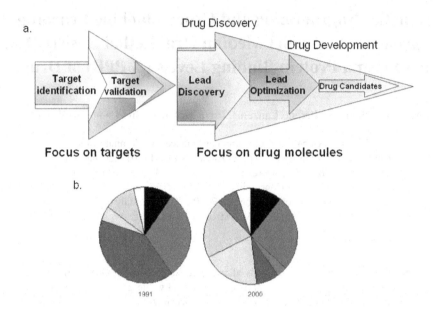

Fig. 1. a) The process of drug discovery from target protein identification to candidate drugs: the identification and validation of the target are followed by lead discovery and optimization. **b)** Reasons for failure in drug development in 1991 and 2000: clinical safety (black), efficacy (red), formulation (green), PK/bioavailability (blue), commercial (yellow), toxicology (gray), cost of goods (purple) and others (white).

development of computational tools applicable for pharmacokinetic profiling, enabling both the generation of reliable predictive models and the management of large and heterogeneous databases of outmost relevance [4] [25]. Reliable prediction tools allow the risk reduction of late-stage research failures, while reducing the number of cavies used in pharmacological research. In this paper, we empirically show that Genetic Programming (GP) [12] is a promising and valuable tool for predicting the values of Median Oral Lethal Dose (LD50) and Plasma Protein Binding (%PPB) levels. LD50 is one of the parameters measuring the toxicity of a given compound. More precisely, LD50 refers to the amount of compound required to kill 50% of the cavies. It is usually expressed as the number of milligrams of drug related to one kilogram of mass of cavies (mg/kg). Depending on the specific organism (rat, mice, dog, monkey and rabbit usually) and on the precise way of supplying (intravenous, subcutaneous, intraperitoneal, oral generally) chosen, it is possible to define a wide spectrum of LD50 experimental protocols. We consider the LD50 measured using rats as model organisms and supplying the compound orally. %PPB corresponds instead to the percentage of the initial drug dose which binds plasma proteins [13]. This measure is fundamental, both because blood circulation is the major vehicle of drug distribution into human body and because only free (unbound) drugs permeate the cellular membranes and reach the targets.

This paper is structured as follows: section 2 describes the mostly employed methods in literature for LD50 and %PPB levels predictions. In section 3, we describe the ML techniques used and our experimental settings including dataset collection and preparation methods, and we pinpoint all the parameters needed for a complete reproduction of our experiments. Section 4 shows the experimental results obtained by using four different implementations of GP and their comparison to the results obtained by employing five non evolutionary ML methods. Finally, in section 5 we discuss the results and we suggest hints for future research.

2   State of the Art and Related Work

Usually, pharmacokinetics prediction tools can be based on two paradigms: *molecular modeling* (using approaches of intensive chemical structure calculations) and *data modeling*. Methods based on data modeling are widely reported in literature and all are based on Quantitative Structure Activity Relationship (also called Q.S.A.R.) models [5]. Such models try to infer a quantitative relationship between the structure of a molecule and its biological activity by means of statistical regression methods. To accomplish this task, it is necessary to collect a training set of drugs for which the biological parameter to be predicted is known experimentally. A spectrum of features, called molecular descriptors [17], is calculated from the structural representation of each compound. Descriptors could be computed by representing molecules as graphs with labeled nodes (atoms) and labeled edges (bonds between atoms). In this sense two main categories of features are used in QSAR procedures: 2D-chemical descriptors, based on a bi-dimensional representation of compounds and 3D-chemical descriptors, whose computation is time-consuming since they are calculated from a tri-dimensional molecular structure. After the computation of the molecular descriptors, it comes natural to look for a mathematical relationship between them and the biological parameter, applying statistical or Machine Learning (ML) procedures. Drug toxicity prediction is a very complex task, because of the complexity of possible interactions between the organism and the pharmacological molecule. i.e. similar compounds may undergo different toxic behaviors). A benchmark of mathematical methods, including recursive partitioning (RP) and Partial Least Square regression (PLS) among others, has been tested in [9] for pharmacokinetic predictions. Multivariate Linear Regression and Artificial Neural Networks have been applied for building a model of LD50 in fathead minnow (*Pimephales promelas*) for 397 organic chemicals [23]. For %PPB, a sigmoidal relationship between the percentage of drug bound and the logD at pH 7.4 has been proposed [6], even though a considerable scatter of data around the predicted trend has been detected. A technique called Genetic Function approximation (GFA) considering 107 molecular descriptors has been proposed in [2]. As suggested in [10], artificial Neural Networks are ubiquitously used for pharmacokinetics parameters estimation. Neural Networks are widely used also in the existing commercial packages, that are usually developed by software vendors traditionally involved in the field of molecular modeling. GP has not been extensively used in pharmacokinetics until now. Four noteworthy exceptions are given by refernce [27] (where GP is used to classify molecules in terms of their Bioavailability), reference [1] (where GP is used to quantitatively predict the fraction of drug absorbed), reference [26] (where GP is used as feature selection technique for QSAR applications) and reference [15] (where multi-objective optimization frameworks are used for QSAR applications).

3   Experimental Protocol Settings

In this section we first describe the procedures used for datasets collection and preparation. We have obtained a set of molecular structures and the corresponding LD50 and %PPB values parsing a public database of FDA approved drugs and drug-like compounds [19]. Chemical structures are all expressed as SMILES code (Simplified Molecular Input Line Entry Specification), i.e. strings codying the 2D molecular structure of a compound in an extremely concise form. The resulting libraries of molecules contained 234 (respectively 662) molecules with measured LD50 (respectively %PPB) values. SMILES strings have been transformed into vectors of 626 molecular descriptors using the on-line DRAGON software [24]. Thus, data have been gathered in matrices composed by 234 (respectively 662) rows and 627 columns. Each row represents a drug; each column a molecular descriptor, except the last one, that contains the known values of LD50 (respectively %PPB). These two datasets, comprehensive of SMILES data structures and molecular descriptors can be downloaded from our laboratory webpage: <omitted to keep anonimity>. Training and test sets have been obtained by randomly splitting the two datasets. For each dataset, 70% of the molecules have been randomly selected with uniform probability and inserted into the training set, while the remaining 30% form the test set.

3.1   Genetic Programming Configurations

The four versions of GP methods proposed in this paper are briefly described below.

"Canonic" (or standard) GP. The first GP setting used also called canonic, or standard, GP and indicated as stdGP, is a deliberately simple version of standard tree-based GP [12]. In particular, we used the parameter setting, sets of functions and terminal symbols as similar as possible to the ones originally adopted in [12] for symbolic regression problems. Each molecular feature has been represented as a floating point number. Potential solutions (GP individuals) have been built by means of the set of functions $F = \{+, *, -, \%\}$ (where % is the protected division, i.e. it returns 1 if the denominator is zero), and the set of terminals T composed by n floating point variables (where n is the number of columns in the training sets, i.e. the number of molecular features of the compounds). The fitness of each individual has been defined as the root mean squared error ($RMSE$) in prediction computed on the training set. Afterwards, for each individual k, we evaluated the $RMSE$ and the correlation coefficient (CC) measured on the test set, and we used them for comparing GP results with the ones of the other employed methods. GP parameters used in our experiments are: population size of 500 individuals; ramped half-and-half initialization; tournament selection of size 10; maximum tree depth equal to 10; subtree crossover [12] rate equal to 0.95; subtree, swap and shrink mutation [12] rates equal to 0.1; maximum number of generations equal to 100. Generational GP has been used with elitism (i.e. copy of the best individual in the next population at each generation).

LS2-GP. The second version of GP differs the previous one for the fitness function, which in this case is obtained executing two steps: (1) application of *linear scaling* to

RMSE (a detailed introduction of this method can be found in [11]). (2) computation of a weighted average between the *RMSE with* linear scaling and the *CC* between expected outputs and the results returned by the GP candidate solution. Weights equal to 0.4 and 0.6 have been respectively assigned to *RMSE* and *CC*, after being normalized into the range $[0, 1]$, thus giving slightly higher importance to *CC*. This values have been chosen empirically through a simple experimentation phase (whose results are not shown here for lack of space) from which resulted that they are the values for which the GP system has the best generalization ability. The idea behind this weighted sum is that optimizing only the *RMSE* on the training set may leads to overfitting and thus to a poor generalization power of GP solutions (i.e. bad results on the test set). If we optimize more than one criterium, GP returns more likely individual which is good for all the criteria even though not optimal for all of them. This GP version will be called "Linear Scaling with 2 criteria" GP or LS2-GP from now on.

LS2-C-GP. The third GP version presented in this paper is similar to LS2-GP with the only difference that a set of ephemeral random constants (*ERCs*) is added to the set of terminal symbols to code GP expressions. These *ERCs* are generated uniformly at random from the range $[m, M]$, where m and M are the minimum and the maximum target values on the training set respectively. In the experiments presented in this paper, a number of *ERCs* equal to the number of floating points variables has been used. This choice has been empirically confirmed to be suitable by a set of GP runs in which different numbers of *ERCs* extracted from different ranges have been used. The results of these experiments are not shown here for lack of space. This version of GP will be called "LS2 with Constants" GP and indicated as LS2-C-GP.

DF-GP. In the fourth version of GP the fitness function changes dynamically changes during the evolution. In particular, the evolution starts with the *CC* used as the only optimization criterium. When at least the 10% of the individuals in the population reaches a value of the *CC* which is largest or equal to 0.6, the fitness function changes, and it becomes equal to the *RMSE* for all those individuals which have a *CC* larger or equal to 0.6; to all the other individuals we artificially assign a *very bad* fitness value. In this way, selection operates as a strong pruning algorithm, giving a chance to survive for mating only to those individuals whose correlation is largest or equal to 0.6. The idea behind this method is that the search space is too large for GP to perform efficiently; furthermore, we hypothesize that individuals with a good, although not optimal, *CC* between outputs and goals have a largest generalization ability and thus should take part in the evolutionary process. Some experiments (whose results are not shown here for lack of space) have empirically confirmed that the threshold value 0.6 for the *CC* is large enough to avoid underfitting and small enough to reduce overfitting. Of course, this value has been tested and has revealed suitable only for the datasets used in this paper and can by no means be interpreted as a general threshold. Nevertheless, the experiments that we have executed to obtain this value are very simple and if we wish to evolve new expressions on new data we could easily replicate them. This GP version has been called Dynamic Fitness GP and will be indicated as DF-GP from now on.

3.2 Non Evolutionary Methods

Various non evolutionary regression models were used for comparison with GP results. They are described below in a synthetic way, since they are well known and well established ML techniques. For more details on these methods and their use, the reader is referred to the respective references quoted below. Furthermore, we adopted two feature selection heuristics: the Correlation based Feature Selection (*CorrFS*) [14] and the Principal Component based Feature Selection (*PCFS*) [8] implemented in weka [28] , which are not described here to save space.

Linear and Least Square Regression. We used the Akaike criterion for model selection (AIC) [3], that has the objective of estimating the Kullback-Leibler information between two densities, corresponding to the fitted model and the true model. The M5 criterion is used for further attribute selection [3]. The least square regression model is founded on the algorithm of Robust regression and outlier detection described in [16], searching for the more plausible linear relationship between outputs and targets.

Artificial Neural Networks. The multilayer Perceptron [20] implementation included in the Weka software distribution [28] was adopted. It uses the Backpropagation algorithm [20] with a learning rate equal to 0,3. All the neurons have a sigmoidal activation function. All the other parameters that we have used have been set to the defaults values proposed by the Weka implementation. A momentum of 0.1 progressively decreasing until 0.0001 has been used to escape local optima.

Support Vector Machines Regression. The Smola and Scholkopf sequential minimal optimization algorithm [21] was adopted for training a Support Vector regression using polynomial kernels. The Weka implementation [28] of this ML method has been used. More precisely we have built two models using polynomial kernels of first and second degree respectively.

4   Experimental Results

Rat Median Oral Lethal Dose. Here we present the experimental results that we have obtained for the prediction of LD50. In table 1 the *RMSE* and *CC* are reported for all the non evolutionary ML techniques. No feature selection procedure has been applied

Table 1. Experimental comparison between different non evolutionary ML techniques for Rat Median Oral Lethal Dose predictions without feature selection

Method	*RMSE* on test set	Correlation coefficient
Linear Regression	9876.5079	0.0557
Least Square Regression	66450.2168	0.1976
Multi layer Perceptron	2412.0279	-0.0133
SVM Regression - first degree polynomial kernel	18533.3505	-0.1723
SVM Regression - second degree polynomial kernel	4815.3241	0.1108

Table 2. Experimental comparison between different non evolutionary ML techniques for Rat Median Oral Lethal Dose predictions using *PCFS*

Method	*RMSE* on test set	Correlation coefficient
Linear Regression	2703.6638	-0.1424
Least Square Regression	3149.4284	-0.1514
Multi layer Perceptron	3836.1868	0.2003
SVM Regression - first degree polynomial kernel	2523.5642	0.0494
SVM Regression - second degree polynomial kernel	3229.7823	-0.1048

here. The best *RMSE* result is returned by the multilayer Perceptron, while the best *CC* has been found by the Least Square Regression. The results returned by the same techniques using the *PCFS* are reported in table 2. These results suggest that the use of this feature selection technique helps to generally improve the performances of all methods. In this case the best *RMSE* result is returned by SVM regression with first degree polynomial kernel, while the best *CC* is found by the multilayer Perceptron. Results obtained by using *CorrFS* istead of *PCFS* are reported in table 3. In this case we can observe a further improvement, and the best *RMSE* is returned by Linear Regression,

Table 3. Experimental comparison between different non evolutionary ML techniques for Rat Median Oral Lethal Dose predictions using *CorrFS*

Method	*RMSE* on test set	Correlation coefficient
Linear Regression	2170.4631	0.2799
Least Square Regression	2753.6281	-0.0358
Multi layer Perceptron	3204.0369	0.1041
SVM Regression - first degree polynomial kernel	2241.0289	0.39
SVM Regression - second degree polynomial kernel	2504.2721	0.115

Table 4. Experimental results of the different GP versions in predicting Rat Oral Median Lethal Dose without feature selection. These results concern the individuals with the best *RMSE* value in all the populations over 20 independent runs.

GP Version	Best *RMSE* individual		Best *CC* individual	
	RMSE	*CC*	*RMSE*	*CC*
stdGP	1776.77	0.0625492	1882.86	0.31329
LS2-GP	1753.58	0.280622	1753.58	0.280622
LS2-C-GP	1779.13	0.224534	1800.07	0.248892
DF-GP	1815.07	0.183483	1879.76	0.201722

while the best *CC* is found by the SVM regression with first degree polynomial kernel. In table 4 we report *RMSE* and *CC* for all the GP versions introduced above without any feature selection method. To have a complete picture of the GP capabilities, we show the performances of both the individual with the best *RMSE* value and the individual with best *CC* contained in the population after 20 independent runs. Although further investigation is needed to confirm these results, for example through the analysis of the ML algorithms with respect to their learning parameters, the GP approaches seems to be promising in that they outperform the other ML methods both for *RMSE* and *CC*. The GP version that has returned the best *RMSE* is LS2-GP, while the best *CC* has been found by stdGP. We also point out that DF-GP does not reach the *CC* value of 0.6 on the test set, even though the selection algorithm prunes all the individuals with a smaller *CC* than 0.6 on the training set. In table 5, we report the performance of the GP versions with *PCFS*. In this case, the number of variables used as terminal symbols to build the

Table 5. Experimental results of the different GP versions in predicting Rat Oral Median Lethal Dose with *PCFS*. These results concern the individuals with the best *RMSE* value in all the populations over 20 independent runs.

	Best *RMSE* individual		Best *CC* individual	
GP Version	*RMSE*	*CC*	*RMSE*	*CC*
stdGP	1883.53	0.00805945	2301.12	0.260822
LS2-GP	1754.87	0.288088	1802.69	0.293959
LS2-C-GP	1775.7	0.249466	1779.93	0.252211
DF-GP	1859.85	0.203944	4340.19	0.223444

trees (and consequently also the number of ERCs in LS2-C-GP, which is always equal to the number of variables in our experiments) is equal to the number of features selected by the *PCFS*. The application of this feature selection strategy does not seem to remarkably enhance performances of any of the GP versions. The same holds for *CorrFS*, results reported in table 6. We remark that when we have explicitly used no feature

Table 6. Experimental results of the different GP versions in predicting Rat Oral Median Lethal Dose with *CorrFS*. These results concern the individuals with the best *RMSE* value in all the populations over 20 independent runs.

	Best *RMSE* individual		Best *CC* individual	
GP Version	*RMSE*	*CC*	*RMSE*	*CC*
stdGP	1789.57	0.200317	1845.63	0.305397
LS2-GP	1841.56	0.106883	2080.02	0.163737
LS2-C-GP	1882.26	0.112631	2644.63	0.240427
DF-GP	1917.21	0.139279	1917.21	0.139279

selection (results reported in 4), GP has, by itself, performed a feature selection, since the best individuals returned by each of the GP versions contain only a very restricted subset of all the possible variable symbols. Thus, we hypothesize that, at least for this training set, GP "prefers" to selects features by its own, rather than considering a set of features filtered by another algorithm.

Plasma-Protein Binding Levels. Here we present the experimental results that we have obtained for the prediction of %PPB. In table 7 the *RMSE* and *CC* for all the non evolutionary ML techniques are reported, without applying any feature selection procedure. Both the best *RMSE* and the *CC* results are returned by the SVM regression with first degree polynomial kernel. The results of the same techniques using

Table 7. Experimental comparison between different non evolutionary ML techniques for Plasma-Protein Binding predictions without feature selection

Method	*RMSE* on test set	Correlation coefficient
Linear Regression	39.8402	0.3802
Least Square Regression	56.2947	0.0323
Multi layer Perceptron	32.6068	0.1398
SVM Regression - first degree polynomial kernel	31.4344	0.5056
SVM Regression - second degree polynomial kernel	53.0044	0.2616

Table 8. Experimental comparison between different non evolutionary ML techniques for Plasma-Protein Binding predictions using *PCFS*

Method	*RMSE* on test set	Correlation coefficient
Linear Regression	27.2677	0.5324
Least Square Regression	36.4912	0.1844
Multi layer Perceptron	56.9232	0.251
SVM Regression - first degree polynomial kernel	29.1552	0.4993
SVM Regression - second degree polynomial kernel	30.8581	0.5116

PCFS are reported in table 8. The usage of this feature selection procedure improves the performances of all the methods, with the exception of multilayer Perceptron that has returned a considerably higher *RMSE*. This time, both the best *RMSE* and the *CC* results are returned by Linear Regression. Results obtained using *CorrFS* are shown in table 9. As in the case of LD50, this kind of feature selection improves the performances of all the non evolutionary ML techniques better than *PCFS*. As for the LD50, the best *RMSE* result is returned by Linear Regression, and the best *CC* is found by the SVM regression with first degree polynomial kernel. In table 10, *RMSE* and *CC* for all

Table 9. Experimental comparison between different non evolutionary ML techniques for Plasma-Protein Binding predictions using *CorrFS*

Method	*RMSE* on test set	Correlation coefficient
Linear Regression	26.7496	0.5512
Least Square Regression	33.7744	0.4711
Multi layer Perceptron	31.7581	0.4379
SVM Regression - first degree polynomial kernel	27.2962	0.5586
SVM Regression - second degree polynomial kernel	28.3638	0.5188

Table 10. Experimental results of the different GP versions in predicting Plasma Protein Binding without feature selection. These results concern the individuals with the best *RMSE* value in all the populations over 20 independent runs.

GP Version	Best *RMSE* individual		Best *CC* individual	
	RMSE	*CC*	*RMSE*	*CC*
stdGP	34.6617	0.142542	36.9151	0.228315
LS2-GP	31.8281	0.319194	31.8281	0.319194
LS2-C-GP	32.4555	0.221073	32.5728	0.252989
DF-GP	32.1279	0.288191	32.245	0.294923

Table 11. Experimental results of the different GP versions in predicting Plasma Protein Binding with *PCFS*. These results concern the individuals with the best *RMSE* value in all the populations over 20 independent runs.

GP Version	Best *RMSE* individual		Best *CC* individual	
	RMSE	*CC*	*RMSE*	*CC*
stdGP	34.9778	0.197143	37.816	0.26463
LS2-GP	31.0209	0.335335	31.2919	0.35923
LS2-C-GP	31.2285	0.329228	31.5918	0.334473
DF-GP	31.0119	0.361486	31.0119	0.361486

the GP versions introduced above are reported. No feature selection has been applied here. Once again, for each GP version, the performances of both the individual with the best *RMSE* value and the individual with best *CC* contained in the population at the end of 20 independent runs are reported. Results obtained in the prediction of %PPB show that the GP approach outperforms non evolutionary methods only when feature selection is not applied. The GP version that turned out to be the best solution, both in terms of *RMSE* and *CC*, is LS2-GP. In table 11, we report the performance of the four GP proposed approaches combined with *PCFS*. Also in this case the application of feature selection does not remarkably enhance their performances. The same holds for *CorrFS*, see table 12.

Table 12. Experimental results of the different GP versions in predicting Plasma Protein Binding with *CorrFS*. These results concern the individuals with the best *RMSE* value in all the populations over 20 independent runs.

GP Version	Best *RMSE* individual		Best *CC* individual	
	RMSE	*CC*	*RMSE*	*CC*
stdGP	33.938	0.231034	36.3262	0.370302
LS2-GP	30.8059	0.394357	30.8059	0.394357
LS2-C-GP	31.2334	0.352454	31.2712	0.35958
DF-GP	31.0393	0.370221	31.0556	0.383304

5   Discussion and Future Work

An empirical study based on various versions of Genetic Programming (GP) and other well known ML techniques to predict Median Oral Lethal Dose (*LD50*) and Plasma Protein Binding (*%PPB*) levels has been presented in this paper. The availability of good prediction tools for pharmacokinetics parameters is critical for reducing the costs of the ADMET experimentation phase. This meets the recent encouragement of UE community contained in the REACH [18] proposal, whose aim is to improve the protection of human health and environment through the better and earlier identification of the properties of chemical substances and, among these, toxicity. In our work the technique that has returned the best results in the estimation of (*LD50*) was GP, while for %PPB GP has been outperformed by Linear Regression and SVM with first degree polynomial kernel. Nevertheless, also considering the good results obtained in literature by GP in different (but similar) applications [1,27], GP can undoubtely be considered one of the most promising techniques for this kind of tasks. We have also shown that feature selection procedures, which generally significantly improve the performance of non evolutionary ML techniques, are not relevant for improving GP prediction capabilities. The reason of this is that GP intrinsically acts as a wrapper, by automatically selecting and evaluating a specific subset of features to be kept inside the population. In some senses, we could say that feature selection is implicit in GP, while for the other ML techniques it has to explicitly be executed in a pre-processing phase. Furthermore, in our experiments we have remarked that the individual characterized by the best *RMSE* is often different from the one characterized by the best *CC* both on the training and test sets. For this reason, we confirm an observation that has already recently been done in [1]: using more (possibly uncorrelated) criteria for evaluating regression models is often beneficial. In fact, in our experiments the best results on the test set have been returned by the GP versions that use both the *RMSE* and the *CC* as optimization criteria on the training set. Nevertheless, in this paper we evaluate individuals by simply performing a weighted average between *RMSE* and *CC*. Future work will focus on testing more sophisticated multi-objective learning algorithms, possibly based on the concept of Pareto front and using more than two criteria. Finally, in order to deeply characterize the ability of GP in the ADMET in silico arena, we are planning to test our methodologies on other datasets, for example in the prediction of Blood Brain Barrier Permeability or Cytochrome P450 interactions.

Acknowledgments

We are grateful to the anonymous reviewers of this paper for their pertinent and useful suggestions and remarks.

References

1. F. Archetti, S. Lanzeni, E. Messina and L. Vanneschi. Genetic programming for human oral bioavailability of drugs. In M. Cattolico, editor, *Proceedings of the 8th annual conference on Genetic and Evolutionary Computation*, pages 255 – 262, Seattle, Washington, USA, July 2006.

2. G. Colmenarejo ,A. Alvarez-Pedraglio and J. L. Lavandera . Chemoinformatic models to predict binding affinities to human serum albumin. *Journal of Medicinal Chemistry*, 44:4370–4378, 2001.

3. H. Akaike. Information theory and an extension of maximum likelihood principle. In , editor, *2nd International Symposium on Information Theory*, Akademia Kiado, June 1973.

4. H. van de Waterbeemd and E. Gifford. ADMET in silico modeling: towards prediction paradise ? *Nature Reviews Drug Discovery*, 2:192–204, 2003.

5. H. Van de Waterbeemd and S. Rose. *In The Practice of Medicinal Chemistry, 2nd edition*. ed. Wermuth, L. G., 1367-1385,Academic Press, 2003.

6. H. Van de Waterbeemd, D. A. Smith and B. C. Jones. Lipophilicity in PK design:methyl, ethyl, futile. *Journal of Computationally aided Molecular Design*, 15:273–286, 2001.

7. I. Kola and J. Landis. Can the pharmaceutical industry reduce attrition rates ? *Nature Reviews Dug Discovery*, 3:711–716, 2004.

8. I.T. Jolliffe. *Principal Component Analysis, Second edition*. Springer series in statistics., 1999.

9. J Feng, L. Lurati, H. Ouyang, T. Robinson, Y. Wang, S. Yuan and S.S. Young. Predictive toxicology: benchmarking molecular descriptors and statistical methods. *Journal of Chemical Information Computer Science*, 43:1463–1470, 2003.

10. J. Zupan and P. Gasteiger. *Neural Networks in chemistry and drug design: an introduction, 2nd edition*. Wiley, 1999.

11. M. Keijzer. Improving symbolic regression with interval arithmetic and linear scaling. In C. Ryan, T. Soule, M. Keijzer, E. Tsang, R. Poli, and E. Costa, editors, *Genetic Programming, Proceedings of the 6th European Conference, EuroGP 2003*, volume 2610 of *LNCS*, pages 71–83, Essex, 2003. Springer, Berlin, Heidelberg, New York.

12. J. R. Koza. *Genetic Programming*. The MIT Press, Cambridge, Massachusetts, 1992.

13. L. M. Berezhkovskiy. Determination of Drug Binding to Plasma Proteins Using Competitive Equilibrium Binding to Dextran-Coated Charcoal. *Journal of Pharmacokinetics and Pharmacodynamics*, 33(5):920–937, 2006.

14. M. A. Hall. *Correlation-based Feature Selection for Machine Learning*. PhD thesis, Hamilton, NZ: Waikato University, Department of Computer Science, 1998.

15. O. Nicolotti and V. J. Gillet and P. J. Fleming and D. V. Green. Multiobjective optimization in quantitative structure-activity relationships: deriving accurate and interpretable QSARs. *Journal Med. Chem.*, 45(23):5069–5080, 2002.

16. Peter J. Rousseeuw and Annick M. Leroy. *Robust regression and outlier detection*. Wiley, New York, 1987.

17. R. Todeschini and V. Consonni. *Handbook of Molecular Descriptors*. Wiley-VCH, Weinheim, 2000.

18. REACH.   Registration, Evaluation and Authorisation of Chemicals, 2006. http://ec.europa.eu/environment/chemicals/reach/reach_intro.htm.

19. S. David, Wishart, C. Knox, A. C. Guo, S. Shrivastava, M. Hassanali,P. Stothard, Z. Chang and J. Woolsey. DrugBank: a comprehensive resource for in silico drug discovery and exploration. *Nucleic Acids Research*, 34:doi:10.1093/nar/gkj067, 2006.

20. S. Haykin. *Neural Networks: a comprehensive foundation*. Prentice Hall, London, 1999.

21. Smola Alex J. and Bernhard Scholkopf. A Tutorial on Support Vector Regression. Technical Report Technical Report Series - NC2-TR-1998-030, NeuroCOLT2, 1999.

22. T. Kennedy. Managing the drug discovery/development interface. *Drug Discovery Today*, 2:436–444, 1997.

23. T. M. Martin and D. M. Young. Prediction of the Acute Toxicity (96-h LC50) of Organic Compounds to the Fathead Minnow (*Pimephales promelas*) Using a Group Contribution Method. *Chemical Research in Toxicology*, 14(10):1378–1385, 2001.

24. Tetko, I. V.; Gasteiger, J.; Todeschini, R.; Mauri, A.; Livingstone, D.; Ertl, P.; Palyulin, V. A.; Radchenko, E. V.; Zefirov, N. S.; Makarenko, A. S.; Tanchuk, V. Y.; Prokopenko, V. V. Virtual computational chemistry laboratory - design and description. *Journal of Computer Aided Molecular Design*, 19:453–63, 2005. see www.vcclab.org.
25. U. Norinder and C. A. S. Bergstrom. Prediction of ADMET properties. *ChemMedChem*, 1:920–937, 2006.
26. V. Venkatraman and A. R. Dalby and Z. R. Yang. Evaluation of mutual information and genetic programming for feature selection in QSAR. *Journal Chem. Inf. Comput. Sci.*, 44(5):1686–1692, 2004.
27. W. B. Langdon and S. J. Barrett. Genetic Programming in data mining for drug discovery. *in Evolutionary computing in data mining*, pages 211–235, 2004.
28. Weka. a multi-task machine learning software developed by Waikato University., 2006. See www.cs.waikato.ac.nz/ml/weka/.

Hypothesis Testing with Classifier Systems for Rule-Based Risk Prediction

Flavio Baronti and Antonina Starita

Dipartimento di Informatica, Università di Pisa
Largo B. Pontecorvo, 3—56127 Pisa, Italy
{baronti, starita}@di.unipi.it

Abstract. Analysis of medical datasets has some specific requirements not always fulfilled by standard Machine Learning methods. In particular, heterogeneous and missing data must be tolerated, the results should be easily interpretable. Moreover, with genetic data, often the combination of two or more attributes leads to non-linear effects not detectable for each attribute on its own. We present a new ML algorithm, HCS, taking inspiration from learning classifier systems, decision trees and statistical hypothesis testing. We show the results of applying this algorithm to a well-known benchmark dataset, and to HNSCC, a dataset studying the connection between smoke and genetic patterns to the development of oral cancer.

1 Introduction

Medical research is shifting its focus from populations to individuals. It was already evident that people react differently to the same stimuli (diseases, therapies). The discovery of DNA and the advent of genetic profiling suggested it was possible to track down these differences to their root causes. The genetic profile of a person should have pinpointed her exact reaction to diseases and medicines; unfortunately, this objective is still very far in the future. We can describe two main aspects of genetic understanding which make the goal difficult to reach. The first one is *genome size*. As everyday experience clearly shows, every person is unique. From the genetic point of view, this can be explained through the huge number of different genes in the human genome: it is very unlikely that two people have exactly the same allelic variants of each gene. In principle, this makes generalization impossible: we will never find "the same situation". This problem can be solved however by understanding the effect each gene has, and by considering only those genes involved in the disease being studied. If for a particular disease we can reduce the number of involved genes to a manageable number, finding the same situation becomes possible. The second difficulty is *gene interaction*. Traditional medical analyses typically assume independence of the causing factors, and linearity of their combined effect. These assumptions appear wrong in genetic research: rarely a single gene variant has a direct effect on the outcome; genes work together, and often only their combined effect shows

E. Marchiori, J.H. Moore, and J.C. Rajapakse (Eds.): EvoBIO 2007, LNCS 4447, pp. 24–34, 2007.
© Springer-Verlag Berlin Heidelberg 2007

a significant impact on the outcome. This observation requires different analysis techniques to be used, able to deal with non-linearity and factors dependence.

In this paper we will present a new algorithm, developed in order to analyze data collected to study the effects of genetic variants on development of Head and Neck Squamous Cell Carcinoma (HNSCC), a kind oral cancer very common among smokers. Smoking is clearly a very important risk factor; observation however shows that there exist more sensitive people, which develop cancer with little to no smoking, and more resistant people, which do not develop cancer although smoking a lot. This difference could be explained by different treatment of carcinogens by the organism, regulated by different genetic variants. The analysis was performed through a new machine learning (ML) algorithm, loosely based upon learning classifier systems (LCS) research [1], called HCS (Hypothesis testing with Classifier Systems). Its main aim is to identify subsets of the whole dataset where the risk factor is significantly different from the global risk. Differently from most ML algorithms, HCS loosens the requirement of accuracy of its prediction, allowing to output a risk value instead of an exact classification. The main focus is instead generality of the prediction: we will show how this shift of goals was beneficial to the importance of results.

2   Problem Description

The data set we analyzed (originally presented in [2]) was designed to explore the influence of genotype on the chance to develop head and neck squamous cell carcinoma (HNSCC). It is already well-known that this kind of cancer is associated with smoking and alcohol-drinking habits, it is more common among males and its incidence increases with age. The individual risk however could be modified by genetic factors; therefore genotype information, regarding eleven genes involved with carcinogen-metabolizing (CCND1, NQO1, EPHX1, CYP2A6, CYP2D6, CYP2E1, NAT1, NAT2, GSTP1) and DNA repair systems (OGG1, XPD) was provided by molecular testing.

Nine of these genes have two allelic variants; let's call them a_1 and a_2. Since the DNA contains two copies of each gene, there exist three possible combinations: a_1a_1, a_2a_2 (the homozygotes) and a_1a_2 (the heterozygote — order does not matter). The homozygotes where represented with values 0 and 2, while the heterozygote with 1. Due to dominance, for the examined genes the heterozygote is equivalent to one of the homozygotes; however, for many of the considered genes this dominant effect is not known. So class 1 is either equivalent to class 0, or to class 2. The remaining two genes (NAT1 and NAT2) have 4 allelic variants, which result in 9 combinations; they were sorted by their activity level, and put on an integer scale from 0 to 8.

The full data consists of 355 records, with 124 positive elements (HNSCC patients) and 231 negative (controls). Each record reports the person's gender, age, total smoke and alcohol consumption, gene values, and a boolean target value which specifies whether he had cancer when the database was compiled or not. The data was collected in different periods between 1997 and 2003; this has led to many missing values among the genotypic information of patients. Actually only 122 elements have complete genotypic description; the remaining 233 have missing values ranging from 1 to 9, with the average being 3.58. As an overall figure, of the $11 \times 355 = 3905$ genotype values, just 3070 are present: 21% of the genotype information is missing.

3   HCS: Fundamentals

HCS (Hypothesis testing with Classifier Systems) is the learning algorithm we developed and employed to analyze the oral cancer dataset.

The underlying idea driving HCS design was to construct a system which could find *interesting* information in a dataset. Formally defining interestingness immediately appeared very difficult, if not impossible at all: it is too much a subjective and domain-dependent quantity, to be practically defined in all situations. We then turned our research for the interesting into a research for the *unexpected*. Starting from an established theory, experiments can be run to obtain a set of data in order to test it. If the data is in accordance with the theory, there is nothing new: nothing of interest. But when an experiment produces unexpected results, then something new has been discovered: the original theory needs to be revised — and this is certainly interesting.

Turning the idea into application: the environment where we will run experiments to test the theory is the dataset. The dataset will have a number of independent variables, also called *attributes*, and a dependent variable, also called *target*. While the attributes can have any type (real, ordinal, binary, ...), the target variable can only be binary — thus having only two values: positive and negative. The theory to test will be the *independence of the target from the attributes*.

The dataset will contain T samples, out of which Q with a positive target variable. We will call ρ the proportion of positive values — that is, $\rho = Q/T$. Any random extraction of samples from the dataset will contain some positive and some negative samples too; their proportion will generally be not too far from the global one.

Keeping the assumption of independence between attributes and target, selecting samples with respect to their attribute values instead of completely at random should again yield a proportion of positive and negative compatible with the global ρ. When this does not happen, we found something unexpected. The most diverse the proportion is from the global one, the more unexpected and interesting is the subset of samples we found. Since the subset is defined by a precise set of conditions on the attributes, the researcher can easily read these conditions and realize what is the distinguishing pattern of the subset.

In order to establish how much a particular subset of the data is in accordance with the global proportion, we employ statistical hypothesis testing. The proper distribution of target values for a random extraction of data would be the hypergeometric distribution. However, our final model will necessarily provide a fixed risk value, which with repeated sampling corresponds to a binomial distribution. We decided then to immediately perform this generalization step, and to state the H_0 hypothesis as that extractions from the dataset will follow a binomial distribution, with ρ probability to obtain a positive sample. With this assumption, the probability to extract a subset $S(q,t)$ of t samples, q of which are positive, is

$$\mathbb{P}(S(q,t)) = \binom{t}{q} \rho^q (1-\rho)^{(t-q)} \tag{1}$$

To calculate the p-value of the $S(q,t)$, we have then to decide a rejection region, containing the sample and every "more extreme" result. Out hypothesis testing the ρ value in both directions (either more or less), so we need a two-tailed test. The rejection region of a subset $S(q,t)$ is defined as

$$R(q,t) = \{S(r,t) | r \in \{0,\dots,t\} \wedge \mathbb{P}(S(r,t)) \le \mathbb{P}(S(q,t))\} \tag{2}$$

The rejection region is then simply the set of all results having probability lower or equal to the examined sample. Figure 1 provides an example of the rejection region. The bigger dot represents the example subset, with 16 positive samples out of 60 total samples. The shaded area highlights all the "more extreme" possibilities, which all together form the rejection region.

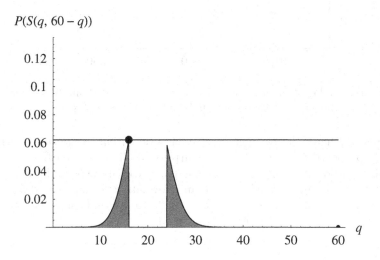

Fig. 1. Rejection region $R(16,60)$ in a setting where $\rho = 1/3$. The x axis shows the possible values for the number of positive samples; the y axis contains the corresponding probability value.

It is then straightforward to calculate the p-value of a subset:

$$p\text{-value}(q,t) = \sum_{S(r,t) \in R(q,t)} \mathbb{P}(S(r,t)) \tag{3}$$

The lower the p-value is, the more our subset is unexpected under the assumption of independence of the target from the attributes. Our algorithm will then search for regions with lowest possible p-value; this will actually happen through maximization of the $-\log$ of the p-value. We will call this quantity $I(q, t)$, to mean *interest*. Figure 2 shows how this value changes in an example setting of 60 extractions and varying amount of positive samples. Clearly the least interesting subset is the one with 20 positives: this is exactly the expected proportion when $\rho = 1/3$. It would be instead very strange to find a subset with all, or almost all positive: this corresponds to a very high interest value.

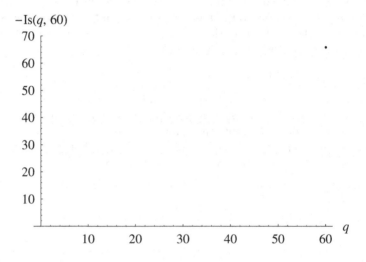

Fig. 2. Negative log of the p-value in a setting where $\rho = 1/3$ and 60 samples. The x axis shows the possible values for the number of positive samples.

Our work shares the basic goal of giving a definition to *unexpected* with the recently proposed Bayesian theory of surprise [3]. This approach starts with a prior distribution over all models, and compares it with the posterior distribution after data has been analyzed through an information theory-related measure of distance. In HCS instead we start with a single model, and check the probability that the data could be obtained by that model.

4   HCS: The Algorithm

Now that a measure of interest of a region has been defined, it is necessary to design an algorithm which can discover such regions. We will pose it as a maximization problem: among all the subsets S of the training data which can be defined with a conjunction of conditions on the attributes, find the one with highest interest value. This problem was solved with an evolutionary algorithm [4], taking inspiration partly from learning classifier systems, and partly from decision trees [5].

Genotype Definition. Each individual has a many genes as attributes in the training set. Each gene expresses a condition on its relative attribute; the gene shape is depending on the data type of its attribute. For instance, boolean attributes have a corresponding gene with three possible values: 0 (matching *false* attribute values), 1 (matching *true*) and # (matching anything). With this respect, genotypes in HCS have the same shape as the ones in other LCS systems, like XCS [6]. However, individuals have no additional data other then their genotype; there is thus no action, performance, accuracy, numerosity, etc.

Phenotype Definition. The phenotype of an individual can be defined as the subset of the training data matched by the individual. Matching occurs if a training sample satisfies all the conditions of the individual. In our implementation, only the largest condition is satisfied by missing data. Taking the boolean example again, only classifiers with a # condition will match data with the boolean value missing. The rationale behind this choice is to avoid taking decisions based on unknown values: if a classifier has a wildcard condition upon an attribute, it will take the same decision regardless of the actual value of the attribute. Other approaches are possible [7].

Fitness Evaluation. The evaluation of fitness is straightforward. The subset of training samples identified by an individual is extracted; the size of the subset is t, and q is the number of positive samples it contains. The fitness is then defined as $I(q, t)$, and will be maximized during the algorithm execution.

4.1   Internal Cycle

The internal cycle starts with an empty population, and evolves for a pre-defined number of steps. The GA employs a steady-state population model: at each step, a new offspring is generated and inserted into the population, removing another one if the population reached its maximum size. Before evolution starts, the proportion of positive samples over all the samples ρ is calculated: this is the number that will be used in subsequent fitness evaluations.

Selection and Reproduction. The selection process employs a niching method inspired to XCS in order to maintain diversity in the classifiers population, which in turn helps to escape local optima.

At each step, a random sample is chosen with replacement from the training set. The current population is scanned to find all the classifiers matching this sample: these classifiers form the *match set*. If the match set is empty, *covering* occurs: a new, random classifier is created, in such a way that it matches at least the extracted sample.

The tournament selection mechanism is used to choose one classifier from the match set. Since the match set has variable size, we use a variable-size tournament, as in [8]. This means that, if n is the match set size, $\tau \times n$ random classifiers are extracted without replacement from the match set, and the one with highest fitness among them is chosen for reproduction.

Reproduction employs the three classical operators. Simple cloning just copies the selected individual to the new population. Mutation puts in the new population a copy of the individual, with small variations. With crossover, a second individual is chosen from the match set with the same method; the two individuals are crossed, and their offspring is copied into the new population. Each reproduction operator will be applied with a probability defined at runtime: χ for crossover, μ for mutation, and cloning otherwise.

The created offspring is finally added to the population. If the population size has reached its limit, a classifier is removed from it with an "inverse" tournament selection: a subset of the whole population is chosen, and the classifier with lowest fitness in the subset is selected for removal. The selection and reproduction cycle is repeated until a pre-determined number of steps has been performed.

4.2 External Cycle

Once the internal cycle is terminated, the classifier with highest fitness in the final population is retrieved and stored. The training set is then split into two subsets: the samples matched by the classifier, and the samples not matched by it. The internal cycle is then recursively repeated on both subsets, creating a tree structure with classifiers at each node. Notice the ρ value is calculated again on each of the two subset.

The external cycle repeats splitting until the classifier fitness is judged too low. This decision is based upon significance testing. When a single hypothesis is tested, the rejection decision is usually based on a single threshold value, called α: the null hypothesis is rejected if the p-value is lower than α. The common-use values of α are generally 0.05 and 0.01. In our setting, we are performing many tests - actually, one for each classifier - increasing the probability to wrongly reject the null hypothesis. To correct this effect, we applied the *Bonferroni correction* [9], dividing the standard α value by the total number of generated classifiers. This approach is often considered too conservative [10]. From another point of view however, the search algorithm is implicitly performing many more tests than it actually does, so the *steps* value is underestimating the real number of tests.

We assumed for simplicity that these two factors cancel out. The external cycle was then allowed to run until the best reported classifier had fitness higher than

$$-\log(\frac{\alpha}{steps}) \tag{4}$$

5   Experiments and Results

We applied the HCS algorithm to the analysis of the Wisconsin Breast Cancer dataset, from the UCI repository [11], and to the HNSCC dataset. Results are compared with Naive Bayes (NB), C4.5, Neural Networks (NN) and XCS [2]. The parameters used for HCS are detailed in table 1.

Table 1. Summary of HCS parameters

Pop. size	300
Steps	45000
μ	0.4
χ	0.5
τ	0.4
S_0	0.75
Mr	0.1
$P_\#$	0.5

The behaviour of the algorithms was assessed with two indexes: accuracy and Brier score. Accuracy is the most used measure of performance for classifiers, whose prediction is a dichotomous value (either 0 or 1). Although it is not well-suited to our model, which predicts risk values, it was used as a reference measure. To calculate accuracy, the risk values predicted by the model were rounded to the closest integer (0 or 1).

The Brier score [12] is a common index applied to evaluate the calibration of a risk prediction model. It ranges from 0 to 1, with lower values corresponding to better models. The formula is

$$B = \frac{1}{T} \sum_{i=1}^{T} (Y_i - r(X_i))^2 \qquad (5)$$

In order to obtain an estimate of accuracy and Brier score, 10-fold stratified cross-validation was applied. The scores were calculated on the reconstructed test set. In order to smooth the variance coming from random fold partitioning, and stochastic algorithms, we repeated each cross-validation experiment 10 times. Tables 2 and 3 show the average of the results for the 10 runs, and their respective standard deviations.

Table 2. WBC dataset: Accuracy and Brier score results on reconstructed test set, 10-fold cross validation. Average(standard deviation) of 10 runs.

	NB	C4.5	NN	HCS
Accuracy	.975 (.001)	.950 (.004)	.965 (.002)	.945 (.005)
Brier	.024 (.001)	.046 (.004)	.028 (.001)	.043 (.004)

On the WBC benchmark dataset, all the tested algorithms have comparable performance. The very good result of Naive Bayes suggests that the attributes can almost independently contribute to the final prediction. HCS on this dataset performs has a very close performance to the other algorithms, which confirms the validity of our approach.

On the HNSCC dataset, traditional algorithms reach unsatisfactory performance levels, with Naive Bayes and C4.5 both with 0.7 accuracy. Neural

Table 3. HNSCC dataset: Accuracy and Brier score results on reconstructed test set, 10-fold cross validation. Average(standard deviation) of 10 runs.

	NB	C4.5	NN	XCS	HCS
Accuracy	.694 (.007)	.70 (.01)	.78 (.02)	.79 (.01)	.838 (.008)
Brier	.212 (.001)	.198 (.008)	.17 (.01)	—	.136 (.006)

Networks substantially improve this figure, thanks to its capability to model non-linear interactions between attributes. HCS has the best performance on this dataset, both from the Brier score and from the accuracy point of view. This performance increase can clearly not be generalized to all datasets. However it shows how the ability to make multiple decisions at each node of the tree, together with hypothesis testing in place of information gain, can find better explanation for complex datasets, where attributes have a relevant effect on the target only when they are combined.

HCS appears to perform better than XCS too. The weak point of XCS when applied to this dataset is its aim to find perfect classifiers, which identify areas containing either all ill or all healthy patients. In the HNSCC dataset (and we believe in most medical datasets), this happens only by chance, in relatively small subsets. The focus on not noisy classification is then hindering the more important search for general classifiers.

After the test performance of HCS was assessed, we executed 10 runs of the algorithm on the whole dataset. We obtained an average accuracy of 0.89(0.02), and an average Brier score of 0.09(0.01). Paired with the cross-validation results, these figures suggest that the algorithm extracts useful information from the data. It is then meaningful to submit the classifiers to the clinicians, for biological validation.

The algorithm is stochastic, so the runs produced slightly different trees, with either one, two or three conditions. Since however all the results had a very similar main condition, and nine out of ten had similar second condition (while the remaining had none), we report a typical two-conditions tree as an example of the algorithm result.

The first classifier extract 77 patients from the dataset, 73 of whom are ill. This is a highly significant area, where the p-value of the null hypothesis is roughly 10^{-28}. The particular genetic combination reported in this rule appears to be favorable to developing cancer: removing *age* and *packyears* from the conditions yielded a subpopulation of 103 patients, 74% of whom were ill — still a high risk value, compared to the baseline of 35%.

The second classifiers, which applies on the people not taken into account by the first one, finds a subset of 22 people, with 18 being ill. Considering that at this point the baseline risk is $(124 - 73)/(355 - 77) \approx 0.18$, this subset is still very interesting (p-value $\approx 10^{-10}$). Again, a genotype causing a higher risk is identified: on the full dataset, the second classifier without *age* and *packyears* conditions extracts a subset of 70 patients, 74% of whom were ill.

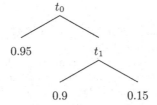

t_0: 73/77 [28] $gender$=Male $\wedge$ $age \in [49, 79]$ $\wedge$ $packyears \geq 4$ $\wedge$ $ogg1 \in \{0, 1\}$ $nat2gen \leq$
        $7 \wedge gstp1 \in \{0, 1\}$
t_1: 18/22 [11] $age \in [35, 82]$ $\wedge$ $packyears \geq 24$ $\wedge$ $ephx1 \in \{1, 2\}$ $\wedge$ $gstp1 \in \{0, 1\}$

Fig. 3. One result for the HNSCC dataset. Conditions are satisfied on the left branch, and unsatisfied on the right branch. Numbers in brackets report rule fitness.

Since this was a case-control study, the dataset is not a representative sample of the global population. It is then important to underline that the identified risk values are not realistic, and do not represent the actual risk of a random person satisfying their conditions. However, the result it still valid with respect to the identified conditions: the genotypes reported by the classifiers will generally have a worse prognosis than the average — although a different study design is necessary to estimate the exact values.

6   Conclusions and Future Work

In this paper we presented a new machine learning algorithm, HCS, aimed at discovering risk prediction rules over a dataset. The algorithm was designed taking into account medical data analysis requirements. It can work seamlessly with different data types, and it is robust to missing data. The algorithm provides readily interpretable rules in an *if-then* format. We applied it to the analysis of the HNSCC dataset, obtaining interesting results both from the accuracy point of view, and from the interpretation point of view. Despite the stochastic nature of genetic algorithms, the final result appeared quite stable; this is important to increase the confidence of the researcher on the relevance of the result.

The algorithm is still in its first draft, as it was specifically designed for the HNSCC dataset over which it was applied. It will be necessary to validate it on the standard benchmark datasets (from the UCI repository [11], for instance). Further work will include testing the effectiveness of algorithm design choices (eg. binomial vs. hypergeometric hypothesis, tournament vs. fitness-proportionate selection, niched vs. standard population). We will investigate the possibility to apply a precise multiple hypothesis testing correction, and empirically evaluate it. An interactive version is planned, where the clinician can choose at each external cycle step which rule he thinks more interesting, and possibly manually tune it to better comply with biological significance. Finally, the algorithm could also be applied to different settings (eg. survival analysis), by using the appropriate statistical tests.

References

1. Holland, J.H.: Adaptation. In Rosen, R., Snell, F.M., eds.: Progress in theoretical biology, 4. New York: Plenum (1976)
2. Passaro, A., Baronti, F., Maggini, V.: Exploring relationships between genotype and oral cancer development through xcs. In Rothlauf, F., ed.: GECCO 2005, Workshop Proceedings, Washington DC, USA, June 25-26, 2005. (2005) 147–151
3. Itti, L., Baldi, P.: Bayesian surprise attracts human attention. In: Advances in Neural Information Processing Systems, Vol. 19 (NIPS*2005), Cambridge, MA, MIT Press (2006) 1–8
4. Goldberg, D.: Genetic Algorithms in Search, Optimization and Machine Learning. Addison-Wesley, Reading, MA (1989)
5. Quinlan, J.R.: C4.5: programs for machine learning. Morgan Kaufmann Publishers Inc. (1993)
6. Wilson, S.W.: Classifier fitness based on accuracy. Ev. Comp. $3(2)$ (1995)
7. Holmes, J.H., Sager, J.A., Bilker, W.B.: Methods for covering missing data in XCS. In Keijzer, M., ed.: Late Breaking Papers at GECCO 2004, Seattle, WA (2004)
8. Butz, M.V., Sastry, K., Goldberg, D.E.: Strong, stable, and reliable fitness pressure in XCS due to tournament selection. Genetic Programming and Evolvable Machines $6(1)$ (2005) 53–77
9. Shaffer, J.P.: Multiple hypothesis testing. Annual Review of Psychology 46 (1995) 561–584
10. Perneger, T.V.: What's wrong with bonferroni adjustments. British Medical Journal 316 (1998) 1236–1238
11. Newman, D.J., Hettich, S., Blake, C.L., Merz, C.J.: UCI repository of machine learning databases [http://www.ics.uci.edu/~mlearn/MLRepository.html] (1998)
12. Brier, G.W.: Verification of forecasts expressed in terms of probability. Montly weather review $78(1)$ (1950) 1–3

Robust Peak Detection and Alignment of nanoLC-FT Mass Spectrometry Data

Marius C. Codrea[1], Connie R. Jiménez[2], Sander Piersma[2], Jaap Heringa[1], and Elena Marchiori[1]

[1] Centre for Integrative Bioinformatics VU (IBIVU)
Department of Computer Science
mcodrea@few.vu.nl, heringa@few.vu.nl, elena@cs.vu.nl
[2] Cancer Center Amsterdam
Vrije Universiteit Amsterdam, The Netherlands
c.jimenez@vumc.nl, S.Piersma@vumc.nl

Abstract. In liquid chromatography-mass spectrometry (LC-MS) based expression proteomics, samples from different groups are analyzed comparatively in order to detect differences that can possibly be caused by the disease under study (potential biomarker detection). To this end, advanced computational techniques are needed. Peak alignment and detection are two key steps in the analysis process of LC-MS datasets. In this paper we propose an algorithm for LC-MS peak detection and alignment. The goal of the algorithm is to group together peaks generated by the same peptide but detected in different samples. It employs clustering with a new weighted similarity measure and automatic selection of the number of clusters. Moreover, it supports parallelization by acting on blocks. Finally, it allows incorporation of available domain knowledge for constraining and refining the search for aligned peaks. Application of the algorithm to a LC-MS dataset generated by a spike-in experiment substantiates the effectiveness of the proposed technique.

1 Introduction

Computational analysis of proteomic datasets is becoming of crucial relevance for discovery of reliable and robust candidate biomarkers. In particular, quantitation of changes in protein abundance and/or state of modification is the most promising, yet most challenging aspect of proteomics. In recent years label-free LC-MS methods that quantify absolute ion abundances of peptides and proteins have emerged as promising approaches for peptide quantitation and profiling of large numbers of clinical samples [1].

Briefly, peptides are subjected to (multi-dimensional) liquid chromatography for separation. Each peptide fraction is then analyzed on an LC-MS system. Each LC-MS run of a sample generates a pattern of very high input dimension consisting of one intensity (relative abundance) measurement for each pair of molecular mass-to-charge ratio (m/z) and retention time (RT) values. Ideally, the same molecules detected in the same LC-MS instrument should have the

E. Marchiori, J.H. Moore, and J.C. Rajapakse (Eds.): EvoBIO 2007, LNCS 4447, pp. 35–46, 2007.

same retention time, molecular weight, and signal intensity. However, in practice this does not happen due to experimental variations. As a consequence, patterns generated by LC-MS runs need to undergo a number of processing steps before they can be comparatively analyzed. Such processing steps include normalization [5,17], background subtraction [8], alignment [4,13,16,17], and peak detection (e.g., [9]). Several tools and algorithms for processing and for difference analysis of LC-MS datasets have been introduced (e.g., [2,3,7,9,11,15,20,17]).

In this paper we focus on peak detection and alignment. As well explained in the overview paper by Listgarten et al [10], alignment algorithms involve either (i) the maximization of an objective function over a parametric set of (generally linear) transformations, or (ii) non-parametric alignment based on dynamic programming, or (iii) combination of these methods like piecewise transformations. They act either on the full pattern or on features (peaks) selected beforehand; they may or may not use the signal intensity and they may or may not incorporate scaling. Most of alignment algorithms require a reference template, to which all time series are aligned. Peak detection is usually performed in an ad-hoc manner [10], involving either a comparison of intensities with neighbours along the m/z axis [19] or detection of coinciding local maxima [18].

Whether one should perform peak detection before [14,17] or after [12] alignment has not been clearly established. In this paper we circumvent this issue by performing both tasks at the same time by means of a novel clustering algorithm. The motivations for a new algorithm rely also on the desire to overcome drawbacks of alignment algorithms, such as the need for a reference template, or the assumption of a given (local or global, usually linear) transformation in RT dimension. The versatility of clustering for simultaneous alignment and peak detection has been already recognized by Tibshirani et al. in [16]. They align MALDI (a technique that generates two dimensional patterns of m/z and intensity values from each sample) data along the m/z axis by applying one-dimensional hierarchical clustering with complete linkage for constructing the dendrogram and a specific cutoff for extracting clusters representing peaks.

The algorithm proposed here, called Peak Detection and Alignment (PDA), acts on blocks of m/z values. Blocks are obtained by splitting the runs along the m/z axis (e.g., in blocks of equal size). Each block is processed individually. The input of PDA is a set of (blocks of) runs described by a list of triplets (m/z, RT, run id) consisting of 'm/z' value, 'RT' value and run identifier. The output of PDA is a set of clusters of (m/z, RT) pairs representing peaks, together with information about their signal intensity.

Novel features of PDA include:

1. a similarity measure for comparing features, where a feature is a triplet (m/z, RT, id). Weights are associated to each of the three attributes to specify their relevance;
2. a cluster merging strategy;
3. a cluster refinement procedure for handling peaks occurring near the boarder of blocks.

Application of PDA to a dataset generated by a spike-in experiment substantiates its effectiveness. The results indicate that PDA provides a flexible, efficient and robust approach for simultaneous peak detection and alignment of high-throughput label-free LC-MS data.

2   Materials and Methods

2.1   nanoLC-FT Mass Spectrometry Proteomic Data

Efficiently identifying and quantifying disease- or treatment-related changes in the abundance of proteins in easy accessible body fluids such as serum is an important area of research for biomarker discovery. Currently, cancer diagnosis and management are hampered by lack of discriminatory and easy obtainable biomarkers. In order to improve disease management more sensitive and specific biomarkers need to be identified. In this light, the simultaneous detection and identification of multiple biomarkers (*molecular signatures*) may be more accurate than single marker detection. Therefore, great promise holds in combining global profiling methods, such as proteomics, with powerful bioinformatics tools that allow for marker identification. The additional dimension of separation provided by coupling nanoliquid chromatography to high-performance Fourier transform mass spectrometry (nanoLC-FTMS) allows for profiling large numbers of peptides and proteins in complex biological samples at great resolution, sensitivity and dynamic range.

In LC-MS, the *chromatographic column* separates peptide mixtures based on one or more physicochemical properties prior to MS. An *ionization source* converts eluting peptides into ions which are separated by the *mass analyzer* on the basis of m/z ratios. The *detector* then registers the relative abundance of ions at discrete m/z values. Therefore, LC-MS yields values indicating that, at a particular time, an ion with a particular m/z value was detected with a particular intensity. In the proteomics community, the unit of measure of the m/z axis is one Dalton (Da), defined as 1/12 of the mass of one atom of carbon-12.

The dataset was generated as follows. Angiotensin 1 (1296.68 Da) was spiked at 1 fmol/μl into a background of a 50 fmol/μl tryptic digest consisting of bovine cytochrome c, hen egg lysozyme, bovine serum albumin, bovine apo-transferrin and Escherichia coli β-galactosidase. 1 μl of the peptide mixture was loaded on column (Pepmap C18 100 Å, 3 mm, 75 μm x 150 mm) and was separated at 250 nl/min in a 50 minute linear gradient of 10-40% buffer B (80% acetonitrile/0.05% formic acid). Eluting peptides were detected using an LTQ-FT hybrid mass spectrometer; mass spectra were acquired every second at resolution 100000. In total 12 samples were generated, consisting of six replicates for each class (spiked, unspiked).

Figure 1 (a) illustrates a sample spectrum comprising the set of intensities (relative abundance of ions) measured at different m/z values across a particular range. Such spectra are measured during the chromatographic period at discrete time points, producing what is often called a 'run'. The chromatographic period is also referred to as elution time or retention time (RT). 2D possible visualizations of runs are given in panels (b and c) of Fig. 1, where the intensity at each RT-m/z location is represented by a gray-level. The spectrum in (a) is the one indicated by the vertical line at 1420 seconds in (b). The panels (d and e) are close-up views within the range where the spiked-in peptide is located.

Each analyte (organic specimen under study) gives rise to more than one peak in the measured spectra. Firstly, an analyte A with mass m can appear with different charges (e.g., A^{+1}, A^{+2}, A^{+3}) and because the actual quantities being measured are m/z ratios, it gives peaks at the m/z values: $m/1, m/2, m/3$. Secondly, because of the natural isotopic occurrence of the chemical elements, different peaks are generated by the same analyte A at the same charge state. For example, about 1.1% of the carbon atoms are carbon-13 isotopes instead of carbon-12. These have an extra neutron in the nucleus and therefore a mass higher by 1 Da. The spacing between the isotopic peaks depends on the charge of the ions. A group of isotopic peaks is shown in Fig. 1(d), where the distance between peaks is 1/3 Da, due to the +3 charge. The number of isotopic peaks and their relative intensity depend on the chemical composition of the analyte and therefore the distribution (pattern) of the isotopic peaks is mass dependent. Note that our algorithm detects and align isotopic peaks. In order to retrieve peptides from isotopic peaks procedures for assembling isotopes into peptides can be used, such as the one implemented in msInspect [3].

2.2   Methods

PDA takes a set of runs as input and outputs a set of (isotopic) peaks. The search strategy for detecting and aligning peaks acts on a three dimensional search space with dimensions given by m/z value, RT and run identifier.

PDA performs the following sequence of steps:

1. *Feature extraction*: Extract features from individual runs (in our experiments we use MZmine [7]).
2. *Block construction*: Split runs into blocks along the m/z axis.
3. Apply the following two steps to each block:
   (a) *Features clustering*: Generate a common set of clusters using weighted hierarchical clustering.
   (b) *Peak construction*: Generate peaks from clusters.
4. *Peak refinement*: Refine peaks located nearby splitting points of blocks.

1. Feature Extraction. We use a routine from MZmine [7] called "Centroid peak detection" for extracting features from runs. A feature in a run is defined here as the (m/z, RT) pair of coordinates of a local signal intensity maximum. These coordinates are computed by first detecting local maxima in each scan (spectrum). Certain features can then be discarded if their intensity is smaller than a chromatographic threshold of the extracted chromatogram (XIC) that contains the local maxima. An XIC is constructed by summing all intensities across RT present within a m/z range (or bin).

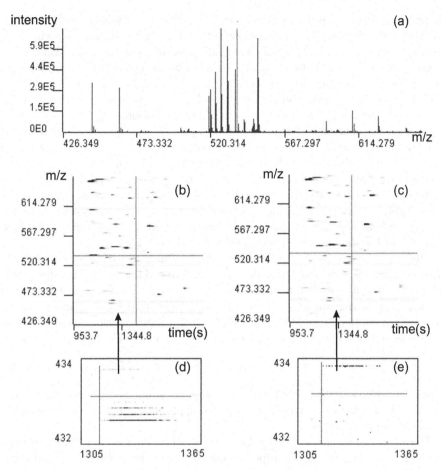

Fig. 1. (a) a sample spectrum (relative abundance of ions) measured at RT=1420s indicated by the vertical line in (b). (b and c) 2D visualizations of two runs from control and spiked-in samples. (d and e) are close-up views of the two runs within the range 1305-1365 s and 432-434 Da where the spiked-in peptide is located.

2. Block Construction. To speed up the processing, runs are split into blocks along the m/z-axis. Clustering and peak extraction are then performed on each block independently, thus (possibly) processed in parallel. The peak refinement step is applied to peaks detected nearby block boarders, in order to deal with the initial splitting of true peaks.

3(a). Features Clustering. This step constitutes the core of PDA. The particularities of the LC-MS measurements lead us to introduce a weighted distance within the clustering algorithm. The following properties of LC-MS measurements were used for setting the values of weights. m/z measurements are rather reproducible across different runs. The RT of the same analyte may, however, greatly differ among distinct runs, even in the order of tens of seconds. It is therefore sensible to assign a larger weight to the m/z coordinate than to the RT, thus enforcing small m/z variations while allowing for discrepancies in time. Moreover, since our aim is also to align peaks over different runs, it is also desired to group features from different runs. To this aim we use also the run identifier as attribute for comparing features.

Our clustering operates on the set of all features, augmented with a 'run id' component. Searching for peaks in this three-dimensional space is performed by hierarchical clustering with average linkage and a novel similarity measure defined below. Since the desired number of peaks is not known a-priori, we have to estimate it. Estimation of the number of clusters can be performed by a model selection criterion, like BIC or AIC (see e.g., [6] for an experimental comparison of the effectiveness of these criteria). Here we adopt a more efficient knowledge-based approach. Starting from a high number of clusters we apply a heuristic for joining clusters which relies on domain knowledge about the minimal distance in m/z and RT dimension between two valid peaks.

Clusters are generated as follows. Firstly, in order to overcome problems related to different m/z and RT scale, the standardized scores of the values (also called z-scores) are used. Informally, for a value x (here m/z or RT value) from a set X, the corresponding z-score measures how far the observation is from the mean in units of standard deviation. The similarity measure used in PDA is defined as follows. For two features $F_1 = (mz1, RT1, r1)$, $F_2 = (mz2, RT2, r2)$

$$sim(F_1, F_2) = sqrt(w1 * (mz1 - mz2)^2 + w2 * (RT1 - RT2)^2) + w3 * I(r1, r2)$$

where $I(r1, r2)$ is 1 if $r1 = r2$ and zero otherwise, and $w1, w2, w3$ are weights associated to the three attributes. We elaborate on the choice of the weights values in the next section.

The dendrogram generated by hierarchical clustering is used for clustering features. A (loose) user-given upper bound M on the desired number of clusters is employed for computing the cutting threshold of the dendrogram. The output of the feature clustering step consists of M 3-dimensional clusters.

3(b). Peak Construction. In this step nearby clusters are merged to form peaks by means of the following algorithm. The distance between the mean m/z values and mean RT values of each pair of centroid clusters is computed and used in an iterative procedure that generate peaks. The procedure starts from a given centroid (selected at random) as initial part of a peak. Then, centroids are incrementally added to that peak if their distance from at least one element of the actual peak is smaller than a threshold in m/z and RT. The construction of a peak terminates when no centroid can be added to it. Then the construction of a new peak can begin using the remaining centroids. The process terminates when the set of centroids becomes empty.

4. Peak Refinement. The peak refinement step processes those peaks whose features result distributed in clusters belonging to neighbour blocks. Because the runs are split into independent, non-overlapping blocks, features that belong to

the same true peak may lie on either side of a certain m/z cutting point. These
features will be assigned to different clusters even if they are close to each other
(corresponding clusters fulfil the peak construction constraints). Therefore, we
introduce a peak refinement step that joins such clusters. This is accomplished
by repeating the peak construction routine on the set of cluster centroids, or
more efficiently, on centroids close to block boarders.

The computational complexity of PDA is dominated by the *Features clustering*
step, that is quadratic in the number of features to be clustered. This number corresponds to the number of intensities occurring in a block. Then on the average,
when blocks are processed in parallel and assuming features are uniformly distributed, PDA (worse case) computational complexity is $O((N/nb)^2 + n_{centroids}^2)$
where N is the number o features, nb the number of blocks, and $n_{centroids}$ the
number of centroids obtained after application of step 3(b).

3   Experimental Analysis

In the feature extraction step with "Centroid peak detection" (MZmine [7]), we
used a fine bin size (0.02 Da) and a loose chromatographic threshold of 40%,
which caused a data reduction from the original measured intensities $\sim 1.8 \times 10^7$
per run to about 4.3×10^5. We split the runs at each 5 Da and produced blocks
containing a relatively small number of features, in the order of thousands. The
upper bound on the number of clusters M is set to one tenth of the total number
of intensities present in each block. Finally, weights of the three attributes m/z,
RT, and 'run id' were set to 2, 0.01 and 0.1, respectively. These values are suitable
for the type of high resolution LC-MS measurements used in this study. Slight
changes of the values did not markedly affect the clustering results. As long as
the the the m/z weight was significantly larger than the RT and the 'run id' weights,
diversity within a cluster was kept moderate and the results were consistent.

PDA results using different m/z and RT weights are illustrated in Fig. 2,
while Fig. 3 shows resulting clusterings based on different values of 'the run'
id weight. For the purpose of illustration, we only show results on small yet
representative m/z, RT regions. Unit weights for the m/z and RT axes (regular
Euclidean distance) lead to undesired split of the peak on the right in Fig. 2(a).
Using the m/z axis alone ($w = (1\ 0\ 0)$) produces clusters that contain peaks of
arbitrary different RTs (b). The RT axis alone is also not sufficient for accurate
peak detection, since peaks are split in favor of small RT deviations (c). The
output of PDA using the weights selected in our experiments is shown in (d).

The 'run id' attribute is used to discourage shifted values (in either m/z or
RT) from joining a cluster if other features from the same run already match
well with features from other runs. However, care should be taken about the
effects on the true peaks. This is illustrated in Fig. 3 (a and b), where different
markers indicate that the features originate from different runs and large markers
indicate cluster centers. A relatively high weight value for 'run ids' can result
in clusters that contain more than one true peak. For example, in panel (b) the
top two and the bottom three peaks are grouped in two clusters, while the five

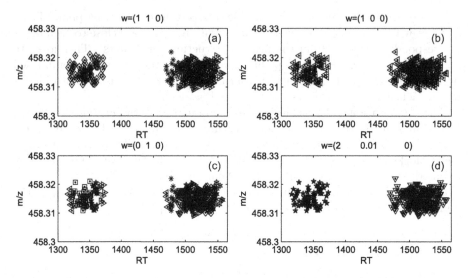

Fig. 2. The influence of the m/z and RT weights on the clustering results (see text for details). Different markers stand for different cluster membership of the features.

correct clusters should appear as in (a). Panels (c) and (d) show a zoomed view on a different region of the run. Note that the features in (c) belong to a cluster that lies beyond the shown RT limit and therefore its center is at about 1282 seconds. This cluster even comprises the peak on the left in Fig. 2 which has similar m/z values but obviously different RTs. A small 'run id' weight value helps in isolating sparse features, as desired (see (d)).

Since the feature clustering step uses an upper bound M on the number of clusters, features that belong to the same true peak can be assigned to different clusters. To fix this problem, the peak construction step 3(b) was used with thresholds for m/z and RT of 0.01 Da and 10 seconds, respectively. The m/z threshold relates to the minimum distance between two different peaks along the m/z axis, and can be calculated from the instrument set-up and the type of measurements. Two distinct isotopic peaks can be 0.01 Da apart if they originate from an analyte with charge +10, which is not practical. The RT threshold is chosen such that within the time window it defines, two different compounds with the same m/z value do not elute simultaneously. Our investigation revealed that for this dataset double peaks are observed within 30 seconds only after increasing the m/z window to 1 Da. It is therefore obvious that at 0.01 Da, all peaks are separated within a time window of 20 seconds.

Results

The spiked-in peptide, Angiotensin 1, has theoretical m/z value 432.90032 Da and RT value of 21.75-22.76 minutes, the latter estimated from the experiment by our domain expert. The peptide gives rise to four isotopes (named here P_i,

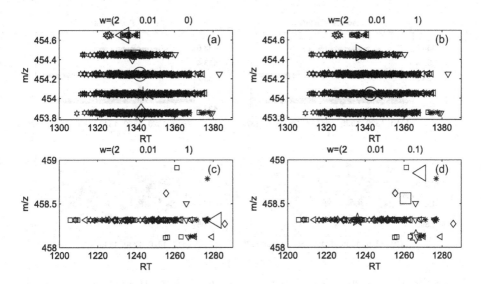

Fig. 3. The influence of the run identifier weight on the clustering results (see text for details). Different markers indicate that the features originate from different runs, large markers indicate cluster centers.

$i = 1 \ldots 4$) with decreasing intensity and duration, located at 432.90032 Da (highest intensity isotope), 433.29, 433.57, 433.91. While $P1, P_2, P_3$ were accurately detected by the PDA, P_4 was missed already by PDA's first step (feature extraction). This means that although the local maxima detection method is simple, care should be taken in the choice of its parameters.

The results of PDA zoomed in a region containing the spiked-in peptide, are shown in Figure 4. The hierarchical clustering produced a large number of clusters belonging to the same isotope (a and c). Because the peak construction step operates on the cluster centroids, their distance may exceed the estimated RT threshold resulting in few peaks belonging to one isotope. More specifically, P_1 and P_2 both resulted represented by two peak clusters, with 31, 159, and 19, 117 features, respectively. P_3 was detected as single peak with 33 features. All clusters contain features from all the six spiked-in runs.

To get an estimate of the actual RT shifts across the runs we computed, for each obtained peak cluster, a value here called 'average time shift'. This value is defined as the mean of the minimum RT (pairwise) distances between features within a cluster belonging to different runs. This measure describes the average shift needed for alignment of features in a given cluster. The minimum pair-wise distance reflects the amount of time required to consider two sets of features (from two different runs) aligned. The behaviour of average time shift for some values of the RT threshold used in the peak construction step is illustrated in Figure 5. The horizontal line within each box represents the median value. The vertical limits of each box are at the lower quartile and upper quartile values.

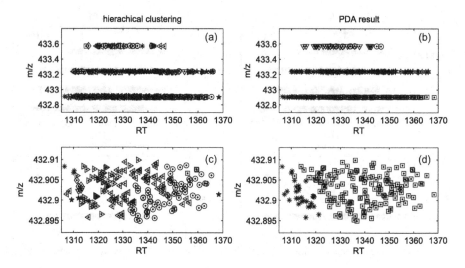

Fig. 4. Results of hierarchical clustering (a, c) and PDA (b, d). Occasionally, isotopic peaks can result in more than one cluster as in (d).

The rest of the observations extend between the lower and upper whiskers. The large amount of outliers (marked with '+') illustrates the difficulty of the alignment problem and substantiates the robustness of PDA. Even large drifts in RT do not obscure the PDA. Different RT threshold settings (between 5 and 30 seconds) do not cause significant changes in 'average time shift', which indicates the consistency of our initial assessment for a suitable threshold value. Moreover the figure indicates the good quality of the experimental data. The observation that 75% of the shift values are within a range of less than 10 seconds suggests that in most cases, local small shifts are sufficient for alignment. Thus local alignment schemes seem to be suited for this type of mass spectrometry data.

Finally, to get a flavour about the performance of other algorithms for LC-MS data alignment and peak detection, we consider two popular LC-MS data processing tools: MZmine [7] and MetAlign [2]. They both perform first peak detection followed by alignment. The nine parameters of MZmine's peak detection routine, as well as the twenty parameters of MetAlign were set by our local MS domain expert. We did not succeed in detecting the spike-in peptide using MetAlign. MZmine detected isotopes only in few of the spiked runs, represented by multiple peaks. In particular, the peak detection routine of MZmine detected isotopes in four of the six spiked-in runs, with detection of only one or two of the four isotopes. This lead to alignment results where P_1 was detected four times, three times represented by one peak from one run and the other time represented by two peaks from different runs. P_2 was identified by two peaks from different runs and P_1 by a single peak from a single run.

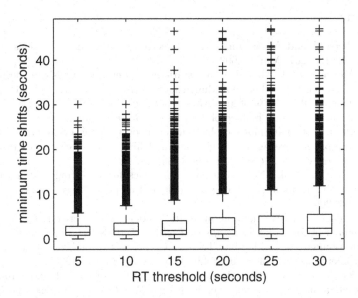

Fig. 5. The behaviour of the minimum shifts among features within each cluster. 75% of the values (within boxes) are smaller than 10 s indicating that local shifts are sufficient for alignment. Different RT thresholds (5 to 30) did not cause significant changes, as the general trend is only slightly increasing. The outliers (marked with '+') shows robustness of PDA.

4   Conclusion

In this paper, we proposed a method for simultaneous peak detection and alignment with nanoLCFT-MS data. Results on a spiked-in dataset indicate the effectiveness of this approach. We are currently working on further refinement steps such as: filtering out small cardinality clusters, irrespective of their locations within runs, removing clusters whose diameter in RT is too big (the elution time range of a peptide is limited), and improving the peak construction step. Moreover, we are planning to incorporate this technique in advanced biomarker detection algorithms based on feature selection. Such a comparative profiling implies the identification of clusters that contain relevant features for discrimination between control and case classes. A comparative analysis of results of the extended method on other datasets will provide more through assessment of its power as a computational tool for disease biomarker detection.

Acknowledgment

Marius C. Codrea was financially supported by an NWO-Bioinformatics Breakthrough Grant (050-71-047), The Netherlands.

References

1. R. Aebersold and M. Mann. Mass spectrometry-based proteomics. *Nature*, 422(6928):198–207, March 2003.
2. A.H. America, et al. Alignment and statistical difference analysis of complex peptide data sets generated by multidimensional lc-ms. *Proteomics*, 6(2):641–53, 2006.
3. M. Bellew, et al. A suite of algorithms for the comprehensive analysis of complex protein mixtures using high-resolution lc-ms. *Bioinformatics*, 2006.
4. D. Bylund, et al. Chromatographic alignment by warping and dynamic programming as a pre-processing tool for parafac modelling of liquid chromatography-mass spectrometry data. *J. Chromatography*, 961:237–244, 2002.
5. S.J. Callister, et al. Normalization approaches for removing systematic biases associated with mass spectrometry and label-free proteomics. *J. Proteome Res.*, 5:277–86, 2006.
6. X. Hu and L. Xu. Investigation on several model selection criteria for determining the number of cluster. *Neural Inform. Proces. - Lett. and Reviews*, 4(1):1–10, 2004.
7. M. Katajamaa, et al. Mzmine: Toolbox for processing and visualization of mass spectrometry based molecular profile data. *Bioinformatics*, 2006.
8. B.M. Kohli, et al. An alternative sampling algorithm for use in liquid chromatography/tandem mass spectrometry experiments. *Rapid Commun. Mass Spectrometry*, 19(5):589–596, 2005.
9. E. Lange, et al. High accuracy peak-picking of proteomics data using wavelet techniques. In *Proc. Pacific Symposium on Biocomputing (PSB-06)*, 243–254, 2006.
10. J. Listgarten and A. Emili. Statistical and computational methods for comparative proteomic profiling using liquid chromatography-tandem mass spectrometry. *Mol. Cell. Proteomics*, 4:419–434, 2005.
11. J. Listgarten, et al. Difference detection in lc-ms data for protein biomarker discovery. *Bioinformatics*, 2006. in print.
12. J. Listgarten, et al. Multiple alignment of continuous time series. In *Advances in Neural Information Processing Systems 2005, (NIPS 2004)*, 2005.
13. N.-P. Vest Nielsen, et al. Aligning of single and multiple wavelength chromatographic profiles for chemometric data analysis using correlation optimised warping. *J. Chromatography*, 805(1-2):17–35, 1998.
14. D. Radulovic, et al. Informatics platform for global proteomic profiling and biomarker discovery using liquid-chromatography-tandem mass spectrometry. *Mol. Cell. Proteomics*, 3(10):984–97, 2004.
15. C.A. Smith, et al. Xcms: Processing mass spectrometry data for metabolite profiling using nonlinear peak alignment, matching, and identification. *Anal. Chem.*, 78:779 – 787, 2006.
16. R. Tibshirani, et al. Sample classification from protein mass spectrometry, by 'peak probability contrasts'. *Bioinformatics*, 20(17):3034–44, 2004.
17. P. Wang, et al. A statistical method for chromatographic alignment of lc-ms data. *Biostatistics*, doi:10.1093/biostatistics/kxl015, 2006.
18. W. Wang, et al. Quantification of proteins and metabolites by mass spectrometry without isotope labeling or spiked standards. *Anal. Chem.*, 75:4818–26, 2003.
19. Y. Yasui, et al. A data-analytic strategy for protein biomarker discovery: profiling of high-dimensional proteomic data for cancer detection. *Biostatistics*, 4(3):449–63, 2003.
20. X. Zhang, et al. Data pre-processing in liquid chromatography-mass spectrometry-based proteomics. *Bioinformatics*, 21(21):4054–9, 2005.

One-Versus-One and One-Versus-All Multiclass SVM-RFE for Gene Selection in Cancer Classification

Kai-Bo Duan[1], Jagath C. Rajapakse[1,2], and Minh N. Nguyen[1]

[1] BioInformatics Research Centre, School of Computer Engineering
Nanyang Technological University, Singapore
[2] Singapore-MIT Alliance, Singapore
{askbduan, asjagath, nmnguyen}@ntu.edu.sg

Abstract. We propose a feature selection method for multiclass classification. The proposed method selects features in backward elimination and computes feature ranking scores at each step from analysis of weight vectors of multiple two-class linear Support Vector Machine classifiers from one-versus-one or one-versus-all decomposition of a multi-class classification problem. We evaluated the proposed method on three gene expression datasets for multiclass cancer classification. For comparison, one filtering feature selection method was included in the numerical study. The study demonstrates the effectiveness of the proposed method in selecting a compact set of genes to ensure a good classification accuracy.

1 Introduction

Large number of genes together with relatively small number of samples is one characteristic of gene expression data available for cancer classification. This characteristic makes feature selection (FS) a necessary procedure for cancer classification with gene expression data to ensure reliable and meaningful classification results along with other benefits such less data storage and computation cost.

One category of multilclass classification methods are based on the combination of two-class (binary) classifiers. One-versus-one (OVO) or one-versus-all (OVA) binary classifiers are constructed to separate one class from another or to separate one class from all other classes respectively. These binary classifiers are then combined to conduct multiclass classification prediction. OVO and OVA are also two ways to extend popular Support Vector Machine (SVM) [1][2] classification method for multiclass classification. For multiclass classification methods that are based on OVO or OVA binary classifiers, one simple feature selection strategy probably is to select feature variables for each of the binary classifiers separately by using existing feature selection methods for two-class classification. In this way, different binary classifiers actually are constructed in different feature sub-spaces. However, combination of these binary classifiers is infeasible if the combination strategy is based on the distance of a sample to decision

E. Marchiori, J.H. Moore, and J.C. Rajapakse (Eds.): EvoBIO 2007, LNCS 4447, pp. 47–56, 2007.

boundaries of different binary classifiers. Unfortunately, the decision function of SVMs (before taking the sign function) corresponds to such a distance measurement and commonly used *Winner-Takes-All* combination strategy [3] for one-versus-all binary SVMs is primarily based on that distance measurement.

Specially for OVO and OVA multiclass SVMs, we will develop in this paper a feature selection method which is based on OVO or OVA binary SVM classifiers and assume all binary classifiers in combination take the same set of features as input variables. The proposed methods can be viewed as a multiclass generalization of SVM-RFE [4] feature selection method for two-class classification. SVM-RFE selects feature in backward elimination and uses coefficients of the weight vector of a binary linear SVM to compute feature ranking scores. The proposed feature selection method selects features using a backward elimination procedure similar to that of SVM-RFE but computes feature ranking scores at each step from analysis of the weight vectors of all OVO or OVA binary linear SVMs from OVO or OVA decomposition of a multiclass classification problem. The method will be evaluated on three multiclass gene expression datasets for cancer classification and will also be compared with a filtering feature selection method.

The rest of the paper is organized as follows. Section 2 gives a short description of two-class SVMs and, OVO and OVA multiclass SVMs. Section 3 describes the SVM-RFE feature selection for two-class classification; Section 4 presents the proposed multiclass SVM-RFE feature selection method; Section 5 is about the numerical study on three multiclass gene expression datasets for cancer classification. Section 6 contains the conclusions and some discussions.

2   SVMs

2.1   Two-Class SVMs

Support Vector Machines (SVMs) [1] [2] are one of the most popular supervised classification methods due to its superior classification performance in many applications. SVMs are also fairly insensitive to the *curse of dimensionality* and efficient in handling classification problems of large scale in both samples and input variables. SVMs was originally developed for two-class classification. First, input sample vectors $\mathbf{x}$ are mapped from input space to a so-called *feature space* via a mapping function $\Phi(\cdot) : \mathbf{z} = \Phi(\mathbf{x})$, where $\mathbf{z}$ denotes a vector in the feature space. Then, in the feature space, an optimal hyperplane discrimination function $f(\mathbf{z}) = \mathbf{w} \cdot \mathbf{z} + b$ is constructed to separate samples from two classes. The mapping function is implicitly defined by a kernel function which computes the inner-product of vectors in the feature space: $\mathcal{K}(\mathbf{x}_i, \mathbf{x}_j) = \mathbf{z}_i \cdot \mathbf{z}_j = \Phi(\mathbf{x}_i) \cdot \Phi(\mathbf{x}_j)$. Commonly used kernel functions are linear, polynomial and Gaussian [1]. SVMs with a linear kernel function are referred to as linear SVMs.

2.2   Multiclass SVMs

Extension of SVMs from two-class to multiclass basically falls into two categories. The first category mutliclass SVMs consider all samples and all classes all together at once and formulate the multiclass classification problem as a single quadratic optimization problem [5][6][7]. The second category of multiclass SVMs are based on combination of several binary SVMs that are from the decomposition of a multiclass problem into several two-class classification problems [8][9][10][11][12][13]. OVO and OVA are two ways to decompose a multiclass problem and construct the binary classifiers for combination. Based on the way that the binary classifiers in combination are constructed, this category of multiclass SVMs are referred to as *OVO-SVM* or *OVA-SVM*. OVO-SVM and OVA-SVM are the basis of the multiclass feature selection method to be presented.

OVO-SVM. Given a classification problem with M classes, OVO-SVM constructs $M(M-1)/2$ binary SVM classifers, each for every distinct pair of classes. Each of binary classifiers takes samples from one class as positive and samples from another class as negative. Max-Wins voting (MWV) [10] is one of the most commonly used combination strategies for OVO-SVM. MWV assigns an instance to a class which has the largest votes from $M(M-1)/2$ OVO binary classifiers.

OVA-SVM. OVA-SVM constructs M binary classifiers and each binary classifier classify one class (positive) versus all other classes (negative). *Winer-Takes-All* (WTA) [3] is the most common combination strategy for OVA-SVM. WTA strategy assigns a sample to the class whose decision function value is largest among all the M OVA binary classifiers.

All OVO-SVMs in this paper are implemented with MWV combination strategy while all OVA-SVMs in this paper are implemented with WTA combination strategy.

3   SVM-RFE

SVM-RFE [4] as a feature selection method for two-class classification was initially developed for gene selection. It also has been successfully applied in other applications, such as in cancer classification with mass spectrometry data [14]. Features are eliminated one by one in a backward selection procedure that is referred to as Recursive Feature Elimination (RFE) in [4]. At each step, coefficients of the weight vector $\mathbf{w}$ of a linear SVM $f(\mathbf{x}) = \mathbf{w} \cdot \mathbf{x} + b$ are used to compute ranking scores for all the remaining features. The feature, say the i-th feature, with the smallest ranking score $(w_i)^2$ is eliminated, where w_i represents the corresponding i-th component of weight vector $\mathbf{w}$. Using $(w_i)^2$ as ranking criterion removes the feature whose removal changes the objective function least [15] [16]. This objective function is $J = \|\mathbf{w}\|^2/2$ in SVM-RFE.

Once a feature is removed, a new linear two-class SVM is trained with all the features left and all the remaining features are ranked again by using weight vector of the new linear SVM. SVM-RFE repeats this procedure until only one feature is left.

4 Multiclass SVM-RFE

Although most of the existing feature selection methods are tailored towards binary classification, feature selection for multiclass classification has begun to draw more and more attention from researchers [17][18]. The new feature selection method that we present in this section can be viewed as a multiclass generalization of SVM-RFE [4] and is inspired by MSVM-RFE [15] feature selection method for two-class classification. MSVM-RFE uses a backward feature selection procedure similar to that of SVM-RFE but, at each recursive step, computes feature ranking scores using weight vectors of multiple linear SVMs trained on different sub samples of the original training data. Inspired by the good performance of MSVM-RFE, we propose the following gene selection method for multiclass classification with feature ranking scores computed from weight vectors of multiple linear binary SVM classifiers from OVO-SVM or OVA-SVM (linear kernel function is used for binary SVMs in OVO-SVM and OVA-SVM).

For a multiclass problem with M classes, suppose T linear binary SVM classifiers are obtained from a OVO-SVM or OVA-SVM multiclass classifier; $T = M(M-1)/2$ for OVO-SVM and $T = M$ for OVA-SVM. Let $\mathbf{w}_j$ be the weight vector of the j-th linear SVM and w_{ji} be the corresponding weight value associated with the i-th feature; let $v_{ij} = (w_{ji})^2$. We can compute feature ranking scores with the following criterion:

$$c_i = \frac{\bar{v}_i}{\sigma_{v_i}} \tag{1}$$

where $\bar{v}_i$ and σ_{v_i} are mean and standard deviation of variable v_i:

$$\bar{v}_i = \frac{1}{T} \sum_{j=1}^{T} v_{ij} \tag{2}$$

$$\sigma_{v_i} = \sqrt{\frac{\sum_{j=1}^{T} (v_{ij} - \bar{v}_i)^2}{T-1}} \tag{3}$$

Before computing the ranking score for each feature, we normalize weight vectors:

$$\mathbf{w}_i = \frac{\mathbf{w}_i}{\|\mathbf{w}_i\|} \tag{4}$$

Note that, Eqs. (1)–(3) used to compute feature ranking scores in multiclass SVM-RFE are same in form as those in MSVM-RFE. The difference is that, the multiple linear SVM classifiers in multiclass SVM-RFE are derived from OVO-SVM or OVA-SVM while the multiple linear SVM classifiers in MSVM-RFE are from training on different sub samples of the original training data for the same two-class problem.

With feature ranking criterion (1), features are selected using the following backward elimination procedure similar to that of SVM-RFE and MSVM-RFE:

1. Initialize:
   Ranked feature set $R = [\ ]$;
   Selected feature subset $S = [1, \cdots, d]$.
2. Repeat until all features are ranked:
   (a) Get T linear SVMs from the training of an OVO-SVM (or OVA-SVM), with features in S as input variables;
   (b) Compute and normalize the weight vectors;
   (c) Compute the ranking scores c_i for features in S using (1);
   (d) Find the feature with the smallest ranking score: $e = \arg\min_i c_i$;
   (e) Update: $R = [e, R]$, $S = S - [e]$;
3. Output: Ranked feature list R

We refer to the above proposed multiclass SVM-RFE with binary classifiers from OVO-SVM and OVA-SVM as *OVO-SVM-RFE* and *OVA-SVM-RFE*, respectively. As in SVM-RFE and MSVM-RFE, all features are ranked from the RFE procedure. Earlier one feature is eliminated, lower is it ranked. Thus, nested feature subsets are obtained similarly as in filtering feature selection methods which however usually rank all features at a single step.

However, there is a risk in using feature ranking criterion (1). In a situation that the standard deviation σ_{v_i} is extremely small for some feature, Eq.(1) will produce an extremely large ranking score even if the mean $\bar{v}_i$ is small. Thus we propose the following alternative ranking criterion by adding a small non-negative constant ϵ to the denominator of ranking criterion (1) to deal with such a situation:

$$c_i = \frac{\bar{v}_i}{\epsilon + \sigma_{v_i}} \tag{5}$$

We set $\epsilon = 1.0$ in this paper. We respectively refer to ranking criteria (1) and (5) as *RC1* and *RC2*.

5   Numerical Study

Effectiveness of the proposed feature selection methods is evaluated on three multiclass gene expression datasets: ALL [19], MLL [20] and SRBCT [21]. ALL dataset is about classification of 6 subtypes of pediatric acute lymphoblastic leukemia. Samples of MLL dataset are from 3 classes of leukemia: lymphoblastic leukemia with MLL translocation, conventional acute lymphoblastic leukemia and acute myelogenous leukemia. SRBCT dataset is about small, round, blue cell tumors of childhood. For all the datasets, expression values of each gene are normalized to zero mean and unit standard deviation, over all samples. During the normalization process, in ALL dataset, we found 2064 genes whose expression values show no variation across all the samples. These genes are removed in our numerical study. Some basic information of the three datasets are summarized in Table 1.

Table 1. Basic information of the datasets

Dataset	# Classes	# Training Samples	# Testing Samples	# Genes
ALL	6	153	85	10524
MLL	3	57	15	12582
SRBCT	4	60	23	2308

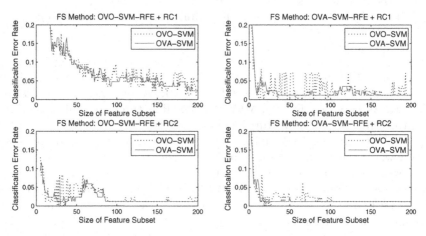

Fig. 1. Classification performance of the feature subsets selected by the proposed methods, on ALL dataset

We also include a filtering feature selection method in our numerical study for comparison. The filtering method ranks genes on the basis of the ratio of their between-classes to within-classes sum of squares [18]. For a gene i, this ratio is:

$$r_i = \frac{BSS(i)}{WSS(i)} = \frac{\sum_{j=1}^{\ell} \sum_{k=1}^{M} I(y_j = k)(\bar{x}_{ki} - \bar{x}_{.i})^2}{\sum_{j=1}^{\ell} \sum_{k=1}^{M} I(y_j = k)(x_{ji} - \bar{x}_{ki})^2} \tag{6}$$

where $\bar{x}_{.i}$ denotes the average expression level of gene i across all samples and $\bar{x}_{ki}$ denotes the average expression level of gene i across samples belongs to the k-th class; $I(.)$ is the indicator function.

We ran all the feature selection methods on the training data only. Classification performance of the nested feature subsets with size varying from 1 to 200 were validated on the test data. For OVO-SVM-RFE, the natural choice of the classification method is OVO-SVM; for OVA-SVM-RFE, OVA-SVM is the natural choice. Even though, we also validated the subsets selected by OVO-SVM-RFE and OVA-SVM-RFE respectively with OVA-SVM and OVO-SVM as the classification methods. For all OVO-SVM and OVA-SVM in both the feature selection phase and in the testing phase, linear kernel function was used for the binary SVMs; OVO-SVM was implemented with MWV combination strategy while OVA-SVM was implemented with WTA combination strategy.

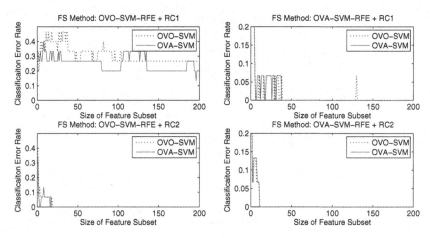

Fig. 2. Classification performance of the feature subsets selected by the proposed methods, on MLL dataset

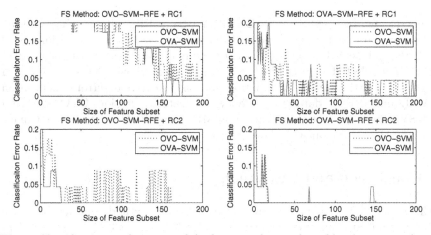

Fig. 3. Classification performance of the feature subsets selected by the proposed methods, on SRBCT dataset

Test error rates versus size of the feature subsets are plotted in Figs. 1–4 for all datasets and for all the feature selection methods, with both OVO-SVM and OVA-SVM respectively as the classification methods. From Figs. 1-3, we can see that, first, OVO-SVM-RFE and OVA-SVM-RFE work better with feature ranking criterion RC2 than with RC1; and second, OVA-SVM-RFE performs better than OVO-SVM-RFE, with either RC1 or RC2. From Fig. 4, it is quite clear that the performance of the filtering method is dataset-dependent. From these figures, we also observe that the classification performance of OVA-SVM overall is better than OVO-SVM, even on feature sets selected by OVO-SVM-RFE.

For a feature selection method, we take the most compact feature subset with smallest number of test errors on as the best feature subset. Best feature

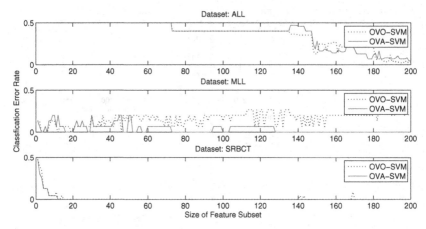

Fig. 4. Classification performance of the feature subsets selected by the filtering method on the three datasets, ALL, MLL and SRBCT

subsets selected by OVO-SVM-RFE and OVA-SVM-RFE with ranking criterion RC2 achieve almost perfect classification results (0 or 1 misclassification error) on all the datasets, but with only a few number of genes (36 at most) Best feature sets selected by the filtering method make 2 (with OVO-SVM) or 4 (with OVA-SVM) errors on ALL dataset although achieve no-error classification on MLL and SRBCT datasets. Without any feature selection, OVO-SVM and OVA-SVM make perfect classification on MLL and SRBCT datasets, but make 2 classification errors on ALL dataset.

6   Conclusion and Discussion

In this paper we have explored new feature selection methods that are based on one-versus-one and one-versus-all multiclass SVMs. The methods select features in a backward elimination procedure similar to that of SVM-RFE and can be viewed as a multiclass generalization of SVM-RFE. Feature ranking scores are computed from weight vectors of multiple binary linear SVMs of OVO-SVM or OVA-SVM, and two feature ranking criteria were proposed and tested. The proposed multiclass feature selection methods were evaluated on three multiclass gene expression data for cancer classification, together with one filtering feature selection method.

The proposed methods, OVO-SVM-RFE and OVA-SVM-RFE, consistently perform better with feature ranking criteria RC2 than with criterion RC1. With only a smalll number of genes (36 at most) selected OVO-SVM-RFE and OVA-SVM-RFE (with RC2 as the feature ranking criteria), the classification performance of OVO-SVM and OVA-SVM are almost perfect on all the datasets, and are better than or equal to the performance of OVO-SVM and OVA-SVM with all the genes as inputs. MSVM-RFE [15] that inspired the work of this paper uses a ranking criterion similar to RC1 in form. Better performance of OVO-SVM-RFE and OVA-SVM-RFE with feature ranking criterion RC2 than with criterion RC1 implies that, even better performance of MSVM-RFE can be expected when ranking criterion similar to RC2 is adopted.

For ranking criterion RC2, if we set $\epsilon = 0$, RC2 will degenerate to RC1. The performance difference of the proposed method with RC1 and RC2 indicates that the non-negative constant ϵ in (5) does affect the gene ranking and gene selection results of the proposed methods. Although RC2 with $\epsilon = 1$ in this paper achieves satisfactory results, better results can be expected if ϵ is optimized. Having said that, optimizing ϵ at each iterative step of the RFE procedure is not an easy task and efficient strategies need to be derived.

We also observed that, OVA-SVM-RFE overall performs better than OVO-SVM-RFE, on all the datasets, with either RC1 or RC2 as the feature ranking criterion. OVA-SVM also performs better than OVO-SVM, even on feature sets selected by OVO-SVM-RFE. Generally, we do not expect OVA-SVM outperforms OVO-SVM, as many researchers have already observed [22][23]. However, for gene expression datasets, the training samples are usually sparse as in our study. OVA-SVM uses all the samples in one-versus-all manner to construct the binary classifiers. We believe this probably explains the better performance of OVA-SVM over OVO-SVM, which further explains the better performance of OVA-SVM-RFE over OVO-SVM-RFE. For cancer classification with gene expression data, especially when the number of available samples is small, OVA-SVM-RFE and OVA-SVM are favorably recommended as the methods of choice respectively for gene selection and classification.

Better performance of OVA-SVM over OVO-SVM even on feature sets selected by OVO-SVM-RFE implies that the classification performance of the features selected by OVO-SVM-RFE or OVA-SVM-RFE probably are not correlated so much to OVO-SVM or OVA-SVM as classification method. The features selected by OVO-SVM-RFE and OVA-SVM-RFE may also work well with classification methods other than OVO-SVM or OVA-SVM and we need to investigate this further.

Acknowledgement

The work is partly supported by a grant to J. C. Rajapakse, by the Biomedical Research Council (grant no. 04/1/22/19/376), of Agency of Science and Technology Research, administered through the National Grid Office, Singapore.

References

1. Vapnik, V.: Statistical Learning Theory. Wiley Interscience (1998)
2. Boser, B.E., Guyon, I., Vapnik, V.: A training algorithm for optimal margin classifiers. In: Computational Learing Theory. (1992) 144–152
3. Friedman, J.H.: Another approach to polychotomous classification. Technical report, Department of Statistics, Stanford University (1996)

4. Guyon, I., Weston, J., Barnhill, S., Vapnik, V.: Gene selection for cancer classification using support vector machines. Machine Learning **46**(1-3) (2002) 389–422
5. Hsu, C.W., Lin, C.J.: A comparison of methods for multi-class support vector machines. IEEE Transactions on Neural Networks **13** (2002) 415– 425
6. Crammer, K., Singer, Y.: On the learnability and design of output codes for multiclass problems. In: Computational Learing Theory. (2000) 35–46
7. Weston, J., Watkins, C.: Support vector machines for multiclass pattern recognition. In: Proceedings of the Seventh European Symposium On Artificial Neural Networks. (1999)
8. Duan, K.B., Keerthi, S.S.: Which is the best multiclass SVM method? An empirical study. In: Multiple Classifier Systems. (2005) 278–285
9. Duan, K., Keerthi, S.S., Chu, W., Shevade, S.K., Poo, A.N.: Multi-category classification by soft-max combination of binary classifiers. In: Multiple Classifier Systems. (2003) 125–134
10. Kreel, U.H.G.: Pairwise classification and support vector machines. In Scholkopf, B., Burges, C.J.C., Smola, A.J., eds.: Advances in Kernel Methods: Support Vector Learning. The MIT Press (2002) 255–268
11. Platt, J., Cristianini, N., Shawe-Taylor, J.: Large margin DAGs for multiclass classification. (2000)
12. Hastie, T., Tibshirani, R.: Classification by pairwise coupling. In Jordan, M.I., Kearns, M.J., Solla, S.A., eds.: Advances in Neural Information Processing Systems. Volume 10., The MIT Press (1998)
13. Dietterich, T.G., Bakiri, G.: Solving multiclass learning problems via error-correcting output codes. J. of Artificial Intelligence Research **2** (1995) 263–286
14. Rajapakse, J.C., Duan, K.B., Yeo, W.K.: Proteomic cancer classification with mass spectrometry data. American Journal of PharmacoGenomics **5**(5) (2005) 281–292
15. Duan, K.B., Rajapakse, J.C.: Multiple SVM-RFE for gene selection in cancer classification with expression data. IEEE Trans Nanobioscience **4**(3) (2005) 228–234
16. Rakotomamonjy, A.: Variable selection using SVM-based criteria. Journal of Machine Learning Research, Special Issue on Variable Selection **3** (2003) 1357–1370
17. Li, G., Yang, J., Liu, G., Xue, L.: Feature selection for multi-class problems using support vector machines. In: Proceedings of 8th Pacific Rim International Conference on Artificial Intelligence(PRICAI-04). (2004) 292–300
18. Dudoit, S., Fridlyand, J., Speed, T.P.: Comparison of discrimination methods for the classification of tumors using gene expression data. Journal of the American Statistical Association **97**(457) (2002) 77–87
19. Yeoh, E.J., Ross, M.E., Shurtleff, S.A., and et al: Classification, subtype discovery, and prediction of outcome in pediatric acute lymphoblastic leukemia by gene expression profiling. Cancer Cell **1** (2002) 133–143
20. Armstrong, S.A., Staunton, J.E., Silverman, L.B., and et al: MLL translocations specify a distinct gene expression profile that distinguishes a unique leukemia. Nature Genetics **30** (2002) 41–47
21. Khan, J., Wei, J.S., Ringn, M., and et al: Classification and diagnostic prediction of cancers using gene expression profiling and artificial neural networks. Nature Medicine **7**(6) (2001) 673–679
22. Rifkin, R., Klautau, A.: In defense of one-vs-all classification. Journal of Maching Learning Research **5** (2004) 101–141
23. Scholkopf, B., Smola, A.J.: Learning with Kernels. MIT Press (2002)

Understanding Signal Sequences with Machine Learning

Jean-Luc Falcone[1,*], Renée Kreuter, Dominique Belin[2], and Bastien Chopard[1]

[1] Département d'informatique, Université de Genève, 1211 Genève 4, Switzerland
[2] Département de Pathologie et d'Immunologie, Université de Genève, Switzerland

Abstract. Protein translocation, the transport of newly synthesized proteins out of the cell, is a fundamental mechanism of life. We are interested in understanding how cells recognize the proteins that are to be exported and how the necessary information is encoded in the so called "Signal Sequences". In this paper, we address these problems by building a physico-chemical model of signal sequence recognition, using experimental data. This model was built using *decision trees*. In a first phase the classifier were built from a set of features derived from the current knowledge about signal sequences. It was then expanded by feature generation with *genetic algorithms*. The resulting predictors are efficient, achieving an accuracy of more than 99% with our wild-type proteins set. Furthermore the generated features can give us a biological insight about the export mechanism. Our tool is freely available through a web interface.

1 Introduction

1.1 Signal Sequences

Proteins synthesized in the cell must be transported to the correct cellular compartment so that they can achieve their role. This process is called protein targeting and is a fundamental aspect of cell protein metabolism [1]. For instance blood plasma proteins and polypeptidic hormones must be delivered to the extracellular space. We are interested in the secretion pathway, which involves the targeting and transport of the proteins out of the cell. The protein complex (called *translocon*) which exports the proteins varies from one species to another.

All the proteins that must be exported, carry a particular region of conserved function, the *signal sequence* (SS) or *signal peptide*, located in N-terminal extremity. The length varies slightly from 10 to 50 amino-acids (AA). The protein is exported before folding and the SS is usually cleaved after the export. The precise location where the cleavage occurs is called the cleavage site.

The most interesting feature of SS is their inter- and intra-species variability. Their sequence as well as their length vary. Thus, they do not carry any systematic consensus. However, three properties have been proposed as distinguishing

---

* Supported by the Swiss National Science Foundation.

E. Marchiori, J.H. Moore, and J.C. Rajapakse (Eds.): EvoBIO 2007, LNCS 4447, pp. 57–67, 2007.

features of SS [2]: (i) They begin with an N-terminal region which includes one or several positively charged lysine or arginine residues. This region is called the *N-Region*. (ii) Following the N-Region, SS contain a stretch of hydrophobic AA forming the so-called *H-Region*. (iii) In the majority of secreted proteins there is a third region, the *C-Region*, located between the H-Region and the cleavage site. It carries a weak consensus recognized by the leader-peptidase.

The above properties are too vague to easily determine whether or not a protein will be secreted. The hypothesis that these three regions, as defined above, characterize an exported protein is based on observations made on known SS. However there are no experiments that confirm that these three properties are sufficient and/or relevant for the recognition process. To address the problem of correctly discriminating secreted proteins from the other ones (cytosolic), artificial intelligence techniques have been considered.

1.2   Computer Predictions of Signal Sequences

Many methods for the recognition of SS have been proposed. For all these methods, the predictors are built by an algorithm based on supervised learning techniques. The SignalP method is currently considered as the best classification method [3] and is the most widely used. It consists of two feed-forward neural networks [4]. The first is trained to recognize the SS itself; the second recognizes the cleavage site.

SignalP cannot help us understand the physico-chemical properties recognized by the translocon. The problem is intrinsic to the nature of classical neural networks which does not allow the retrieval of high-level symbolic rules. Another weakness resides in the fact that SignalP uses the existence of a valid cleavage site as a strong classification criterion. Although it is true that most SS include such a site, there are proteins like ovalbumin which are exported but lack a cleavage site. Furthermore mutated SS are poorly recognized by existing predictors. Therefore there is a need to develop new approaches with better prediction scores on such proteins and which gives better insight into the mechanisms at work.

1.3   Decision Trees

In this paper we propose a novel approach based on *decision tree* classifiers to understand how SS are recognized. Decision trees are classification programs that can classify objects according to properties (or features). Different algorithms exist to build such trees from a list of properties and a set of example objects representative of the different classes. As it is the case with neural networks, the learning strategy is supervised, i.e. based on a training set of sequences. The output of this algorithm is a tree in which the non-terminal nodes are evaluations based on the properties characterizing the objects being classified. The leaves of the tree (the terminal nodes) are the possible classes.

An important advantage of decision tree building algorithms is that only the properties necessary for the classification are retained. The most discriminant properties appear at the root of the tree and the less discriminant ones are near the leaves. This allows us to identify the classification mechanisms explicitly, from the most relevant to the least important one.

1.4   Machine Learning to Investigate Signal Sequences

Our goal was to apply the decision tree method to the problem of SS recognition. We started our investigation using only the N- and H-regions, because the C-region is recognized by the cleavage enzyme after export but probably not by the translocon itself.

Note that since these regions were only described qualitatively, it is necessary to specify a procedure to measure them from a given protein sequence. We propose in Sec. 2.2 a way to compute them.

The purpose of building a decision tree is not only the construction of a new and better SS predictor, but also to show if one or more of the properties mentioned above are relevant and sufficient. If we can reach good efficiency using only these properties we can conclude that they are indeed sufficient. If not, an extended set of features needs to be considered.

Hence, to generate these extra features (or to refine the parameters of existing ones) we have considered feature generation with genetic algorithms. The evolution will optimize individuals whose genes are potential new features. Those features are mathematical functions modeling the various physico-chemical interactions between translocon and the N-terminus of protein. For this purpose, they compute a score for each protein according to the physico-chemical properties of the AA. To evaluate the fitness of the individuals, we add its gene (the new features) to the existing features described above. The fitness is then the performance of the resulting decision tree. This method can lead us not only to improve our efficiency but to suggest new biological criteria and new wet-lab experiments to confirm them.

In a first phase we focus our research on *E. coli*. This choice is motivated by the large number of experimentally known secreted proteins and the existence of a collection of mutant SS.

2   Decision Trees for Signal Sequences

2.1   Datasets

We used four datasets: (i) 104 wild-type *E. Coli* proven signal sequence proteins dataset ; (ii) 160 wild-type *E. Coli* cytoplasmic proteins; (iii) a collection of 17 signal-defective mutants of the *phoA*, *malE* and *rbsB* genes; (iv) a collection of 145 maspin "gain-of-function" mutants and derived constructions.

The datasets (i) and (ii) were obtained from the UniRef100 database as follows: the set (i) was composed only the proteins with proven signal sequence annotations after removal putative TAT proteins with the TatP predictor [5]; the set (ii) was composed of proteins containing the line: CC -!- SUBCELLULAR LOCATION: Cytoplasm. The sequences of all datasets were truncated to the first

70 AA. The final dataset was made by pooling together the four datasets for a total of 427 instances (130 SS and 297 cytoplasmic proteins).

We lack space to describe properly the mutants collection, but all the datasets, their descriptions and references are available online on [6].

2.2   Computation of the Signal Sequence Features

In order to build our decision tree and to assign specific values to the SS features (properties), it is necessary to find a way to define the N- and H-regions and to attribute them a score.

H-Region Features: The H-region score quantify the presence of a stretch of hydrophobic AA. The difficulties in scoring the H-region are (i) to identify the relevant stretch, (ii) to choose the most appropriate hydrophobicity scale and (iii) to find the proper way to combine the hydrophobicity of each residue along the stretch.

We compute four different H-region features, mixing two different hydrophobicity scales with two different calculations. The two scales are (i) AA hydropathy [7]; (ii) percentage of buried residues [8]. The first is a composite index obtained from different measurements and modified to match biological insight. This scale is the most widely used in bioinformatics experiments. The second derives from statistical structural data and only takes the distribution of a residue at the surface or in the core of a set of globular proteins into account.

These two scales can be used in two different ways to define the H-region feature score: we scan the sequence searching for the most hydrophobic stretch of AA, or the longest one. A stretch is defined as a contiguous substring in which all AA have an hydrophobic score larger than a given threshold (with respect to the chosen scale). The threshold used is given by the average hydrophobicity of the 20 AA minus the standard deviation. Among all the stretches, we select the best one. In the first calculation method we compute the sum of the hydrophobicity of all its AA for each stretch. The H-region is then defined as the stretch with highest sum. The final score is this largest sum. In the second calculation method, the H-region is defined as the longest stretch. The final score is its length.

We chose the following terminology for these four features: H_{LK} is the score based on the Kyte-Doolittle scale and returning the longest stretch. H_{VK} is based on the same scale but returns the value of the most hydrophobic stretch. We use the same notation scheme for the score derived from the percentage of buried residues: H_{LB} and H_{VB}.

N-Region Features: In order to determine the N-region, we used the fact that the N-region terminates where the H-region starts. Thus, the H-region had to be computed first, as explained above. We observe that the four possible ways to extract the H-Region give the same N-region. This is due to the fact that the two hydrophobicity scales discussed above select the same AA as hydrophobic, although they return different final values.

Once the N-region has been identified, we compute its score. We again propose two different features. The first one is the average charge $N_Q(p)$ of protein p and the second one is the lysine-arginine density $N_R(p)$:

$$N_Q(p) = \frac{1}{n(p)} \sum_{i=2}^{n(p)+1} q(p_i), \qquad N_R(p) = \frac{1}{n(p)} \sum_{i=2}^{n(p)+1} r(p_i)$$

where $n(p)$ is the length of the N-Region, p_i the ith AA of the protein and $q(p_i)$ is the charge of AA p_i. $r(p_i)$ is equal to 1 if p_i is a lysine or arginine, 0 otherwise. The difference between these two features is that charged aspartate and glutamate residues are neglected in N_R.

2.3   Tree Building Algorithm and Test Procedure

We want to be able to classify a protein according to its location inside or outside the cell, from the six features described above. Here we consider the *C4.5* tree building algorithm [9] because of the good open-source Weka 3 [10] implementation. Several parameters of the *C4.5* algorithm were tested, but the best results were obtained by setting the *pruning confidence* parameter to 0.15 and the *minimum number of instances* parameter to 2.

We were mainly interested in the accuracy of the classifier: $A = \frac{a+d}{N}$ where A is the accuracy, a the number of true positives, d the number of true negatives and N the total number of instances.

To detect over training we used stratified cross-validation tests. This technique works as follows: the dataset (both signal and cytosolic sequences) is split in 10 subsets $S_1,...,S_{10}$ using random sampling preserving the ratio between positive and negative instances. At each step $1 \leq i \leq 10$, a tree T_i is built with all S_j such that $j \neq i$. The accuracy A_i of step i is computed by applying tree T_i to S_i. The final score is the average of A_i. Note that the final tree is the one obtained with the full dataset. Thus, the stratified cross-validation method tests the robustness of the tree building algorithm with respect to the dataset.

2.4   Results

The resulting tree, called SigTree, was built according to all properties discussed above, and is presented in fig. 1. After a first score-node corresponding to the N_Q score, the tree splits in two main branches. The first one (low N_Q score) corresponds to 91.2% of all 297 cytosolic sequences, whereas the second branch, corresponds to 77.7% of all 130 SS.

SigTree achieves an accuracy of 98.2% on the whole dataset, and a cross-validation accuracy of 89.9% which is significant. We can now test separately the wild type proteins (set i and ii) and the two mutants collections on the resulting tree. The results are summarized on tab. 1.

As we said above, the detection of a cleavage site may not be a good indicator of protein secretion. To verify this assumption, we built another tree by adding a cleavage site feature as described in [11]. The results were almost identical to

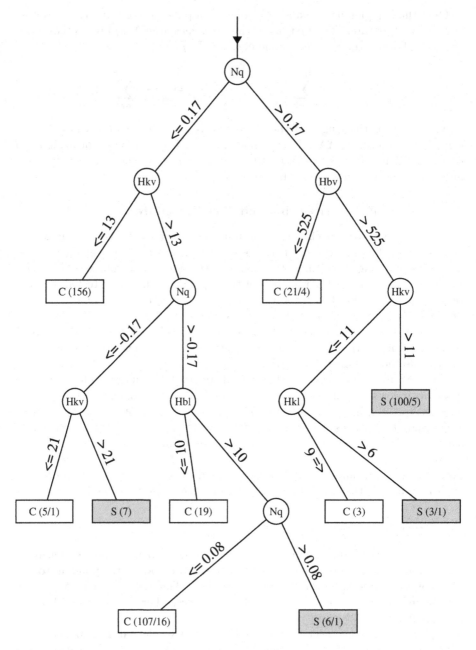

Fig. 1. SigTree. The circle nodes represent the features defined above. The square leaves are the predicted classes. The first number between parenthesis represents the total number of instances classified in the leaves; the second number is the number of dataset sequences wrongly classified (it is omitted if all the sequences are well classified).

Table 1. SigTree and SigTreeGA accuracy on the different datasets (computed on the entire sets)

Dataset	SigTree	SigTreeGA
i+ii	97.4%	99.2%
iii	64.7%	70.6%
iv	89.7%	97.2%

the tree presented above. The cross-validation was only slightly better: 90.4% instead of 89.9%. Therefore we decided to omit this feature in our predictor.

3   Features Generation with Genetic Algorithms

We now want to extend our set of six initial features using Genetic Algorithms (GA). An individual in the GA population will code for a set of new features. These features represent measures of interactions of the N-terminus of proteins with the translocon. Since we do not know how many new features are necessary to improve the classifiers, we used variable size individuals. Those indivuals ranged between one and seven genes and were initialized with four random genes.

As those interactions are depending on the AA physico-chemical properties we consider for each AA the values of its physico-chemical properties. However the large number of AA properties compiled in the database *AAIndex* [12] (more than 500) is impracticable. Thus we have reduced the existing set with the method described in [13] to a set of 45 clusters. This set is available on-line on [6]. A feature is now defined by two components: (i) one of the 45 physico-chemical properties P; (ii) a reduction function which combines the value of P for the AA of the given protein. Each reduction function represents a possible type of interaction between a newly synthesized protein and the translocon.

3.1   Genetic Algorithm Components

Genes: Each gene of the individuals is a possible feature. When a feature is applied on a protein, we first generate a numeric array by replacing each AA with its corresponding value of the coded property. The the reduction function is then applied on this array and returns one number: the feature score for the protein.

We have designed four types of reduction functions defined by an algorithm and structural parameters. Those parameters are optimized by the GA evolution. The reduction functions we chose and their associated parameters are:

Sliding Windows: A sliding window reduction returns the minimum or the maximum sum of all substring of l AA along the protein.

Parameters: the window size l, a Boolean indicating whether the max or min of the sums is chosen.

Stretch operator: This reduction works as the *H-region* feature of Sect. 2.2 but now with an adjustable threshold value.

Parameters: a number representing the threshold, a Boolean indicating whether the highest stretch or the lowest one is returned.

Helix Window: This function assumes that the SS has an helical structure (not necessarily an α-helix) of period λ AA. For all λ possible offsets, it computes the sum of the AA along the SS, with stride λ.

Parameters: an integer λ representing the helix period in AA, a Boolean indicating whether the max or min of all the sums is chosen.

Sliding Helix Window: This reduction function is a combination of the *sliding window* and the *helix window*. It proceeds like the latter but sums up only l AA at time. For example, for $\lambda = 3$ and $l = 4$, we will first add AA with indices $0, 3, 6, 9$, then AA with indices $1, 4, 7, 10$ and so on.

Parameters: an integer λ representing the helix period, the window size l, a Boolean indicating whether the max or the min is taken.

Individuals and Fitness: The fitness is computed by first including the features of the individual to the previous set of six features described above and then by training a decision tree with this extended feature set. The cross-validation of the resulting classifier is computed on the whole dataset and returned. Note that if the same feature appears more than once, the tree building algorithm will use only one.

GA Operators: We chose the classical one-point crossover and point mutation operators but adapted for variable size individuals. For the crossover we simply draw a different crossing point for each mate. For the mutation, we choose randomly for each individual either to add a single gene or to delete an existing one. We used the n-genes, evolutionnary computing environment described in [14] and available on [15].

3.2 Results

New Features. Different runs of our system converge to the same 3 new features. The first two are interpretable: (i) F_1: a Sliding Helix Window which minimizes an hydrophobicity property with $l = 4$ and $\lambda = 4$; (ii) F_2: an Helical window of period 3 which maximizes a turn/coil-propensity. We point out that the periods found are good approximation of the α-helix period (3.6 AA). These two new features are consistent with previous claims about the secondary-structure of the signal-sequence [16].

The third feature, F_3, is a stretch operator which minimizes a property which results from a cluster (see Sect. 3) whose meaning is difficult to understand.

Resulting Tree. The resulting tree (SigTreeGA) is similar to SigTree in its structure and the distribution of the instances. The performance is significantly improved with a cross validation of 93.9%. The value on the specific datasets

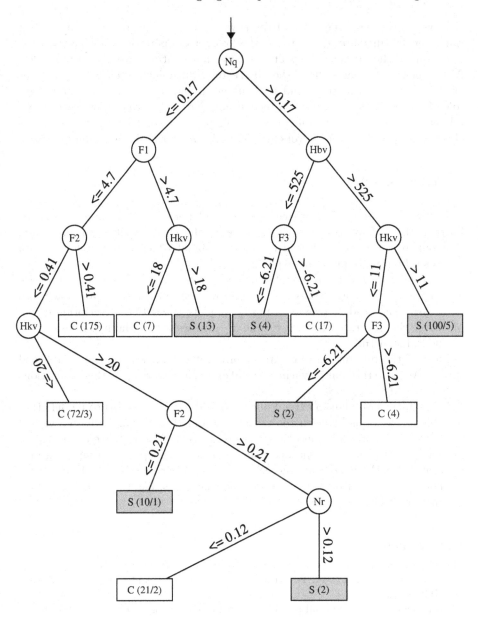

Fig. 2. SigTreeGA. The circle nodes represent the features defined above. The square leaves are the predicted classes. The first number between parenthesis represents the total number of instances classified in the leaves; the second number is the number of dataset sequences wrongly classified (it is omitted if all the sequences are well classified).

are better too. On the entire wild-type protein dataset, we reach an impressive accuracy score of 99.2% and the recognition of mutants is improved. The results are summarized in tab. 1. This predictor is available on [6].

If we analyze now the place of the new features in the tree, we can see that these new features are placed below the first set of operators, allowing to refine the result of the first classifier. The first two new features are placed in the low N_Q branch where the bulk of the non-secreted proteins are classified (271 out of 297). The extra features and their usage in the tree indicate an important role of secondary structure in SS recognition. Not only two of new features are related to helical motives, but also the role of F_2 in the tree can be viewed as eliminating proteins whose propensity of coil/turn formation is to high.

4   Discussion

The first tree produced (SigTree) shows good performances on our datasets. Further, the fact we can omit detection of a valid cleavage site is an indication that our approach is not fooled by another signal which is only relevant to a process taking place after the export.

However, the addition of new features generated by GA allow us to produce a better classifier (SigTreeGA). This improvement justifies our approach and emphasizes the need to provide a more complete theoretical description of SS. Moreover the generated feature F_1 and F_2 give weight to the hypothesis of SS (α-)helical structure.

Most of the mutants in our datasets are not explainable by the established theory. We hope that our new features will eventually reveal how these mutations have changed the export.

In the future we plan to extend our set of reduction functions to model other possible interactions between the translocon and the SS. Furthermore, we will train decision trees on taxonomic groups other than *E. coli*. Since our trees are in fact biological models, the decisions trees of other species/groups can give us insight into the biological differences between their translocons. Finally, this method can be applied to other proteomic challenges, like the prediction of N-terminal acetylation.

References

1. Stryer, L.: Biochemistry. Fourth edn. W. H. Freeman and Company, New-York (1995)
2. von Heijne, G.: The signal peptide. J. Membrane Biology **115** (1990) 195–201
3. Menne, K.M., Hermjakob, H., Apweiler, R.: A comparison of signal sequence prediction methods using a test set of signal peptides. Bioinformatics **16**(8) (2000) 741–2
4. Nielsen, H., Engelbrecht, J., Brunak, S., von Heijne, G.: Identification of prokaryotic and eukaryotic signal peptides and prediction of their cleavage sites. Protein Engineering **10**(1) (1997) 1–6
5. Bendtsen, J.D., Nielsen, H., Widdick, D., Palmer, T., Brunak, S.: Prediction of twin-arginine signal peptides. BMC bioinformatics **6**(167) (2005)
6. Falcone, J.L.: SigTree website. http://cui.unige.ch/spc/tools/sigtree/ (2007)

7. Kyte, J., Doolittle, R.F.: A simple method for displaying the hydropathic character of a protein. J Mol Biol **157**(1) (1982) 105–32

8. Janin, J., Chothia, C.: Role of hydrophobicity in the binding of coenzymes. appendix. translational and rotational contribution to the free energy of dissociation. Biochemistry **17**(15) (1978) 2943–8

9. Quinlan, J.R.: C4.5: Programs for Machine Learning. Morgan Kaufmann San Mateo (1993)

10. Witten, I.H., Frank, E.: Data Mining: Practical machine learning tools with Java implementations. Morgan Kaufmann, San Francisco (2000)

11. von Heijne, G.: A new method for predicting signal sequence clevage site. Nucleic Acid Res. **14** (1986) 4683–4690

12. Kawashima, S., Ogata, H., Kanehisa, M.: Aaindex: amino acid index database. Nucleic Acids Res. (27) (1999) 368–369

13. Falcone, J.L., Albuquerque, P.: Agrégation des propriétés physico-chimiques des acides aminés. IEEE Proc. of CCECE'04 **4** (2004) 1881–1884

14. Falcone, J.L.: Decoding the Signal Sequence. PhD thesis, University of Geneva, Switzerland (2007) (to be published).

15. Fontignie, J., Falcone, J.L.: n-genes website. `http://cui.unige.ch/spc/tools/n-genes/` (2005)

16. Izard, J., Kendall, D.: Signal peptides: exquisitely designed transport promoters. Mol Microbiol. **13**(5) (1994) 765–773

Targeting Differentially Co-regulated Genes by Multiobjective and Multimodal Optimization

Oscar Harari[1], Cristina Rubio-Escudero[1], and Igor Zwir[1,2]

[1] Dept. Computer Science and Artificial Intelligence, University of Granada,
E-18071, Spain
[2] Howard Hughes Medical Institute, Department of Molecular Microbiology,
Washington University School of Medicine, St. Louis, MO 63110-1093, USA
oharari@decsai.ugr.es, crubio@decsai.ugr.es,
zwir@borcim.wustl.edu

Abstract. A critical challenge of the postgenomic era is to understand how genes are differentially regulated in and between genetic networks. The fact that such co-regulated genes may be differentially regulated suggests that subtle differences in the shared *cis*-acting regulatory elements are likely significant, however it is unknown which of these features increase or reduce expression of genes. In principle, this expression can be measured by microarray experiments, though they incorporate systematic errors, and moreover produce a limited classification (e.g. up/down regulated genes). In this work, we present an unsupervised machine learning method to tackle the complexities governing gene expression, which considers gene expression data as one feature among many. It analyzes features concurrently, recognizes dynamic relations and generates profiles, which are groups of promoters sharing common features. The method makes use of multiobjective techniques to evaluate the performance of profiles, and has a multimodal approach to produce alternative descriptions of same expression target. We apply this method to probe the regulatory networks governed by the PhoP/PhoQ two-component system in the enteric bacteria *Escherichia coli* and *Salmonella enterica*. Our analysis uncovered profiles that were experimentally validated, suggesting correlations between promoter regulatory features and gene expression kinetics measured by green fluorescent protein (GFP) assays.

1 Introduction

Genetic and genomic approaches have been successfully used to assign genes to distinct regulatory networks. However, little is known about the differential expression of genes within a regulon. At its simplest, genes within a regulon are controlled by a common transcriptional regulator in response to the same inducing signal. Moreover it is suggested that subtle differences in the shared *cis*-acting regulatory elements are probably significant in the genes expression. However, it is not known which of these features, independently or collectively, can set expression patterns apart. Indeed, similar expression patterns can be generated from different or a mixture of multiple underlying features, thus, making it more difficult to discern the causes of analogous regulatory effects.

E. Marchiori, J.H. Moore, and J.C. Rajapakse (Eds.): EvoBIO 2007, LNCS 4447, pp. 68 – 77, 2007.

The material required for analyzing the promoter features governing bacterial gene expression is widely available. It consists of genome sequences, transcription data, and biological databases containing examples of preciously explored cases. In principle, genes could be differentiated by incorporating into the analysis quantitative and kinetic measurements of gene expression [1] and/or considering the participation of other transcription factors [2-4]. However, there are constraints in such analyses due to systematic errors in microarray experiments, the extra work required to obtain kinetic data and the missing information about additional signals impacting on gene expression. These constraints hitherto allow a relatively crude classification of gene expression patterns into a limited number of classes (e.g., up- and down-regulated genes [5, 6]), thus concealing distinctions among expression features, such as those that characterize the temporal order of genes or their levels of intensity

Here we describe an unsupervised machine learning method that discriminates among co-regulated promoters by simultaneously considering both cis-acting regulatory features and gene expression. By virtue of being an unsupervised method, it is neither constrained by a dependent variable [2, 7], such as expression data, which would restrict the classification to the dual expression classes reported by microarray experiments; nor it requires pre-existing kinetic data. Our method treats each of the promoter features with equal weight, because it is not known beforehand which features are important. Thus, it explores all of the possible aggregations of features; and applies multiobjective and multimodal techniques [8, 9] to identify alternative optimal solutions that describe target sets of genes from different perspectives.

We applied our methodology to the investigation of genes regulated by the PhoP protein of *Escherichia coli* and *Salmonella enterica* serovar Typhimurium. We recovered several profiles that were experimentally validated [10] to establish that PhoP uses different configurations of promoter to regulate genes. We finally correlated these groups with more accurate independent experiments that measure gene expression over time by using GFP assays.

2 Methods

The purpose of this method is to identify all of the possible substructures, here termed profiles (i.e., groups of promoters sharing a common set of features), that characterize sets of genes. These common attributes can ultimately clarify the key *cis*-features that produce distinct kinetic patterns, shedding light in the transcriptional mechanisms that the cell employs to differentially regulate genes belonging to a regulon.

The identification of the promoter features that determine the distinct expression behavior of co-regulated genes is a challenging task because (i) the difficulty in ascertaining the role of the differences in the shared *cis*-acting regulatory elements of co-regulated promoters; (ii) detailed kinetic data that would help the classification of expression patterns is not always available, or it is available for a limited subset of genes; and (iii) the limited extent of genes regulated by a transcriptional factor. To circumvent these constrains, our method explores all of the possible *cis*-feature aggregations, looking for those that better characterize different subset of genes; uses an unsupervised approach, where pre-existing classes are not required; and allows a fuzzy incorporation of promoters to refined hypothesis which enables a same instance to support more than one hypothesis.

Our method represents, learns and infers from structural data by following four main phases: (1) *Database conformation*; (2) *Profile learning*; (3) *Profile evaluation* (4) *Evaluation of external classes*.

2.1 Database Conformation

Biological Model. Multiple independent and interrelated attributes of promoters, naturally encoded into diverse data types, should be considered to perform an integrated analysis of promoter regulatory features. We focus on four types of features for describing our set of co-regulated promoters [2, 3, 10, 11]: *"submotifs"*, fix-length DNA motifs from transcriptional regulator binding sites, represented by position weight matrices *[12]* (Fig 1.a). We used these matrices to prototype DNA sequences, where its elements are the weights used to score a test sequence to measure how close that sequence word matches the pattern described by the matrix; *"orientation"*, which characterizes the binding boxes as either in direct or opposite orientation relative to the open reading frame; *"RNA pol sites"*, represents the RNA polymerase: their location in the chromosome is studied as a distribution and encoded into fuzzy sets (*close, medium,* and *remote*). It also models the class of sigma 70 promoter [13]: *class I* promoters bind to upstream locations (Fig 1.b). By contrast *class II* promoters bind to sites that overlap the promoter region. [14](Fig 1.c); and *"expression"*, which considers gene expression from multiple experiments represented as vector patterns. See [15] for a detail description of the learning process of these features.

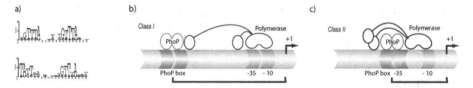

Fig. 1. Different *cis*-features participating in the regulation scheme. a) PhoP binding box modeled as position weight matrices shown as logos: The characters representing the sequence are stacked on top of each other for each position in the aligned sequences. The height of each letter is made proportional to its frequency.. **b-c)**Two transcription factors had binded to a DNA strain and recruited RNA polymerase (*Class I/II* respectively). A PhoP box might be located in the same strain as the polymerase (b) or in the opposite direction (c).

Representation Model. We use fuzzy sets as a common framework to represent the domain independent features. We cluster promoters considering each feature independently by using fuzzy C-means clustering (FCM) method and a validity index [16] to estimate the number of clusters, as an unsupervised discretization of the features [9, 17]. For example, we obtained three clusters for the *"expression"* feature (E_1^1 : strong evidence of upregulation; E_2^1 : mild evidence of upregulation; and E_3^1 : evidence of downregulation). As a result of this process, we obtain initial prototypes of profiles, and are able to account for the variability of the data by treating these features as fuzzy (i.e., not precisely defined) instead of categorical entities. Thus, our database is conformed by the membership of each promoter to each of the cluster of every feature.

2.2 Profile Learning

Our method uses a conceptual clustering approach to incrementally find significant characterization of promoters (profiles) while exploring the features space [18-20]. Initial profiles are aggregated to create compound higher level profiles (i.e. offspring profiles) by using the fuzzy intersection[1]. In a hierarchical process, the number of features shared by a profile is increased, resulting in a lattice of profiles. Level n profiles are built by aggregating level n-1 profiles (Fig. 2). This is because the method re-discretizes the original features:

$$V_{fj} = \square_{k=1}^{n} \mu_{jk} x_{fk} / \square_{k=1}^{n} \mu_{jk} \tag{1}$$

where μ_{jk} is the membership of the promoter k to cluster j; and x_{kf} is the original raw data for feature f. This allows to the prototypes of the profiles to be dynamically adapted to the promoters recovered by it. In account of these new prototypes, the membership of the entire database of promoters is re-evaluated:

$$\mu_{fij}(x) = \left[1 + \left(\left\| x_{if} - V_{fj} \right\|_f^2 / w_{fj} \right)^{1/m-1} \right]^{-1} \tag{2}$$

where w_{fj} is the "bandwith" of the fuzzy set V_{fj} [16]. This allows re-assignations of observations between sibling profiles [21], which is especially useful to gain support to hypothesis in problems, such as ours, that have a reduced number of samples.

2.3 Profile Evaluation

We applied multiobjective and multimodal techniques to evaluate the performance of the profiles [8, 9, 22], considering the conflicting criteria of the extent of the profile, and the quality of matching among its members and the corresponding features.

The extent of the profile is calculated by using the hypergeometric distribution that gives the probability of intersection (PI) of an offspring profile and its parents:

$$PI(V_{i,j}) = 1 - \sum_{q=0}^{p} \binom{h}{q}\binom{q-h}{n-q} / \binom{g}{h} \tag{3}$$

where V_i is an alpha-cut of the offspring profile, of size h; V_j is an alpha-cut of the union of its parents, of size n; p is the number of promoters of the intersection; and g is the number of candidates. The PI is an adaptive measure that is sensitive to small sets of examples, while it retains specificity with large datasets [23].

---

[1] Fuzzy logic-based operations, such as T-norm/T-conorm, include operators which are used as basic logic operators, such as AND or OR, [16]. In this work we used the MINIMUN and MAXIMUM as T-norm and T-conorm, respectively.

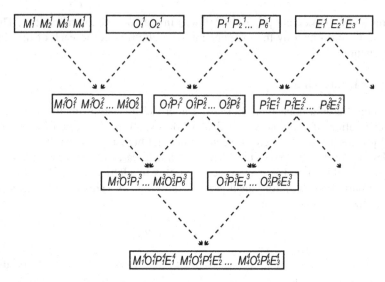

Fig. 2. Schematic view of the method. The method navigates through the feature-space lattice generating and evaluating profiles. Hierarchically, profiles of one level are combined to generate the profiles of the following one. Observations can migrate from parental to offspring clusters (i.e., hierarchical clustering), and among sibling clusters (i.e., optimization clustering).

The quality of matching between promoters and features of a profile (i.e., similarity of intersection (SI)) is calculated using the equation (4), where U_α is an alpha-cut of the profile i and n_α is its number of elements.

$$SI(V_i) = \left(1 - \sum\nolimits_{k \in U_\alpha} \mu_{ik} \Big/ n_\alpha\right)\Big/ f \ \ U_\alpha = \left\{\mu_{ik} : \mu_{ik} > \alpha\right\} \tag{4}$$

The tradeoff between the opposing objectives (i.e., PI and SI) is estimated by selecting a set of solutions that are non-dominated, in the sense that there is no other solution that is superior to them in all objectives (i.e., Pareto optimal frontier) [8, 9]. The dominance relationship in a minimization problem is defined by:

$$a \prec b \ iif \ \forall i \ O_i(a) \le O_i(b) \ \exists j O_j(a) < O_j(b) \tag{5}$$

where the O_i and O_j are either PI or SI. This approach is less biased than weighting the objectives because it identifies the profiles lying in the Pareto optimal frontier [8, 9], which is the collection of local multiobjective optima in the sense that its members are not worse than (i.e. dominated by) the other profiles in any of the objectives being considered.

Another objective indirectly considered is the profile diversity, which consists of maintaining a distributed set of solutions in the Pareto frontier, and thus, identifying clusters that describe objects from alternative regulatory scenarios. Therefore, our approach applies the non-dominance relationship locally, that is, it identifies all non-dominated optimal profiles that have no better solution in the local neighborhood

[8, 9]. We evaluate niches by applying equation (3) to every pair of solution and establish a small threshold value as boundaries of neighborhoods.

2.4 Evaluation of External Classes

This proposed unsupervised method, in contrast to supervised approaches, does not need the specification of output classes. Consequently, the discovered profiles can be used for independently explain external classes as a process often termed labeling [7]

Instead of choosing a single profile to characterize an external target set, the method selects all of the profiles that are correlated enough to the query set. To find its classes of equivalence it applies equation (3) to the target set and the entire collection of profiles previously produced. In this way, the method can recover all of the alternative profiles that match the external class, including the most specific and general solutions.

3 Results

We investigated the utility of our approach by exploring the regulatory targets of the PhoP protein in *E. coli* and *S. enterica*, which is at the top of a highly connected network that controls transcription of dozens of genes mediating virulence and the adaptation to low Mg^{2+} environments [24]. As little is known about the mechanism by which *cis*-regulatory features govern gene expression, we searched through the space of all potential hypotheses; evaluated them, by considering both their extent and similarity of the recovered promoters; and obtained alternative descriptions for target set of genes. Moreover, to tackle constrains of the crude classification obtained by microarray experiments -which would not have allowed finding detail topologies of promoters- in an unsupervised approach we modeled gene expression as one feature among many.

We demonstrated that our method makes predictions at two levels: it detects new candidate promoter for a regulatory protein; and it indicates alternative possible configurations by which genes previously identified as controlled by a regulator are differentially expressed. We recovered several optimally evaluated profiles, thus, revealing distinct putative profiles that can describe the PhoP regulation process:

One profile ($O_1^4 E_2^4 M_3^4 P_2^4$: PI=1.57E-4, SI=0.002) corresponds to canonical PhoP-regulted promoters (e.g., those of the *phoP, mgtA, rstA, slyB, yobG, ybjX, ompX, PagP, pdgL, pipD, and pmrD* genes) characterized by a class II RNA polymerase sites situated close to the PhoP boxes, high expression patterns and a typical PhoP box submotif in a direct orientation. Notably, this profile recovers promoters previously not known to be directly regulated by PhoP. The method was also able to describe this target by using other profiles, being the most general ones composed of only two features (Fig 3.a)

Another profile ($E_3^4 M_1^4 O_2^4 P_1^4$:PI=3.53E-4, SI=0.032) includes promoters (e.g., those of the *mgtC, mig-14, pagC, pagK,* and *virK* genes of *Salmonella*) that share PhoP boxes in the opposite orientation of the canonical PhoP-regulated promoters, as well as class I RNA polymerase sites situated at medium distances from the PhoP boxes. As expected, the method was able to identify this target set by more general hypothesis that aggregates again only two features (Fig 3.b).

Finally, another profile ($E_2^3 O_2^3 P_4^3$: PI=6.48E-06, SI=0.070), which is slightly different from the former, includes promoters (e.g., those of the *ompT* gene of *E. coli* and the *pipD, ugtL* and *ybjX* genes of *Salmonella*) is defined by a PhoP binding site in the opposite orientation, the RNA polymerase of the canonical PhoP regulated promoters and a mild evidence of upregulation. The method was also able to characterize this target by a specific Phop box submotif and the same type of RNA polymerase.

The above profiles differ in the number of features because our method uses a multivariate environment, where feature selection is locally performed for each profile, as not every feature is relevant for all profiles. The predictions made by our method were experimentally validated [10] to establish that the PhoP protein uses multiple mechanisms to control gene transcription.

Furthermore, as these profiles can be used to effectively explain the different kinetic behavior of co-regulated genes, we measured the promoter activity and growth kinetics for GFP reporter strains with high-temporal resolution (Fig. 4); and obtained independent target sets by clustering them by using FCM. We found that the cluster that recovers those promoters that expressed earlier rise times and higher levels of transcription (e.g. mgtA, ompX, pagP, phoP, pmrD, rstA, slyB, ybjX, yobG) is correlated to profile $O_1^4 E_2^4 M_3^4 P_2^4$ (*p*-value < 0.03) (Fig. 3.a). Another target set includes those promoters that expressed the latest rise time and lowest levels of transcription (e.g. mgtC, mig-14, pagC, pagK, pipD, ugtL, virK, pagD); and it is correlated to profile $E_3^4 M_1^4 O_2^4 P_1^4$ (*p*-value < 0.013) (Fig. 3.b). The cluster which contains the promoters that showed intermediate values (e.g., those of the *ompT* gene of *E. coli* and the *pipD, ugtL* and *ybjX* genes of *Salmonella*) is correlated to profile $E_2^3 O_2^3 P_4^3$ (*p*-value < 0.025)

This detailed analysis of the gene expression behavior would not be possible to be obtained by applying a supervised machine learning approach because of the lack of kinetic data for some promoters.

4 Discussion

We showed that our method can make precise mechanistic predictions even with incomplete input dataset and high levels of uncertainty; making use of several characteristics that contribute to its power: (i) it considers crude gene expression as one feature among many (unsupervised approach), thereby allowing classification of promoters even in its absence; (ii) it has a multimodal nature that allows alternative descriptions of a system by providing several adequate solutions [9] that characterize a target set of genes; (iii) it allows promoters to be members of more than one profile by using fuzzy clustering thus explicitly treating the profiles as hypotheses, which are tested and refined during the analysis; and (iv) it is particularly useful for knowledge discovery in environments with reduced datasets and high levels of uncertainty.

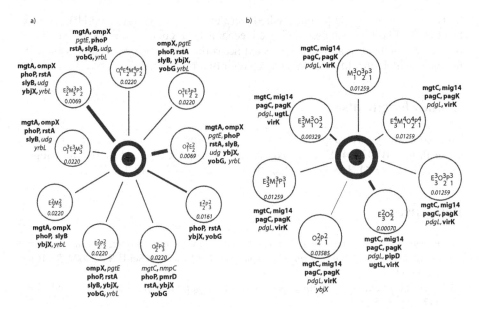

Fig. 3. Chart of Correlated Profiles. Targets are display at the center of each chart, surrounded by the profiles that hit them. Optimal profiles are situated closer to the targets. For each profile it is displayed the features that characterizes it, the promoters that recovers (bold-face belonging to the target, and italic not belonging to it) and the correlation to the target set. E stands for *"Expression"*, P for *"RNA Pol. Sites"*, O for *"Orientation"* and *"M"* for *"Submotif"*; subscripts denote the cluster and superscripts the re-discretized level.

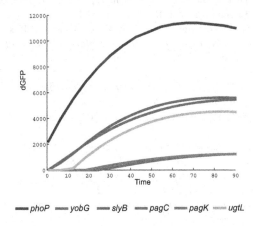

Fig. 4. Rise time and levels of transcription. Transcriptional activity of wild-type *Salmonella* harboring plasmids with a transcriptional fusion between a promoterless *gfp* gene and the *Salmonella* promoters. The activity of each promoter is proportional to the number of GFP molecules produced per unit time per cell $[dG_i(t)/dt]/OD_i(t)]$, where $G_i(t)$ is GFP fluorescence from wild-type *Salmonella* strain 14028s, and $OD_i(t)$ is the optical density. The activity signal was smoothed by a polynomial fit (sixth order). Details about genetic experiments can be found in http://www.pnas.org/ and about GFP assays available under requirements to the authors.

The predictions made by our method were experimentally validated [10] to establish that the PhoP protein uses multiple mechanisms to control gene transcription, and is a central element in a highly connected network. These profiles can be used to effectively explain the different kinetic behavior of co-regulated genes.

Acknowledgments

This work was partly supported by the Spanish Ministry of Science and Technology under Project BIO2004-0270-E, and I.Z. is also supported by and by Howard Hughes Medical Institute. O.H. acknowledges the doctoral MAEC- AECI fellowship.

References

1. Ronen, M., et al., *Assigning numbers to the arrows: parameterizing a gene regulation network by using accurate expression kinetics.* Proc Natl Acad Sci U S A, 2002. **99**(16): p. 10555-60.
2. Beer, M.A. and S. Tavazoie, *Predicting gene expression from sequence.* Cell, 2004. **117**(2): p. 185-98.
3. Bar-Joseph, Z., et al., *Computational discovery of gene modules and regulatory networks.* Nat Biotechnol, 2003. **21**(11): p. 1337-42.
4. Conlon, E.M., et al., *Integrating regulatory motif discovery and genome-wide expression analysis.* Proc Natl Acad Sci U S A, 2003. **100**(6): p. 3339-44.
5. Oshima, T., et al., *Transcriptome analysis of all two-component regulatory system mutants of Escherichia coli K-12.* Mol Microbiol, 2002. **46**(1): p. 281-91.
6. Tucker, D.L., N. Tucker, and T. Conway, *Gene expression profiling of the pH response in Escherichia coli.* J Bacteriol, 2002. **184**(23): p. 6551-8.
7. Mitchell, T.M., *Machine learning.* 1997, New York: McGraw-Hill. xvii, 414.
8. Deb, K., *Multi-objective optimization using evolutionary algorithms.* 1st ed. Wiley-Interscience series in systems and optimization. 2001, Chichester ; New York: John Wiley & Sons. xix, 497.
9. Ruspini, E.H. and I. Zwir, *Automated generation of qualitative representations of complex objects by hybrid soft-computing methods*, in *Pattern recognition : from classical to modern approaches*, S.K. Pal and A. Pal, Editors. 2002, World Scientific: New Jersey. p. 454-474.
10. Zwir, I., et al., *Dissecting the PhoP regulatory network of Escherichia coli and Salmonella enterica.* Proc Natl Acad Sci U S A, 2005. **102**(8): p. 2862-7.
11. Li, H., et al., *Identification of the binding sites of regulatory proteins in bacterial genomes.* Proc Natl Acad Sci U S A, 2002. **99**(18): p. 11772-7.
12. Stormo, G.D., *DNA binding sites: representation and discovery.* Bioinformatics, 2000. **16**(1): p. 16-23.
13. Romero Zaliz, R., I. Zwir, and E.H. Ruspini, *Generalized analysis of promoters: a method for DNA sequence description*, in *Applications of Multi-Objective Evolutionary Algorithms*, C.a.L. Coello Coello, G., Editor. 2004, World Scientific: Singapore. p. 427-450.
14. Salgado, H., et al., *RegulonDB (version 4.0): transcriptional regulation, operon organization and growth conditions in Escherichia coli K-12.* Nucleic Acids Res, 2004. **32**(Database issue): p. D303-6.

15. Zwir, I., H. Huang, and E.A. Groisman, *Analysis of differentially-regulated genes within a regulatory network by GPS genome navigation 10.1093/bioinformatics/bti672*. Bioinformatics, 2005. **21**(22): p. 4073-4083.

16. Bezdek, J.C., *Pattern Analysis*, in *Handbook of Fuzzy Computation*, W. Pedrycz, P.P. Bonissone, and E.H. Ruspini, Editors. 1998, Institute of Physics: Bristol. p. F6.1.1-F6.6.20.

17. Kohavi, R. and G.H. John, *Wrappers for feature subset selection*. Artificial Intelligence, 1997. **97**(1-2): p. 273-324.

18. Cheeseman, P. and R.W. Oldford, *Selecting models from data : artificial intelligence and statistics IV*. 1994, New York: Springer-Verlag. x, 487.

19. Cook, D.J., et al., *Structural mining of molecular biology data*. IEEE Eng Med Biol Mag, 2001. **20**(4): p. 67-74.

20. Cooper, G.F. and E. Herskovits, *A Bayesian Method for the Induction of Probabilistic Networks from Data*. Machine Learning, 1992. **9**(4): p. 309-347.

21. Falkenauer, E., *Genetic Algorithms and Grouping Problems*. 1998, New York: John Wiley & Sons.

22. Rissanen, J., *Stochastic complexity in statistical inquiry*. World scientific series in computer science, 15. 1989, Singapore: World Scientific. 177.

23. Tavazoie, S., et al., *Systematic determination of genetic network architecture*. Nat Genet, 1999. **22**(3): p. 281-5.

24. Groisman, E.A., *The pleiotropic two-component regulatory system PhoP-PhoQ*. J Bacteriol, 2001. **183**(6): p. 1835-42.

Modeling Genetic Networks:
Comparison of Static and Dynamic Models

Cristina Rubio-Escudero[1], Oscar Harari[1], Oscar Cordón[1,2], and Igor Zwir[1,3]

[1] Department of Computer Science and Artificial Intelligence,
[2] European Center for Soft Computing, Mieres, Spain.
[3] Howard Hughes Medical Institute, Washington University School of Medicine, St. Louis, MO.
{crubio, oharari, ocordon, zwir}@decsai.ugr.es

Abstract. Biomedical research has been revolutionized by high-throughput techniques and the enormous amount of biological data they are able to generate. The interest shown over network models and systems biology is rapidly raising. Genetic networks arise as an essential task to mine these data since they explain the function of genes in terms of how they influence other genes. Many modeling approaches have been proposed for building genetic networks up. However, it is not clear what the advantages and disadvantages of each model are. There are several ways to discriminate network building models, being one of the most important whether the data being mined presents a static or dynamic fashion. In this work we compare static and dynamic models over a problem related to the inflammation and the host response to injury. We show how both models provide complementary information and cross-validate the obtained results.

1 Introduction

Advances in molecular biology and computational techniques permit the systematical study of molecular processes that underlie biological systems (Durbin *et al.*, 1998). One of the challenges of this post-genomic era is to know when, how and for how long a gene is turned *on/off*. Microarray technology has revolutionized modern biomedical research in this sense by its capacity to monitor the behavior of thousands of genes simultaneously (Brown *et al.*, 1999; Tamames *et al.*, 2002). The reconstruction of genetic networks is becoming an essential task to understand data generated by microarray techniques (Gregory, 2005). The enormous amount of information generated by this high-throughput technique is raising the interest in network models to represent and understand biological systems.

Systems biology research arises at this point as the field to explore the life regulation processes in a cohesive way making use of the new technologies. Proteins have a main role in the regulation of genes (Rice and Stolovitzky, 2004), but unfortunately, for the vast majority or biological datasets available, there is no information about the level of protein activity. Therefore, we use the expression level of the genes as an indicator of the activity of proteins they generate.

Gene networks represent these gene interactions. A gene network can be described as a set of nodes which usually represent genes, proteins or other biochemical entities. Node interaction is represented with edges corresponding to biologic relations.

E. Marchiori, J.H. Moore, and J.C. Rajapakse (Eds.): EvoBIO 2007, LNCS 4447, pp. 78–89, 2007.
© Springer-Verlag Berlin Heidelberg 2007

There is a wide range of models available to build genetic networks up. One of the differences between such models is whether they represent static or dynamic relations. Static modeling explains causal interactions by searching for mutual dependencies between the gene expression profiles of different genes (van Someren *et al.*, 2002). Clustering techniques are widely applied for static genetic network, since they group genes that exhibit similar expression levels.

In dynamic modeling, the expression of a node A in the network at time t_{+1} can be given as the result of the expression of the nodes in the network with edges related to A at time t (van Someren *et al.*, 2002). The understanding of the relations helps to describe all the relations occurring in a given organism we would be able to know the behavior of such organism throughout time.

The question arises as which network model is the most appropriate given a set of data. In the present work we have applied both static (*K*-means clustering method, (Duda and Hart, 1973)) and dynamic network models (a Boolean method, described in (D'onia *et al.*, 2003) and implemented in (Velarde, 2006) and a graphic Gaussian method (GGM) (Schäfer and Strimmer, 2005)) to a set of data derived from an experiment on inflammation and the host response to injury (Calvano *et al.*, 2005). The results show how dynamic models are capable to recover temporal dependencies that static models are not able to find. Temporal studies are becoming widely used in biomedical research. In fact, over 30% of published expression data sets are time series (Simon *et al.*, 2005).

2 Problem Description

In this work we compare the behavior of static vs. dynamic modeling in a problem derived from the inflammation and the host response to injury. On the one hand, static modeling searches for relations between the expression levels of genes throughout time. The relation found by static methods might not only be similar behavior throughout time (direct correlation), but an inverse correlation (two genes having exactly opposite profiles over time), a proximity on the expression values (distance measures such as Euclidean Distance or City block distance) (see Fig. 1). On the other hand, dynamic modeling retrieves temporal dependencies among genes, i.e., it detects dependencies of a gene at time t_{+1} related to some other(s) gene at time t (see Fig. 1).

To compare the performance of these two models, we have applied them to a data set derived from an experiment over inflammation and the host response to injury as part of a Large-scale Collaborative Research Project sponsored by the National Institute of General Medical Sciences (www.gluegrant.org) (Calvano *et al.*, 2005). Human volunteers have been treated with intravenous endotoxin and compared to placebo, obtaining longitudinal blood expression profiles. Analysis of the set of gene expression profiles obtained from this experiment is complex, given the number of samples taken and variance due to treatment, time, and subject phenotype. The data were acquired from blood samples collected from eight human volunteers, four treated with intravenous endotoxin (i.e., patients 1 to 4) and four with placebo (i.e., patients 5 to 8). Complementary RNA was generated from circulating leukocytes at 0, 2, 4, 6, 9 and 24 hours after the and hybridized with GeneChips® HG-U133A v2.0 from Affymetrix Inc., which contains 22216 probe sets, analyzing the expression level of 18400 transcripts and variants, including 14500 well-characterized genes.

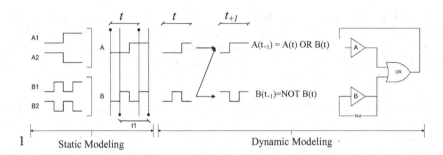

Fig. 1. The static modeling captures the relation (inverse correlation) between A_1 and A_2 (profile A) and between B_1 and B_1 (profile B). However, it does not capture the relation between A and B describing profile A at time t_{+1}. This relation is only captured by the dynamic model.

3 Genetic Network Construction

We have applied both static and dynamic models to the set of data just described. As said in Section 1, clustering techniques are widely applied for static genetic network, so we have used a classic clustering algorithm based on Euclidean distance, the *K-means* (Duda and Hart, 1973) which is a very popular clustering algorithm widely used with data from microarray experiments (Guiller *et al.*, 2006). Two dynamic methods have been applied as well: a Boolean method, described in (D'onia *et al.*, 2003) and implemented in (Velarde, 2006) and a graphic Gaussian method (GGM) (Schäfer and Strimmer, 2005). These two methods have been chosen as representation of discrete and continuous models respectively, the two big families in which dynamic models can be divided (van Someren *et al.*, 2002). We now describe each of these methods.

Classification of gene expression patterns to explore shared functions and regulation can be accomplished using clustering methods (D'haeseleer *et al.*, 2000). We have applied a classic clustering algorithm based on Euclidean distance, the *K-*means algorithm (Duda and Hart, 1973). The number of resulting clusters k is estimated by application of the Davies-Bouldin validity index (Davies and Bouldin, 1979). The groupings obtained using this method, i.e., gene expression profiles, are expected to be functionally cohesive since genes sharing the same expression profiles are likely to be involved in the same regulatory process (D'haeseleer *et al.*, 2000). This can be proved applying the EMO-CC algorithm (Romero-Záliz *et al.*, 2006), which validates the gene groupings obtained using external information from the Gene Ontology database, which provides a controlled vocabulary to describe gene and gene product attributes in any organism (Ashburner *et al.*, 2000).

3.1 Dynamic Discrete Modeling : Boolean Networks

A Boolean network is composed by a set of nodes n which represent genes, proteins or other biochemical entities. These nodes can take *on/off* values. The net is determined by a set of at maximum n Boolean functions, each of them having the state of k specific nodes as input, where k depends on each node. Therefore, each node has its own Boolean function which determines the next state based on the actual state of the input nodes. The changes in the net are assumed to occur at discrete time intervals.

The algorithm applied to build the Boolean network with our data is the GeneYapay (D'Onia *et al.*, 2003). It performs an exhaustive search of Boolean functions over the data, where a number of nodes, less or equal then k, univocally determines the output of some other gene. All possible subsets of 1, 2, ..., k elements are visited calculating the number of inconsistencies of the Boolean functions in relation to the output value of each gene. The algorithm stops the search for each node when a subset of nodes is found which defines the expression profile. The implementation applied (Velarde, 2006) only uses the NAND function since all other Boolean function -AND, OR, NOT- can be expressed using NAND (see Table 1).

Table 1. Boolean functions obtained only using the NAND function

NOT A ≡ A NAND A
A AND B ≡ (A NAND B) NAND (A NAND B)
A OR B ≡ (A NAND A) NAND (B NAND B)

3.2 Dynamic Continuous Modeling : Graphic Gaussian Network

The graphical gaussian models were first proposed by Kishino and Waddell (2000) for the association structure among genes. GGMs are similar to Bayesian networks in that they allow to distinguish direct from indirect interactions (i.e. whether gene A acts on gene B directly or through a third gene C). As any graphical model, they also provide a notion of conditional independence of two genes. However, in contrast to Bayesian networks, GGMs contain only undirected rather than directed edges. This makes graphical Gaussian interaction modeling on the one hand conceptually simpler, and on the other hand more widely applicable (e.g. there are no problems with feedback loops as in Bayesian networks).

The GGM applied in this work has been developed by Schäfer and Strimmer, (2005) and is based on (1) improved (regularized) small-sample point estimates of partial correlation, (2) an exact test of edge inclusion with adaptive estimation of the degree of freedom and (3) a heuristic network search based on false discovery rate multiple testing.

4 Results

High-throughput techniques provide great amounts of data that need to be processed before being used to build genetic networks up. The first step is the identification of genes relevant for the problem under study. We have applied the methodology described in Rubio-Escudero *et al.,* (2005): a process based on the meta analysis of microarray data. The proliferation of related microarray studies by independent groups, and therefore, different methods, has lead to the natural step of combination of results (Gosh *et al.,* 2003). Thus, a battery of analysis methods has been applied (Student's T-Tests (Li and Wong, 2003), Permutation Tests (Tusher *et al.,* 2001), Analysis of

Variance (Park *et al.*, 2003) and Repeated Measures ANOVA (Der and Everitt, 2001)). A total of 2155 genes have been identified as relevant for the problem under study. For this particular problem the number of genes retrieved is very high compared to other microarray experiments, since the problem under study, inflammation and host response to injury, is a process that affects the human system in a global manner, hence altering the behavior of a large number of genes (Calvano *et al.*, 2005).

At the view of these, we decide to use the expression profiles of the genes as the input for the genetic network building algorithms, since the number of genes involved in the problem is unfeasible for both building and analyzing the genetic networks. The set of profiles used is the one obtained from the static model applied, the K-*means* algorithm.

4.1 Static Modeling: *K*-Means Clustering

We apply a clustering method, the *K*-means algorithm, as described in section 3.1. We have identified 24 expression profiles (Rubio-Escudero *et al.*, 2005) (see Fig. 2). These profiles have been proved as functionally cohesive by application of the EMO-CC algorithm (Romero-Záliz *et al.*, 2006). For instance, the majority of the genes exhibiting profile #22 are related to the inflammatory response (GO:0006954) and are annotated as intracellular (GO:0005622). Another sample is profile #16, with genes sharing the apoptosis (GO:0006915) and integral to plasma membrane (GO:0005887) annotations.

The functional identification of the 24 profiles resulting from the clustering method represents a further analysis of the data behind the identification of the genes relevant for the problem.

4.2 Dynamic Discrete Modeling: Boolean Network

Boolean building network algorithms use discrete data which take two possible values: *on* or *off*, i.e., 1 or 0. Therefore, the set of 24 differential profiles obtained in the inflammation and host response to injury problem (Calvano *et al.*, 2005) needs to be transformed to fit the binary scheme. First of all, each of the profiles will be scaled in the [0, 1] interval according to the maximum value scored in the expression level of such profile throughout the six time points stored. The individual scaling has been used instead of a global one (scaling the 24 profiles according to the global maximum) since the profiles fluctuate in different levels of expression. For instance, profile #1 takes values between 1224.2 and 1724.4, while profile #24 changes between 13632 and 16436. If we scaled all values together, the variations between the expression values in profile #1 would result to small to be traceable, although they could be significative. In Table 2 (A) the expression levels before scaling are shown.

Once the values are scaled in the [0, 1] interval we have assign them [0-1] values. The simplest approach is to establish a threshold value, for instance 0.5, and to set each time point value depending whether they are over/under the threshold. The obvious problem with this approach is the "border value", such as 0.45 or 0.55. These will be set 0 and 1 respectively, while they are so close to each other that they should take the same value. Our approach consists in setting the value based on the

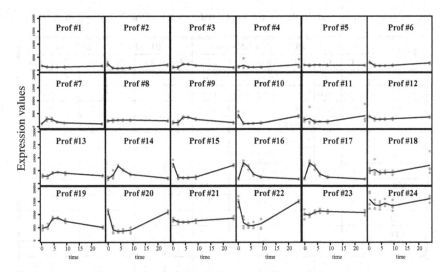

Fig. 2. Set of 24 expression profiles obtained from the inflammation and host response to injury problem

proximity to the expression level in the previous time point, which solves the previously described problem and captures the behavior of the profile over time. The scheme used to set the values is:

$$
\begin{cases}
if\ (|t-t_{+1}|<\delta)\quad then\ t_{+1}=t \\[2ex]
if\ (|t-t_{+1}|>\delta)\quad then\ \begin{cases} if\ (|t-t_{+1}|<0)\ then\ t_{+1}=0 \\ if\ (|t-t_{+1}|>0)\ then\ t_{+1}=1 \end{cases}
\end{cases}
$$

where t_{+1} is the gene value to be set and t is the gene value in the previous time point. Table 2 (B) shows the obtained Boolean values for the 24 profiles in our problem.

The resulting Boolean network is shown in Fig. 3. This net is the result of an exhaustive search of Boolean functions over the data which univocally determines the output of the other genes. We see that some nodes represent more than one expression profile. This is due to the processing the data has to undergo. The scaling of the data to the [0, 1] interval, makes profiles at different levels of expression end up sharing a common Boolean profile. A sample of this in our particular problem is the one represented by profiles #9, #13 and #19. These three expression profiles share similar behaviour throughout time at different levels of expression (see Fig. 4). The net shows valuable information about relation between profiles. For instance, the relation established between profiles #7 and #17 with profiles #3 and #14 is confirmed when searching in the KEGG database (Kanehisa *et al.*, 2004), a metabolic pathway database. Genes exhibiting profiles #7 and #17 are in the same pathway and regulate

Table 2. Continuous and Boolean values obtained for each of the 24 profiles in the data set

PROFILES	CONTINUOUS VALUES (A)						BOOLEAN VALUES (B)					
	T0	T2	T4	T6	T9	T24	T0	T2	T4	T6	T9	T24
#1	1724.4	1316.4	1224.2	1236.9	1327.5	1666	1	0	0	0	0	1
#2	2546.2	734.44	700.28	737.5	867.44	2107.8	1	0	0	0	0	1
#3	1108.8	1027.9	2403.2	2376	1843.3	1069.6	0	0	1	1	0	0
#4	1323.6	2001.9	1089.4	1139.8	1192.7	2230.8	0	1	0	0	0	1
#5	1933.1	1829.8	1970.5	1983.6	1966.4	1907.5	1	0	1	1	1	1
#6	3146	1694.2	1669.1	1746.3	1889.8	2872.3	1	0	0	0	0	1
#7	1265.8	3551.7	3079	2008.1	1656.4	1160.3	0	1	1	0	0	0
#8	2396.3	2577.6	2721.5	2726.6	2712	2412.9	0	1	1	1	1	0
#9	1614.2	1619	3756.4	3972.6	3116.5	1676.8	0	0	1	1	1	0
#10	4844.2	1278.3	1248.4	1316.9	1468.1	4240.1	1	0	0	0	0	1
#11	2730.3	3351.4	1921.3	2114.9	2146.3	4459.3	0	1	0	0	0	1
#12	4176	2984.1	2974	3068.7	3265.5	4021.8	1	0	0	0	0	1
#13	3022.8	2898.1	4262.2	4666.1	4329.1	3150.8	0	0	1	1	1	0
#14	2117.6	3289.7	7298.8	5871.3	4036.8	2229.4	0	0	1	1	0	0
#15	7849.5	2328	2297.4	2450	2738.6	7171.7	1	0	0	0	0	1
#16	4836.6	4220.5	5085.4	5398.3	5356.3	4829.7	1	0	1	1	1	0
#17	1950.7	9001.6	7946	4268.8	2804	1787.1	0	1	1	0	0	0
#18	5238.2	5734.5	4445.8	4654.6	4665.7	7584.4	1	0	0	0	0	1
#19	4935.7	5335.4	9034.5	9171	7858	5285.3	0	0	1	1	1	0
#20	11615	4161.2	3578.6	3760.8	4149.9	11344	1	0	0	0	0	1
#21	8358.3	7308.8	7244.2	7652.2	8139.2	8913.8	1	0	0	0	0	1
#22	15442	7021.5	5798.9	5918.8	6632.3	15605	1	0	0	0	0	1
#23	10473	10132	11396	11871	11531	10980	0	0	1	1	1	1
#24	16095	13749	13632	14364	13741	16436	1	0	0	1	0	1

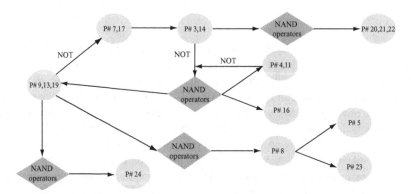

Fig. 3. Genetic network obtained using the Boolean model. The round nodes represent the gene expression profiles (groups of genes with a common behavior) and the diamond shape nodes represent the Boolean function based on the NAND operator. Note that some nodes represent more than one expression profile.

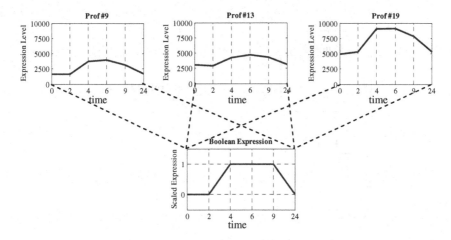

Fig. 4. Profiles at different levels of expression but sharing a common behavior throughout time share the same Boolean profile

genes exhibiting profile #14 (See Fig. 5). That is the case of gene IL1RN (prof. #17, Interleukin-1 receptor antagonist protein precursor), related to the immune response (GO:0006955) and gene IL1R2 (prof. #14, Interleukin-1 receptor type II), also related to the immune response. We can see in Fig. 5(A) more examples of gene relations found in KEGG and present in the Boolean network obtained.

4.3 Dynamic Continuous Modeling: Graphical Gaussian Model

We have applied a Graphic Gaussian algorithm (Schäfer and Strimmer, 2005), which takes as input continuous data that can be in longitudinal format (Opgen-Rhein and Strimmer, 2006), very convenient for microarray time course experiments since it deals with repeated measurements, irregular sampling, and unequal temporal spacing of the time points. To select the edges, and thus the nodes, we have used the local false discovery rate (fdr) (expected proportion of false positives among the proposed edges), an empirical Bayes estimator of the false discovery rate (Efron, 2005). An edge is considered *present* or *significant* if its local fdr is smaller than 0.2 (Efron, 2005). Three independent networks are found (see Fig. 6). Network (B) confirms the information provided by the Boolean network about profiles #7, #14 and #17. In network (A) there is a relation established between profiles #11, #23 and #16 that is confirmed when searching in the KEGG database (see Fig. 5(B)). That is the case of gene RACK (Reversion-inducing cysteine-rich protein with Kazal motifs), which exhibits profile #11 and is related to gene MMP9 (Matrix metalloproteinase-9), which exhibits profile #23. Both genes are related to the inflammation problem. Another relation is found between a gene exhibiting profile #23, CEBPB (CCAAT/enhancer-binding protein beta), related to the immune response (GO:0006955) and to the; inflammatory response (GO:0006954) and a gene exhibiting profile #16, CASP1 (Caspase-1) related to apoptosis (GO:0006915).

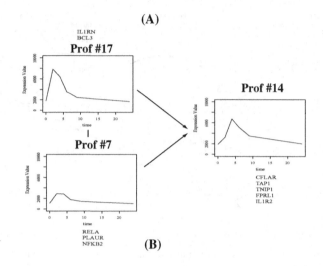

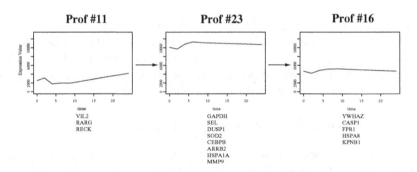

Fig. 5. Gene relations detected by the network building algorithms and confirmed in the KEGG database. (A) has been found by both the Boolean algorithms and GGM while (B) has only been found by GGM. The genes regulate other genes with the same color.

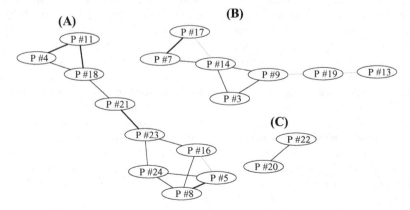

Fig. 6. Three independent networks found by the GGM algorithm

5 Discussion

We have applied both static and dynamic methods for the analysis of a data set derived from the inflammation and the host response to injury (Calvano et al., 2005). The static method has been the K-means clustering algorithm, and the dynamic methods have been a discrete one, Boolean model described in (D'Onia et al., 2003) and implemented by (Velarde, 2006) , and a continuous one, Graphic Gaussian Model developed by (Schäfer and Strimmer, 2005). We have already described some of the findings these methods have made on the dataset: the static method is capable of grouping the genes based on their behaviour throughout time and these groupings are cohesive in biological functionality. The dynamic models provide temporal relations between the genes, or in this case, between the profiles they exhibit, organizing them in regulatory networks that are validated using the KEGG database. These temporal relations would not have been found only applying static models.

When comparing the two dynamic models, we see that they cross-validate in general their results i.e., the profiles involved and the relations between those profiles are concordant with one another. The Boolean algorithm and GGM show different and complementary information about the problem under study. In a GGM network the relation between nodes is based on the levels of correlation but the time dependency is not so clearly pointed out as in Boolean networks. For instance, in our GGM net we see that profiles #5, #8 and #23 are related since they are in the same subnet, but the Boolean network specifically describes the behavior of those profiles: #8 determines the behavior of both #5 and #23 (see Fig. 7), since the behavior shown by profile #8 is shifted over time in profiles #5 and #23. This kind of information is only available in network models which strongly stress the temporal dependencies, as it is the case with Boolean networks.

However, Boolean algorithms lack the capacity to distinguish among expression profiles with similar behaviour throughout time at different levels of expression (see Fig. 4). For instance, the Boolean algorithm considers profiles #9, #13 and #19 as only one node. GGM uses continuous values solving this problem and taking advantage of the diversity or the data, but it misses some information. The network (C) provided by GGM covers profiles #20 and #22. In the Boolean network they are considered as one single profile along with #21, since their Boolean representation is the same. GGM has not been able to capture the similarity between these three profiles, only between two of them, #20 and #22. However, the Boolean model considers them as the same node, so any temporal relation between them is impossible to capture. In fact, when searching in KEGG (Kanehisa et al., 2004), we see that one of the genes that exhibit profile #20 is NFKB2 (nuclear factor of kappa light polypeptide gene enhancer in B-cells 2) and one of the genes exhibiting profile #22 is TNIP1 (TNFAIP3-interacting protein 1). When searching for information about these two genes, which are related in their behavior, we see they are also functionally related since TNIP1 interacts with zinc finger protein A20/TNFAIP3 and inhibits TNF-induced NF-kappa-B-dependent gene expression (NFKB2). This valuable information is only prone to be found with network models such as GGM which permit the representation of temporal dependencies among strongly correlated profiles.

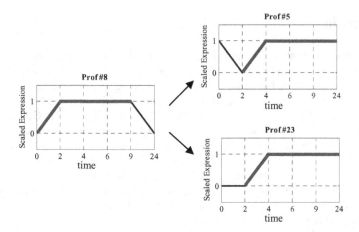

Fig. 7. Time relations found by the Boolean algorithm. Profile #8 determines the behavior of profiles #6 and #23.

The evaluation of static and dynamic models over the inflarmmation and host response to injury problem allows us to conclude that static models provide very valuable information but a step further is needed to get a deeper knowledge of the problem under study. Dynamic models provide information of the temporal dependencies in the data what is very valuable especially for time-course experiments, which are becoming very popular used in biomedical research. Dynamic discrete models miss valuable information when discretizing the data, while the continuous models do not suffer this problem. However, dynamic continuous models are not capable to find some of the dependencies that discrete model discover and vice versa. Therefore, they are complementary methods and it is a recommendable practice to apply both models to extract the maximum information possible from experiments.

References

Ashburner, M., Ball, C.A., Blake, J.A., Botstein, D., Butler, H., Cherry, J.M., Davis, A.P., Dolinski, K., Dwight, S.S., Eppig, J.T., Harris, M.A., Hill, D.P., Issel-Tarver, L., Kasarskis, A., Lewis, S., Matese, J.C., Richardson, M., Rubin, G.M. and Sherlock, G. (2000) Gene ontology: tool for the unification of biology. The Gene Ontology Consortium, *Nat Genet*, 25: 25-29.

Brown,P. and Botstein,D. (1999) Exploring the new world of the genome with DNA microarrays. *Nature Genet.*, 21 (Suppl.), 33-37.

Calvano,S.E., Xiao,W., Richards,D.R., Feliciano,R.M., Baker, H.V., Cho, R.J., Chen, R.O., Brownstein,B.H., Cobb,J.P., Tschoeke,S.K., Miller-Graziano,C., Moldawer,L.L., Mindrinos, M.N., Davis, R.W., Tompkins,R.G. and Lowry,S.F. (2005) The Inflammation and Host Response to Injury Large Scale Collaborative Research Program. A Network-Based Analy-sis of Systemic Inflammation in Humans. *Nature*, 13:437(7061):1032-7.

Davies, D. L. and Bouldin, W. (1979) A cluster separation measure. IEEE PAMI, 1, 224-227.

D'haeseleer P., Liang S. and Somogyi R. (2000) Genetic network inference: from co-expression clustering to reverse engineering. *Bioinformatics*, 16(8):707-726.

D'Onia D., Tam L., Cobb J. P., and Zwir I. A hierarchical reverse-forward methodology for learning complex genetic networks. *Proceedings of the 3rd International Conference on Systems Biology (ICSB)*, Stockholm Sweden.

Duda, R. O., and Hart, P. E. (1973) Pattern Classification and Scene Analysis. John Wiley & Sons, New York, USA.

Durbin,R., Eddy,S., Krogh,A. and Mitchison,G. (1998) Biological Sequence Analysis: Probabilistic Models of Proteins and Nucleic Acids. Cambridge University Press.

Efron, B. (2005). Local false discovery rates. Preprint, Dept. of Statistics, Stanford University.

Dutilh B. (1999) Analysis of data from microarray experiments, the state of the art in gene network reconstruction. *Report Binf.1999.11.01, Bioinformatics*, Utrecht University.

Gregory W. (2005) Inferring network interactions within a cell. *Bioinformatics*, 6: 380-389.

Guiller A., Bellido A., Coutelle A. and Madec L. (2006) Spatial genetic pattern in the land Helix aspersa inferred from a 'centre-based clustering' procedure. Genet Res., 88(1):27-44.

Kanehisa M., Goto S., Kawashima S., Okuno Y. and Hattori M. (2004) The KEGG resource for deciphering the genome. Nucleic Acids Res, 1;32(Database issue):D277-80.

Karp P.D., Ouzounis C.A., Moore-Kochlacs C., Goldovsky L., Kaipa P., Ahren D., Tsoka S., Darzentas N., Kunin V. and Lopez-Bigas N. Expansion of the BioCyc collection of pathway/genome databases to 160 genomes. Nucleic Acids Research, 19:6083-89 2005.

Kishino,H. and Waddell,P.J. (2000) Correspondence analysis of genes and tissue types and finding genetic links from microarray data. Genome Informatics, 11:83–95.

McAdams H. H. and Arkin A. (1998) Simulation Of Prokaryotic Genetic Circuits. Ann. Rev. Biophys. Biom. Struct., 27:199-224.

Opgen-Rhein R.and Strimmer K. (2006) Inferring gene dependency networks from genomic longitudinal data: a functional data approach. REVSTAT 4:53-65.

Rice J.J. and Stolovitzky G. (2004) Making the most of it: Pathway reconstruction and integrative simulation using the data at hand. Biosilico 2(2):70-7.

Romero-Záliz R., Rubio-Escudero C., Cordón O., Harare O., del Val C. and Zwir I. (2006) Mining Structural Databases: An Evolutionary Multi-Objective Conceptual Clustering Methodology. Proceedings of the 4th European Workshop on Evolutionary Computation and Machine Learning in Bioinformatics. Budapest, Hungary.

Rubio-Escudero C., Romero-Záliz R., Cordón O., Harari O., del Val C., Zwir I. (2005) Optimal Selection of Microarray Analysis Methods using a Conceptual Clustering Algorithm. Proceedings of the 4th European Workshop on Evolutionary Computation and Machine Learning in Bioinformatics. Budapest, Hungary.

Schäfer J. and Strimmer K. (2005) An empirical Bayes approach to inferring large-scale gene association networks. Bioinformatics 21: 754-764.

Silvescu A. and Honavar V. (2001) Temporal Boolean Network Models of Genetic Networks and Their Inference from Gene Expression Time Series. Complex Systems, 13(1):54-75.

Simon I., Siegfried Z., Ernst J. and Bar Z. (2005) Combined static and dynamic analysis for determining the quality of time-series expression profiles. Nat. Biotechnol., 23(12):1503-8.

Velarde, Cyntia. (2006) Master Thesis in Comp Sci, University of Buenos Aires, Argentina.

A Genetic Embedded Approach for Gene Selection and Classification of Microarray Data

Jose Crispin Hernandez Hernandez, Béatrice Duval, and Jin-Kao Hao

LERIA, Université d'Angers,
2 Boulevard Lavoisier, 49045 Angers, France
{josehh,bd,hao}@info.univ-angers.fr

Abstract. Classification of microarray data requires the selection of subsets of relevant genes in order to achieve good classification performance. This article presents a genetic embedded approach that performs the selection task for a SVM classifier. The main feature of the proposed approach concerns the highly specialized crossover and mutation operators that take into account gene ranking information provided by the SVM classifier. The effectiveness of our approach is assessed using three well-known benchmark data sets from the literature, showing highly competitive results.

Keywords: Microarray gene expression, Feature selection, Genetic Algorithms, Support vector machines.

1 Introduction

Recent advances in DNA microarray technologies enable to consider molecular cancer diagnosis based on gene expression. Classification of tissue samples from gene expression levels aims to distinguish between normal and tumor samples, or to recognize particular kinds of tumors [9,2]. Gene expression levels are obtained by cDNA microarrays and high density oligonucleotide chips, that allow to monitor and measure simultaneously gene expressions for thousands of genes in a sample. So, data that are currently available in this field concern a very large number of variables (thousands of gene expressions) relative to a small number of observations (typically under one hundred samples). This characteristic, known as the "curse of dimensionality", is a difficult problem for classification methods and requires special techniques to reduce the data dimensionality in order to obtain reliable predictive results.

Feature selection aims at selecting a (small) subset of informative features from the initial data in order to obtain high classification accuracy [11]. In the literature there are two main approaches to solve this problem: the filter approach and the wrapper approach [11]. In the filter approach, feature selection is performed without taking into account the classification algorithm that will be applied to the selected features. So a filter algorithm generally relies on a relevance measure that evaluates the importance of each feature for the classification task. A feasible approach to filter selection is to rank all the features

E. Marchiori, J.H. Moore, and J.C. Rajapakse (Eds.): EvoBIO 2007, LNCS 4447, pp. 90–101, 2007.
© Springer-Verlag Berlin Heidelberg 2007

according to their interestingness for the classification problem and to select the top ranked features. The feature score can be obtained independently for each feature, as it is done in [9] which relies on correlation coefficients between the class and each feature. The drawback of such a method is to score each feature independently while ignoring the relations between the features.

In contrast, the wrapper approach selects a subset of features that is "optimized" by a given classification algorithm, e.g. a SVM classifier [5]. The classification algorithm, that is considered as a black box, is run many times on different candidate subsets, and each time, the quality of the candidate subset is evaluated by the performance of the classification algorithm trained on this subset. The wrapper approach conducts thus a search in the space of candidate subsets. For this search problem, genetic algorithms have been used in a number of studies [15,14,6,4].

More recently, the literature also introduced embedded methods for feature selection. Similar to wrapper methods, embedded methods carry out feature selection as a part of the training process, so the learning algorithm is no more a simple black box. One example of an embedded method is proposed in [10] with recursive feature elimination using SVM (SVM-RFE).

In this paper, we present a novel embedded approach for gene selection and classification which is composed of two main phases. For a given data set, we carry out first a pre-selection of genes based on filtering criteria, leading to a reduced gene subset space. This reduced space is then searched to identify even smaller subsets of predictive genes which are able to classify with high accuracy new samples. This search task is ensured by a specialized Genetic Algorithm which uses (among other things) a SVM classifier to evaluate the fitness of the candidate gene subsets and problem specific genetic operators. Using SVM to evaluate the fitness of the individuals (gene subsets) is not a new idea. Our main contribution consists in the design of semantically meaningful crossover and mutation operators which are fully based on useful ranking information provided by the SVM classifier. As we show in the experimentation section, this approach allows us to obtain highly competitive results on three well-known data sets.

In the next Section, we recall three existing filtering criteria that are used in our pre-selection phase and SVM that is used in our GA. In Section 3, we describe our specialized GA for gene selection and classification. Experimental results and comparisons are presented in Section 4 before conclusions are given in Section 5.

2   Basic Concepts

2.1   Filtering Criteria for Pre-selection

As explained above, microarray data generally concern several thousands of gene expressions. It is thus necessary to pre-select a smaller number of genes before applying other search methods. This pre-selection can be performed by using simply a classical filter method that we recall in this section. The following filtering or relevance criteria assign to each gene a numerical weight that is used to rank all the genes and then to select top ranked genes.

In the rest of the paper, we shall use the following notations. The matrix of gene expression is denoted by $D = \{(X_i, y_i) \mid i = 1, ..., n\}$, where each (X_i, y_i) is a labeled sample. The labels $y_1, ..., y_n$ are taken from a set of labels Y which represent the different classes (for a two class problem $Y = \{-1, 1\}$). Each $X_i = \{x_{i,1}, ..., x_{i,d}\}$ describes the expression values of the d genes for sample i.

The **BW ratio**, introduced by Dudoit et al. [7], is the ratio of between-group to within-group sums of squares. For a gene j, the ratio is formally defined by:

$$BW(j) = \frac{\sum_i \sum_k I(y_i = k)(\bar{x}_{kj} - \bar{x}_j)^2}{\sum_i \sum_k I(y_i = k)(x_{ij} - \bar{x}_{kj})^2} \tag{1}$$

where $I(.)$ denotes the indicator function, equaling 1 if the condition in parentheses is true, and 0 otherwise. $\bar{x}_j$ and $\bar{x}_{kj}$ denote respectively the average expression level of the gene j across all samples and across samples belonging to class k only.

The **Correlation between a gene and a class distinction**, proposed by Golub et al. [9], is defined as follows.

$$P(j) = \frac{\bar{x}_{1j} - \bar{x}_{2j}}{s_{1j} + s_{2j}} \tag{2}$$

where $\bar{x}_{1j}, s_{1j}$ and $\bar{x}_{2j}, s_{2j}$ denote the mean and standard deviation of the gene expression values of gene j for the samples in class 1 and class 2. This measure identifies for a two-class problem informative genes based on their correlation with the class distinction and emphasizes the signal-to-noise ratio by using the gene as a predictor. $P(j)$ reflects the difference between the classes relative to the standard deviation within the classes. Large values of $|P(j)|$ indicate a strong correlation between the gene expression and the class distinction.

The **Fisher's discriminant criterion** [8] is defined by:

$$P(j) = \frac{(\bar{x}_{1j} - \bar{x}_{2j})^2}{((s_{1j})^2 + (s_{2j})^2)} \tag{3}$$

where $\bar{x}_{1j}, (s_{1j})^2$ and $\bar{x}_{2j}, (s_{2j})^2$ denote respectively the mean and variance of the gene expression values of gene j across the class 1 and across the class 2. It gives higher values to features whose means differ greatly between the two classes, relative to their variances.

Any of the above criteria can be used to select a subset G_p of p top ranked genes. In our case, we shall compare, in Section 4, these three criteria and retain the best one to be combined with our genetic embedded approach for gene selection and classification.

2.2   Support Vector Machines (SVMs) and Feature Ranking

In our genetic embedded approach, a SVM classifier is used to evaluate the fitness of a given candidate gene subset. Let us recall briefly the basic concept of

SVM. For a given training set of labeled samples, SVM determines an optimal hyperplane that divides the positively and the negative labeled samples with the maximum margin of separation.

Formally, given a training set belonging to two classes, $\{X_i, y_i\}$ where $\{X_i\}$ are the n training samples with their class labels y_i, a soft-margin linear SVM classifier aims at solving the following optimization problem:

$$\min_{w,b,\xi_i} \frac{1}{2} \|w\|^2 + C \sum_{i=1}^{n} \xi_i \tag{4}$$

subject to $y_i (w \cdot X_i + b) \geq 1 - \xi_i$ and $\xi_i \geq 0$, $i = 1, ..., n$.

C is a given penalty term that controls the cost of misclassification errors.

To solve the optimization problem, it is convenient to consider the dual formulation [5]:

$$\min_{\alpha_i} \frac{1}{2} \sum_{i=1}^{n} \sum_{l=1}^{n} \alpha_i \alpha_l y_i y_l X_i \cdot X_l - \sum_{i=1}^{n} \alpha_i \tag{5}$$

subject to $\sum_{i=1}^{n} y_i \alpha_i = 0$ and $0 \leq \alpha_i \leq C$.

The decision function for the linear SVM classifier with input vector X is given by $f(X) = w \cdot X + b$ with $w = \sum_{i=1}^{n} \alpha_i y_i X_i$ and $b = y_i - w \cdot X_i$.

The weight vector w is a linear combination of training samples. Most weights α_i are zero. The training samples with non-zero weights are support vectors.

In order to select informative genes, the orientation of the separating hyperplane found by a linear SVM can be used, see [10]. If the plane is orthogonal to a particular gene dimension, then that gene is informative, and vice versa. Specially, given a SVM with weight vector w the ranking coefficient vector c is given by:

$$\forall i, c_i = (w_i)^2 \tag{6}$$

For our classification task, we will use such a linear SVM classifier with our genetic algorithm which is presented in the next section. Finally, let us mention that SVM has been successfully used for gene selection and classification [16,17,10,13].

3 Gene Selection and Classification by GA

As explained in the introduction, our genetic embedded approach begins with a filtering based pre-selection, leading to a gene subset G_p of p genes (typically with $p < 100$). From this reduced subset, we will determine an even smaller set of the most informative genes (typically < 10) which allows to give the highest classification accuracy. To achieve this goal, we developed a highly specialized Genetic Algorithm which integrates, in its genetic operators, specific knowledges on our gene selection and classification problem and uses a SVM classifier as one key element of its fitness function. In what follows, we present the different elements of this GA, focusing on the most important and original ingredients:

problem encoding, SVM based fitness evaluation, specialized crossover and mutation operators.

3.1   Problem Encoding

An individual $I = <I^x, I^y>$ is composed of two parts I^x and I^y called respectively *gene subset vector* and *ranking coefficient vector*. The first part, I^x, is a binary vector of fixed length p. Each bit I_i^x ($i = 1...p$) corresponds to a particular gene and indicates whether or not the gene is selected. The second part, I^y, is a positive real vector of fixed length p and corresponds to the ranking coefficient vector c (Equation 6) of the linear SVM classifier. I^y indicates thus for each selected gene the importance of this gene for the SVM classifier.

Therefore, an individual represents a candidate subset of genes with additional information on each selected gene with respect to the SVM classifier. The gene subset vector of an individual will be evaluated by a linear SVM classifier while the ranking coefficients obtained during this evaluation provide useful information for our specialized crossover and mutation operators.

3.2   SVM Based Fitness Evaluation

Given an individual $I = <I^x, I^y>$, the gene subset part I^x, is evaluated by two criteria: the classification accuracy obtained with the linear SVM classifier trained on this subset and the number of genes contained in this subset. More formally, the fitness function is defined as follows:

$$f(I) = \frac{CA_{SVM}(I^x) + \left(1 - \frac{|I^x|}{p}\right)}{2} \tag{7}$$

The first term of the fitness function $(CA_{SVM}(I^x))$ is the classification accuracy measured by the SVM classifier via 10-fold cross-validation. The second term ensures that for two gene subsets having an equal classification accuracy, the smaller one is preferred.

For a given individual I, this fitness function leads to a positive real fitness value $f(I)$ (higher values are better). At the same time, the c vector obtained from the SVM classifier is calculated and copied in I^y which is later used by the crossover and mutation operators.

3.3   Specialized Crossover Operator

Crossover is one of the key evolution operators for any effective GA and needs a particularly careful design. For our search problem, we want to obtain small subsets of selected genes with a high classification accuracy. Going with this goal, we have designed a highly specialized crossover operator which is based on the following two fundamental principles: 1) to conserve the genes shared by both parents and 2) to preserve "high quality" genes from each parent even if they are not shared by both parents. The notion of "quality" of a gene here is defined by the corresponding ranking coefficient in c. Notice that applying the first principle will have as main effect of getting smaller and smaller gene subsets while applying the second principle allows us to keep up good genes along the search process.

More precisely, let $I = < I^x, I^y >$ and $J = < J^x, J^y >$ be two selected individuals (parents), we combine I and J to obtain a single child $K = < K^x, K^y >$ by carrying out the following steps:

1. We use the boolean logic AND operator ($\otimes$) to extract the subset of genes shared by both parents and arrange them in an intermediary gene subset vector F.

$$F = I^x \otimes J^x$$

2. For the subset of genes obtained from the first step, we extract the maximum coefficients max_I and max_J accordingly from their original ranking vectors I^y and J^y.

$$max_I = max \{I^y_i \mid i \text{ such that } F_i = 1\}$$

and

$$max_J = max \{J^y_i \mid i \text{ such that } F_i = 1\}$$

3. This step aims to transmit high quality genes from each parent I and J which are not retained by the logic AND operator in the first step. These are genes with a ranking coefficient greater than max_I and max_J. The genes selected from I and J are stored in two intermediary vectors AI and AJ

$$AI_i = \begin{cases} 1 & if \ I^x_i = 1 \ and \ F_i = 0 \ and \ I^y_i > max_I \\ 0 & otherwise \end{cases}$$

and

$$AJ_i = \begin{cases} 1 & if \ J^x_i = 1 \ and \ F_i = 0 \ and \ J^y_i > max_J \\ 0 & otherwise \end{cases}$$

4. The gene subset vector K^x of the offspring K is then obtained by grouping all the genes of F, AI and AJ using the logical "OR" operator ($\oplus$).

$$K^x = F \oplus AI \oplus AJ$$

The ranking coefficient vector K^y will be filled up when the individual K is evaluated by the SVM based fitness function.

3.4   Specialized Mutation Operator

As for the above crossover operator, we design a mutation operator which is semantically meaningful with respect to our gene selection and classification problem. The basic idea is to eliminate some "mediocre" genes and at the same time introduce randomly other genes to keep some degree of diversity in the GA population.

Given an individual $I = < I^x, I^y >$, applying the mutation operator to I consists in carrying out the following steps.

1. The first step calculates the average ranking coefficient of a gene in the individual I.

$$\bar{c} = \frac{\sum_{k=1}^{p} I_k^y}{p}$$

2. The second step eliminates (with a probability) "mediocre" genes (*i.e.* inferior to the average) and for each deleted gene introduces randomly a new gene. $\forall I_i^x = 1$ and $I_i^y < \bar{c}$ $(i = 1...p)$, mutate I_i^x with probability p^m. If a mutation does occur, take randomly a I_j^x such that $I_j^x = 0$ and set I_j^x to 1.

3.5 The General GA and Its Other Components

An initial population P is randomly generated such that the number of genes by each individual varies between p and $p/2$ genes. From this population, the fitness of each individual I is evaluated using the function defined by the formula 7. The ranking coefficient vector c of the SVM classifier is then copied to I^y.

To obtain a new population, a temporary population P' is used. To fill up P', the top 40% individuals of P are first copied to P' (elitism). The rest of P' is completed with individuals obtained by crossover and mutation. Precisely, Stochastic Universal Selection is applied to P to generate a pool of $|P|$ candidat individuals. From this pool, crossover is applied $0.4 * |P|$ times to pairs of randomly taken individuals, each new resulting individual being inserted in P'. Similarly, mutation is applied $0.2 * |P|$ times to randomly taken individuals to fill up P'. Once P' is filled up, it replaces P to become the current population. The GA stops when a fixed number of generations is reached.

4   Experimental Results

4.1 Data Sets

We applied our approach on three well-known data sets that concern leukemia, colon cancer and lymphoma.

The leukemia data set consists of 72 tissue samples, each with 7129 gene expression values. The samples include 47 acute lymphoblastic leukemia (ALL) and 25 acute myeloid leukemia (AML). The original data are divided into a training set of 38 samples and a test set of 34 samples. The data were produced from Affymetrix gene chips. The data set was first used in [9] and is available at http://www-genome.wi.mit.edu/cancer/.

The colon cancer data set contains 62 tissue samples, each with 2000 gene expression values. The tissue samples include 22 normal and 40 colon cancer cases. The data set is available at http://www.molbio.princeton.edu/colondata and was first studied in [2].

The lymphoma data set is based on 4096 variables describing 96 observations (62 and 34 of which are respectively considered as abnormal and normal). The data set was first analyzed in [1]. This data set has already been used for benchmarking feature selection algorithms, for instance in [17,16]. The data set is available at http://www.kyb.tuebingen.mpg.de/bs/people/spider/.

Prior to running our method, we apply a linear normalization procedure to each data set to transform the gene expressions to mean value 0 and standard deviation 1.

4.2   Experimentation Setup

The following sub-sections present and analyze the different experiments that we carried out in order to compare our approach with other selection methods. We present here the general context that we adopt for our experimentations.

Accuracy evaluation is realized by cross validation, as it is commonly done when few samples are available. To avoid the problem of selection bias that is pointed out in [3] and following the protocol suggested in the same study, we use a cross-validation process that is external to the selection process. At each iteration, the data set is split into two subsets, a training set and a test set. Our method of selection is applied on the training set and the accuracy of the classifier is evaluated on the test set (which is not used in the selection process). 50 independent runs are performed, with a new split of the data into a training set and a test set each time. We report in the following the average results (accuracy, number of genes) obtained on these 50 runs. This experimental setup is used in many other works, even if the number of runs may be different. Let us note that our GA also requires an internal cross-validation to estimate the classifier accuracy during the selection process [11].

For the genetic algorithm, both the population size and the number of generations are fixed at 100 for all the experimentations presented in this section. The crossover and mutation operators are applied as explained in Section 3.5.

4.3   Comparison of Pre-selection Criteria

The first experiment aims to determine the best filtering criterion that will be used in our pre-selection phase to obtain an initial and reduced gene subset G_p. To compare the three criteria presented in Section 2, we apply each criterion to each data set to pre-select the p top ranked genes and then we apply our genetic algorithm to these p genes to seek the most informative ones. We experiment with different values of p ($p=50\ldots150$) and we observe that large values of p does not affect greatly the classification accuracy, but necessarily increase the computation times. So we decide to pre-select $p=50$ genes.

In order to compare the three filtering criteria, we report in Table 1 the final number of selected genes, the classification accuracy evaluated on the training set and on the test set. From Table 1, one observes that the best results are obtained with the BW ratio measure. Therefore our following experiments are carried out with the BW ratio criterion and the number p of pre-selected genes is always fixed at 50.

4.4   Comparison with Other Selection Methods

In this section, we show two comparative studies. The first compares our method with two well known SVM based selection approaches reported in [16,17]. We

Table 1. Comparison of three pre-selection criteria. NG is the mean and standard deviation of the number of selected genes, AcTr (resp. AcTe) is the average classification rate (%) on training set (resp. on test set).

Dataset	BW ratio criteria			Correlation criteria			Fisher's Criterion		
	NG	$AcTr$	$AcTe$	NG	$AcTr$	$AcTe$	NG	$AcTr$	$AcTe$
Leukemia	3.93±1.16	98.27	89.05	5.07±1.98	94.40	85.59	4.71±1.44	96.59	86.95
Colon	8.05±1.57	90.62	78.81	10.43±2.77	85.47	76.32	9.17±2.03	87.16	76.59
Lymphoma	5.96±1.31	96.53	88.27	8.01±1.94	92.90	84.47	7.13±1.86	93.82	86.02

also carry out a comparison with two highly effective GA-based gene selection approaches [12,14]. Unlike some other studies, these studies are based on the same experimental protocol as ours that avoids the selection bias problem pointed out in [3].

Comparison with SVM-Based Selection Approaches. In [16], the author reports an experimental evaluation of several SVM-based selection methods. For comparison purpose, we adopt the same experimental methodology. In particular, we fix the number of selected genes and adapt our GA to this constraint (the fitness is then determined directly by the classification accuracy of the classifier, c.f. Equation 7). In Table 2, we report the classification accuracies when the number of genes is fixed at 20 and we compare the best results reported in [16] and our results for the data sets concerning the colon cancer and the lymphoma ([16] does not give information about the leukemia data set).

In [17], the authors propose a method for feature selection using the zero-norm (**A**pproximation of the ze**ro**-norm **M**inimization, AROM), and also gives results concerning the colon and lymphoma data sets that we report in Table 2.

Table 2. A comparison of SVM-based selection methods and our method. The columns indicate: the mean and standard deviation of classification rates on test set (Ac), the number of trials (NT), and the number of samples in the test set (NSa).

Dataset	[17]			[16]			Our method		
	Ac (%)	NT	NSa	Ac (%)	NT	NSa	Ac (%)	NT	NSa
Colon	85.83 ±2.0	30	12	82.33 ±9	100	12	82.52 ±8.68	50	12
Lymphoma	91.57 ±0.9	30	12	92.28 ±4	100	36	93.05 ±2.85	50	36

From Table 2, one observes that for the lymphoma data set, our method obtains a better classification accuracy (higher is better). For the colon data set, our result is between the two reference methods. Notice that in this experiment we restrict our method since the number of selected genes is arbitrarily fixed while our method is able to select dynamically subsets of informative genes. The following comparison provides a more interesting experimentation where the number of genes will be determined by the genetic search.

Comparison with Other Genetic Approaches. In [12], the authors propose a multiobjective evolutionary algorithm (MOEA), where the fitness function evaluates simultaneously the misclassification rate of the classifier, the difference in error rate among classes and the number of selected genes. The classifier used in this work was the weighted voting classifier proposed by [9].

In [14], the authors present a probabilistic model building genetic algorithm (PMBGA) as a gene selection algorithm. The Naive-Bayes classifier and the weighted voting classifier are used to evaluate the selection method in a Leave-One-Out-Cross-Validation process.

Table 3 shows our results on the three data sets together with those reported in [12] and [14]. One can observe that our method gives better results than [12], in the sense that the number of selected genes is smaller and the accuracy is higher. Concerning [14], our results are quite comparable.

Table 3. Comparison of other genetic approaches and our method. The columns indicate: the mean and standard deviation of the number of selected genes (NG), the mean and standard deviation of classification rates on test set (Ac). We also report the number (in two cases) or the percentage of samples that form the test set (NSa) for the experiments.

	[12]			[14]			Our method		
Dataset	*NG*	*Ac (%)*	*NSa*	*NG*	*Ac (%)*	*NSa*	*NG*	*Ac (%)*	*NSa*
Leukemia	15.2±4.54	90 ±7.0	30%	3.16±1.00	90 ±6	34	3.17±1.16	91.5 ±5.9	34
Colon	11.4±4.27	80 ±8.3	30%	4.44±1.74	81 ±8	50%	7.05±1.07	84.6 ±6.6	50%
Lymphoma	12.9±4.40	90 ±3.4	30%	4.42±2.46	93 ±4	50%	5.29±1.31	93.3 ±3.1	50%

We must mention that we report the average results obtained by a 10-fold cross validation, but we observe that in some experiments, our method achieves a perfect classification (100% accuracy). Finally, let us comment that these results are comparable to those reported in [6] and better than those of [15].

5   Conclusions and Future Work

In this paper, we have presented a genetic embedded method for gene selection and classification of Microarray data. The proposed method is composed of a pre-selection phase according to a filtering criterion and a genetic search phase to determine the best gene subset for classification. While the pre-selection phase is conventional, our genetic algorithm is characterized by its highly specialized crossover and mutation operators. Indeed, these genetic operators are designed in such a way that they integrate gene ranking information provided by the SVM classifier during the fitness evaluation process. In particular, the crossover operator not only conserves the genes shared by both parents but also uses SVM ranking information to preserve highly ranked genes even if they are not shared by the parents. Similarly, the gene ranking information is incorporated into the mutation operator to eliminate "mediocre" genes.

Using an experimental protocol that avoids the selection bias problem, our method is experimentally assessed on three well-known data sets (colon, leukemia and lymphoma) and compared with several state of the art gene selection and classification algorithms. The experimental results show that our method competes very favorably with the reference methods in terms of the classification accuracy and the number of selected genes.

This study confirms once again that genetic algorithms constitute a general and valuable approach for gene selection and classification of microarray data. Its effectiveness depends strongly on how semantic information of the given problem is integrated in its genetic operators such as crossover and mutation. The role of an appropriate fitness function should not be underestimated. Finally, it is clear that the genetic approach can favorably be combined with other ranking and classification methods.

Our ongoing works include experimentations of the proposed method on more data sets, studies of alternative fitness functions and searches for other semantic information that can be used in the design of new genetic operators.

Acknowledgments. The authors would like to thank the referees for their helpful suggestions which helped to improve the presentation of this paper. This work is partially supported by the French Ouest Genopole®. The first author of the paper is supported by a Mexicain COSNET scholarship.

References

1. A. Alizadeh, M.B. Eisen, E. Davis, C. Ma, I. Lossos, A. Rosenwald, J. Boldrick, H. Sabet, T. Tran, X. Yu, J.I. Powell, L. Yang, G.E. Marti, J. Hudson Jr, L. Lu, D.B. Lewis, R. Tibshirani, G. Sherlock, W.C. Chan, T.C. Greiner, D.D. Weisenburger, J.O. Armitage, R. Warnke, R. Levy, W. Wilson, M.R. Grever, J.C. Byrd, D. Botstein, P.O. Brown, and L.M. Staudt. Distinct types of diffuse large B–cell lymphoma identified by gene expression profiling. *Nature*, 403:503–511, February 2000.
2. U. Alon, N. Barkai, D. A. Notterman, K. Gish, S. Ybarra, D. Mack, and A. J. Levine. Broad patterns of gene expression revealed by clustering analysis of tumor and normal colon tissues probed by oligonucleotide arrays. *Proc Natl Acad Sci USA*, 96:6745–6750, 1999.
3. C. Ambroise and G.J. McLachlan. Selection bias in gene extraction on the basis of microarray gene-expression data. *Proc Natl Acad Sci USA*, 99(10):6562–6566, 2002.
4. E. Bonilla Huerta, B. Duval, and J.-K. Hao. A hybrid ga/svm approach for gene selection and classification of microarray data. *Lecture Notes in Computer Science*, 3907:34–44, Springer, 2006.
5. B. E. Boser, I. Guyon, and V. Vapnik. A training algorithm for optimal margin classifiers. In *Proceedings of the Fifth Annual Workshop on Computational Learning Theory*, pages 144–152, ACM Press, 1992.
6. K. Deb and A. R. Reddy. Reliable classification of two-class cancer data using evolutionary algorithms. *Biosystems*, 72(1-2):111–29, Nov 2003.

7. S. Dudoit, J. Fridlyand, and T. P. Speed. Comparison of discrimination methods for the classification of tumors using gene expression data. *Journal of the American Statistical Association*, 97(457):77–87, 2002.

8. R. O. Duda and P. E. Hart. *Pattern Classification and scene analysis*. Wiley, 1973.

9. T. R. Golub, D. K. Slonim, P. Tamayo, C. Huard, M. Gaasenbeek, J. P. Mesirov, H. Coller, M. L. Loh, J. R. Downing M. A. Caligiuri, C. D. Bloomfield, and E. S. Lander. Molecular classification of cancer: Class discovery and class prediction by gene expression monitoring. *Science*, 286:531–537, 1999.

10. I. Guyon, J. Weston, S. Barnhill, and V. Vapnik. Gene selection for cancer classification using support vector machines. *Machine Learning*, 46(1-3):389–422, 2002.

11. R. Kohavi and G.H. John. Wrappers for feature subset selection. *Artificial Intelligence*, 97(1-2):273–324, 1997.

12. J. Liu and H. Iba. Selecting informative genes using a multiobjective evolutionary algorithm. In *Proceedings of the 2002 Congress on Evolutionary Computation*, pages 297–302, IEEE Press, 2002.

13. E. Marchiori, C. R. Jimenez, M. West-Nielsen, and N. H. H. Heegaard. Robust svm-based biomarker selection with noisy mass spectrometric proteomic data. *Lecture Notes in Computer Science*, 3907:79–90, Springer, 2006.

14. T.K. Paul and H. Iba. Selection of the most useful subset of genes for gene expression-based classification. *Proceedings of the 2004 Congress on Evolutionary Computation*, pages 2076–2083, IEEE Press, 2004.

15. S. Peng, Q. Xu, X.B. Ling, X. Peng, W. Du, and L. Chen. Molecular classification of cancer types from microarray data using the combination of genetic algorithms and support vector machines. *FEBS Letters*, 555(2):358–362, 2003.

16. A. Rakotomamonjy. Variable selection using svm-based criteria. *Journal of Machine Learning Research*, 3:1357–1370, 2003.

17. J. Weston, A. Elisseeff, B. Scholkopf, and M. Tipping. The use of zero-norm with linear models and kernel methods. *Journal of Machine Learning Research*, 3(7-8):1439–1461, 2003.

Modeling the Shoot Apical Meristem in *A. thaliana*: Parameter Estimation for Spatial Pattern Formation

Tim Hohm and Eckart Zitzler

Computer Engineering (TIK), ETH Zurich
{hohm,zitzler}@tik.ee.ethz.ch
http://www.tik.ee.ethz.ch/sop/

Abstract. Understanding the self-regulatory mechanisms controlling the spatial and temporal structure of multicellular organisms represents one of the major challenges in molecular biology. In the context of plants, shoot apical meristems (SAMs), which are populations of dividing, undifferentiated cells that generate organs at the tips of stems and branches throughout the life of a plant, are of particular interest and currently studied intensively. Here, one key goal is to identify the genetic regulatory network organizing the structure of a SAM and generating the corresponding spatial gene expression patterns.

This paper addresses one step in the design of SAM models based on ordinary differential equations (ODEs): parameter estimation for spatial pattern formation. We assume that the topology of the genetic regulatory network is given, while the parameters of an ODE system need to be determined such that a particular stable pattern over the SAM cell population emerges. To this end, we propose an evolutionary algorithm-based approach and investigate different ways to improve the efficiency of the search process. Preliminary results are presented for the Brusselator, a well-known reaction-diffusion system.

1 Motivation

Ordinary differential equations (ODEs) represent a common approach to model genetic regulatory networks [1]. Such models are on the one hand used to quantitatively understand the interactions of multiple genes controlling specific cellular processes and on the other hand applied to make predictions about the cell behavior. One important and challenging problem in this context is the determination of the model parameters that lead to the desired temporal dynamics. For single cell networks, there has been a lot of work on parameter estimation using analytical as well as heuristic methods [13]; in particular, several studies make use of evolutionary algorithms to find suitable parameter settings [8,9,11].

This paper considers a slightly different problem where the focus is on multicellular systems, in particular the shoot apical meristems (SAMs) in the plant *Arabidopsis thaliana*. The main goal is to identify an ODE system that is capable of producing an (experimentally observed) spatial gene expression pattern

E. Marchiori, J.H. Moore, and J.C. Rajapakse (Eds.): EvoBIO 2007, LNCS 4447, pp. 102–113, 2007.

across the cell population, assuming that gene products can cross cell borders via diffusion. Starting with a given set of gene interactions in terms of an ODE system, we address the problem of model parameter determination for such a spatial scenario. In comparison to previous studies on parameter estimation, there are several differences with respect to the scenario under investigation:

- Instead of a single cell, multiple interacting cells are considered which requires a prespecified spatial cell structure and a cell interaction model;
- Instead of achieving a particular temporal behavior, we are interested in obtaining a stable, i.e., non-oscillating system state in which a particular gene expression pattern emerges over the spatial cell structure;
- Instead of considering absolute gene product concentrations as target values, the gene expression patterns are rather defined qualitatively since quantitative measurements in space are scarcely available.

It is an open question of how to efficiently search for model parameters in such a scenario and how to formalize spatial patterns in terms of an objective function.

In the following, we present a preliminary study for this problem where a more general goal is taken as a basis: we do not assume a given target pattern, but aim at finding parameter settings that produce arbitrary, non-chaotic patterns. We first propose a general modeling framework which allows to simulate genetic regulatory networks within multicellular systems. Secondly, for a simple reaction-diffusion system with two genes that has been part of a previously published model for the shoot apical meristem by Jönsson et al. [6], we investigate the issue of parameter estimation. To this end, we introduce and apply an evolutionary approach based on the Covariance Matrix Adaption Evolution Strategy (CMA-ES) [3,4] and investigate different ways to improve the efficiency of the search.

2   Background

2.1   The Shoot Apical Meristem (SAM)

A shoot apical meristem (SAM) consists of multiple dividing, undifferentiated cells and is located at the tips of stems or branches of a plant. It is responsible for generating organs throughout the life of a plant and determines the number, type and position of the resulting lateral organs. A SAM has a particular internal organization that is preserved through its existence and its position at the tip of the stem or a branch remains fixed, although the plant is growing. Therefore, a fundamental question in meristem research is what this structure looks like and how it is maintained.

In various experimental studies, a number of genes and gene interactions have been identified that are involved in the organization of a SAM. At the heart of preserving the organization and functioning of a SAM is a negative feedback loop with two critical elements, the transcription factor gene WUS and the CLAVATA (CLV) genes, which encode components of a ligand/receptor complex.

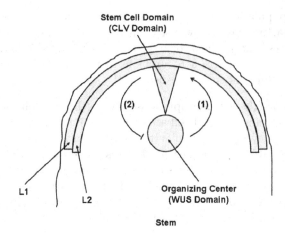

Fig. 1. A sketch of the SAM summarizing the known structural constituents and showing the WUS-CLV feedback loop. In the middle of the model, the organizing center is located. Directly on top of this domain the triangular shaped CLV stem cell domain begins and stretches up to the outermost cell layers L_1 and L_2. This setup remains stable throughout the life of the plant. During growth, cells from the CLV stem cell area move laterally, differentiate and thereby contribute to the plants growth. In regulating the maintenance of this spatial pattern the WUS-CLV feedback loop indicated by (1) and (2) plays a central role. Starting from the organizing center it promotes its own growth and the regeneration of the CLV domain (1) which lost cells due to differentiation. To prevent the system from over stimulating growth in CLV domain and organizing center, in turn CLV3 produced in the topmost layers of the CLV domain (L_1, L_2) gives negative feedback to the WUS organizing center (2). As a result of this interplay both, CLV domain and organizing center can maintain a stable size.

This negative-feedback loop elegantly corrects transient aberrations in stem-cell number. Besides these relatively well-characterized regulators, a range of other elements has been identified. In many cases their function in the meristem is unclear and, so far, there is no overall picture of the genetic regulatory networks in a SAM. Fig. 1 schematically summarizes the main constituents of a SAM that are currently known.

A current limitation in meristem research is the resolution of the measurements. Ideally, for each gene the gene product concentration within each cell of the meristem separately would be known, but it is obvious that such type of measurements are utopian for the near future. For this reason, data-driven modeling approaches where genetic regulatory networks are inferred from quantitative data are currently infeasible. Instead, a knowledge-driven approach is pursued where the topology of the network is determined by hand based on previous knowledge, and novel hypotheses are tested by slightly modifying the existing network and validating it with regard to phenotypic data.

2.2   Pattern Formation

One approach to study the regulation mechanisms enabling plants to maintain these spatial SAM patterns is to use reaction-diffusion systems, a well understood system to produce spatial patterns in general, dating back to work by Turing [12]. He investigated the influence of diffusion as a spatial component on systems described by coupled non-linear differential equations. In contrast to the predominant opinion, he found out that systems which converge to a homogeneous steady state without diffusion can be perturbed in such a way that they form either spatially stable patterns over time or temporally stable patterns in the spatial domain. Using similar systems many pattern forming dynamics in sea shells [7], development of animals like hydra [2] or drosophila [5] have been investigated.

In the context of SAM modeling, Jönsson et al. [6] employed reaction-diffusion systems to simulate the domain formation and maintenance in the SAM. In their work, the authors used a two dimensional model only considering the WUS-CLV feedback loop extended by an additional activator substance; the reported results in simulating phenotypic observations in SAM development and maintenance are promising. As to model parameter determination, their model consisted only of few constituents and therefore it was possible to tune the parameters by hand. Considering the fact that these systems are sensitive to either start conditions and parameters like coupling constants, degradation rates and production rates, it is likely that tuning more complex models by hand becomes intractable. Therefore we here present a method which, using a model similar to the one from Jönsson et al., (1) optimizes parameters of the system in such a way that spatial patterns are formed and (2) thereby can be used to explore the pattern formation capabilities of that given setup.

3   A SAM Modeling Framework

In the following, we present a modeling framework for multicellular systems in general and SAMs in particular that serves two goals: hypothesis testing and hypothesis exploration. On the one hand, it should be testable whether a given system of interacting factors can form certain spatial patterns by finding the necessary parameter settings and simulating the system. On the other hand, based on the parameter optimization, predictions on the possible patterns of novel interactions resulting from novel intracellular and intercellular interactions shall be made.

3.1   Model Structure

The model proposed here is defined by the following core components:

Cells: The model consists of spatially discrete units, the cells. They are used as autonomous units. We assume that all cells are similar to each other in design, in particular regarding the underlying genetic regulatory network, and only differ in their states.

Gene products: The state of a cell v_i is characterized by the concentrations of the gene products produced in the cell. The gene product concentrations are represented by a real valued vector. The term 'gene product' in this case not only refers the the product but has to be understood synonymous for gene products, gene expression levels and all processes on the way from gene to gene product. Since there exists a mapping between expression levels and the resulting amount of gene products, the gene product concentrations are representing the gene expression levels.

Cell structure: The cells are grouped according to a spatial neighborhood defining which cells share common cell surface areas. In this model only a two dimensional horizontal cut through the SAM is considered. We assume that the cells are hexagonal and the cell plane is arranged in rings around a central cell. A schematic picture of the plane is given in Fig. 2. Internally the cell neighborhood is represented by a graph $G(V, E)$ consisting of a set of cells V. Contacts or interaction pathways between the cells are represented by edges $e_{i,j} \in E$ between two cells v_i and v_j.

Cell communication: To form spatial heterogeneous patterns, spatial interactions, namely diffusion, between the constituting components are mandatory. In this model diffusive interactions are possible along the edges between the cells. Therefore implicitly zero flux boundary conditions are used on the boundaries of the cell plane.

The framework is implemented in Java and for the graph representation the JUNG library is used.

3.2 Model Dynamics

During the simulation process the states of the cells change according to (1) intracellular interaction between genes or gene products and (2) the intercellular diffusion. In a formal description the state change of a cell v_i follows a transition function $\delta(q_i, N(v_i))$ depending on the current state q_i of the cell and the states of its interaction partners given by the neighboring vertices $N(v_i)$. Each iteration in the simulation corresponds to calculating the transitions made for every cell based on the status quo.

The reaction equations describing the intracellular interactions can easily be transformed to ordinary differential equations, using the reaction rates from the reaction equations as parameters. The time course of ODE systems can be simulated by numerical integration. Since the intracellular interactions are already represented by ODEs, it is convenient to express the diffusion by ODEs as well. The used ODE approximation for diffusion is given in Eq. 1,

$$\frac{dx_{i,j}}{dt} = \sum_{k \in N(v_i)} D_j(x_{k,j} - x_{i,j}) \tag{1}$$

where $x_{i,j}$ is the concentration of gene product j in cell v_i, $N(v_i)$ encompasses all cells in contact with v_i, and D_j is the diffusion constant for the type of gene

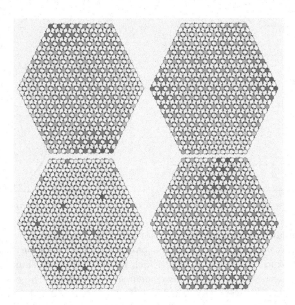

Fig. 2. Activator inhibitor patterns resulting from the Brusselator reaction-diffusion system for two different parameter sets on two dimensional cell planes with a hexagonal lattice. Each vertex represents a cell, each edge indicates an interaction pathway between cells. The cells are colored according to the concentrations of a single gene product. The activator patterns are shown in the left column and the corresponding inhibitor patterns in the right column. Gene product concentration levels are relative and range from low (light color) to high (dark color). The first row shows the patterns simulated using the parameters recorded by Jönsson et al. [6] and the second row shows patterns resulting from parameter optimization using our framework. The difference in size of the patches with high activator concentrations between both parameter sets stems from the difference in the activator diffusion constant D_A. For the optimized parameter set it is smaller and therefore the activator peaks are more local.

product. For our two dimensional meristem simulations, the system is integrated for 5000 steps using a fourth order Runge Kutta integrator with fixed step size $\Delta_t = 0.1$.

Additionally to reaction rates and diffusion constants, the starting conditions or initial gene product concentrations can be considered as a third group of parameters. Due to the non-linearity of the considered system, already slight changes in any of the parameter settings can result in drastic changes in the system behavior whilst the system can be highly robust with respect to other variations. To illustrate this fact, in Fig. 3 two simulation runs of a one dimensional reaction-diffusion system with slightly varying parameter settings are shown. This system, namely the Brusselator, was introduced in 1968 by Prigogine and Lefever [10] and ranges among the best studied reaction-diffusion systems. It is defined by the two equations:

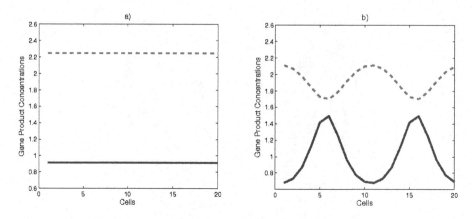

Fig. 3. Two simulations using the Brusselator activator (solid lines) inhihibitor (dashed lines) reaction-diffusion system with similar parameters but one. One diffusion constant (D_b) is changed from 0.69 (a) to 0.7 (b) which results in a state change of the system from spatial homogeneous to spatial heterogeneous waves for both gene products. The other parameter settings were: $a = 0.1$, $b = 0.2$, $\beta = 0.1$, $c = 0.1$, $D_a = 0.1$.

$$\frac{dA}{dt} = a - (b + \beta)A + cA^2B + D_a\nabla^2 A \qquad (2)$$

$$\frac{dB}{dt} = bA - cA^2B + D_b\nabla^2 B \qquad (3)$$

3.3   A Clavata-Wuschel Model

As mentioned in Section 2.1, the feedback loop between CLV produced in the L_1 layer and WUS produced in the organizing center is one of the key regulation mechanisms for maintaining a stable SAM. Jönsson et al. [6] simulated this feed-back loop complemented by a an activator inhibitor reaction-diffusion system. They decided to use the Brusselator model explained in Sec. 3.2 for this task and following this suggestion we use the same system for this study. With help of this system, Jönsson and coworkers were able to reproduce similar pattern formation for the considered horizontal cut through the SAM when compared to the *in vivo* SAM either unperturbed or after laser ablation of the WUS producing or-ganizing center (cf. Fig. 1). Since in our study we are only interested in pattern formation in general, we reduce their model to the Brusselator equations.

4   Model Parameter Estimation

This study is concerned with investigating ways to optimize parameters for the SAM model based on reaction-diffusion systems. The considered optimization problem can be summarized by the following design parameters:

- Search space $X \subseteq \mathbb{R}^n$, for the Brusselator $n = 6$,
- objective space $Z = \mathbb{R}$,
- objective function $f(x) : \mathbb{R}^n \rightarrow \mathbb{R}$ evaluating the resulting patterns for the given parameter set considering the two aspects (1) stability of the pattern over time and (2) significance of the heterogeneity of the resulting pattern.

In the following, we present two types of objective functions used during optimization with the Covariance Matrix Adaption Evolution Strategy (CMA-ES) developed by Hansen and Ostermeier in 1996 [3,4] – a state of the art stochastic optimization method already successfully applied to several real valued optimization problems. The first type of objective functions is designed to avoid using any domain knowledge. Therefore it represents a baseline approach for optimizing reaction-diffusion systems. Secondly, we consider a set of methods to incorporate domain knowledge into the objective function in order to improve the quality of the patterns found.

4.1   Baseline Approach

Method. For the baseline approach we used both spatial heterogeneous gene product concentration distribution and convergence of the gene product concentrations over time and aggregated them into a single objective $f(x)$ as follows:

$$f(x) = \sum_{i \in gp} \left(\max(\delta_t - \Delta_{s_i}, 0) + \Delta_{t_i} \right), \tag{4}$$

where gp are all gene products, Δ_{s_i} is the maximal difference in gene product i measured over all cells at the end of the simulation and δ_t is a threshold value which is used to decide if a given spatial heterogeneity is significant. For our simulations $\delta_t = 0.5$ was used. Δ_{t_i} is the largest change in gene product concentration i in the last integration step. The first term in the fitness function can be seen as a penalty term on parameter settings that fail to generate a stable pattern. In effect, the second term penalizes settings for which the simulation does not converge within the given number of integration steps.

Results. Using the described fitness function we made eleven optimization runs using the CMA-ES. Due to runtime constraint, one optimization run took up to 4 hours, for each variant only eleven runs were conducted. The used (4, 9)CMA-ES parameter values are given in Tab. 1 and the results are shown in Fig. 4.

The undertaken optimization runs failed to converge to an optimum within 1000 objective function evaluations. After investigating which parts of the parameter space had been explored during the optimization runs, it turned out that only 3 percent of the tested settings had relations between the activator diffusion constant D_a and the diffusion constant of the inhibitor D_b of $\frac{D_a}{d_b} \leq \frac{1}{7}$. Although it is known from literature that pattern formation using reaction-diffusion systems only takes place if for the relation of the diffusion constants $\frac{D_a}{d_b} \leq \frac{1}{7}$ holds, the idea behind this base approach was to avoid using domain knowledge and thereby testing the feasibility of our parameter optimization approach on general ODE systems.

Table 1. Parameter settings for CMA-ES

Parameter	Value
Initial ODE Parameters $[a, b, \beta, c, D_a, D_b]$	$[0.45, 0.45, 0.45, 0.45, 0.8, 0.8]$
Initial Standard deviation for the Parameters	$[0.25, 0.25, 0.25, 0.25, 0.7, 0.7]$
Maximal Number of Objective function Evaluations	1000

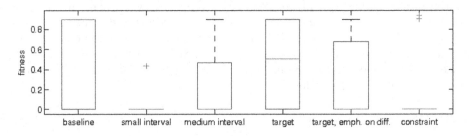

Fig. 4. The results of all conducted runs are shown in boxplots. 'baseline' refers to the first variant not incorporating any domain knowledge, 'interval small' and 'interval medium' both refer to the variant were the optimization process operated on a pre-defined search interval, 'target' refers to the variant where the domain knowledge was integrated by guiding the search to the vicinity of a known solution, 'target, emph. on diff.' denotes the runs using a target setting with an emphasize on the diffusion relation and finally 'constraint' refers to the variant where knowledge about the dependency of the diffusion constants was used.

4.2   Integration of Domain Knowledge

Method. Considering the difficulties in optimizing the parameters without domain knowledge, we decided to include domain knowledge into the optimization process. We tested three different approaches:

1. Restricting the initial search interval of the CMA to a smaller interval which is known or suspected to contain good parameter settings,
2. introducing a term pointing to a region that it is known to be good and thereby generating bias towards this region,
3. constraining the parameters considering known dependencies between parameters like the relation between diffusion constants.

The first approach is trying to increase the probability of identifying a good solution by simply regarding a smaller search space.

Since a study using a sampling grid on the diffusion constants and fixing all other parameters showed that the fitness landscape for the screened part of the parameter space consists of mainly two plateaus, a small sink containing the pattern forming settings and a large plateau of settings for which the system converges to a spatially homogeneous state (cf. Fig. 5), we introduced the latter

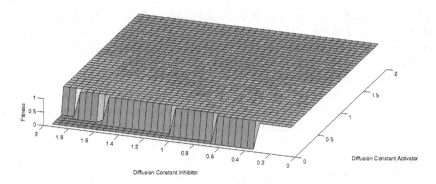

Fig. 5. Slice of the fitness landscape resulting from the objective function given in Eq. 4: For this slice only two of the six parameters of the Brusselator reaction-diffusion system (cp. Eq. 3) are considered, namely the diffusion constants D_A and D_B. They were sampled in the interval $[0.05, 1.95]$ using 0.05 steps while the other parameters are fixed. It can be seen that only for small activator diffusion constants there are pattern forming sets and therefore for most of the tested settings no pattern formation takes place. Further on it can be seen that the transition between pattern forming settings and non-pattern forming settings is ridge like.

two approaches. Both aiming at reshaping the fitness landscapes to become easier to optimize. By integrating new terms in the fitness function, higher plateaus are slightly inclined to point at pattern forming parameter regions. The second approach to this end uses a term penalizing distance to a known promising region. The resulting objective function reads as follows:

$$f(x) = \begin{cases} \sum_{i \in gp}(\max(\delta_t - \Delta_{s_i}, 0) + \Delta_{t_i}) + \|x - x_t\| & \text{if } \|x - x_t\| > \delta_d, \\ \sum_{i \in gp}(\max(\delta_t - \Delta_{s_i}, 0) + \Delta_{t_i}) & \text{else,} \end{cases} \quad (5)$$

where x_t is the target parameter vector and δ_d is a minimal length of the difference vector of x and x_t.

The third approach follows a more general idea: It exploits the knowledge about the necessary relation between the two diffusion constants D_A and D_B. Whenever the relation between both constants exceeds a threshold of 0.1, the actual relation is added to the function value. In effect, the search space is constraint and the resulting objective function reads as follows:

$$f(x) = \begin{cases} \sum_{i \in gp}(\max(\delta_t - \Delta_{s_i}, 0) + \Delta_{t_i}) + \frac{D_A}{D_B} & \text{if } \frac{D_A}{D_B} > 0.1, \\ \sum_{i \in gp}(\max(\delta_t - \Delta_{s_i}, 0) + \Delta_{t_i}) & \text{else.} \end{cases} \quad (6)$$

Results. For all mentioned approaches we did eleven optimizations runs each. Using the two different interval sizes around a parameter set found using the optimization framework ($a = 0.3, b = 0.05, \beta = 0.05, c = 0.25, D_A = 0.075, D_B = 1.525$) and σ settings for the corresponding search distribution in the CMA-ES of $\sigma \in \{0.1, 0.3\}$, for the small σ reproducibly good solutions were found whereas already for the medium σ the results became significantly worse. Therefore the successes have to be contributed to the small size of the explored search space rather than to an effective optimization.

Using the parameter vector ($a = 0.1, b = 0.2, \beta = 0.1, c = 0.1, D_A = 0.1, D_B = 1.5$) as a target vector generating search direction towards a good region in parameter space (for the CMA-ES the parameter settings in Tab. 1 were used), the obtained results were reasonable but still the runs did not converge to a setting with an objective value below $1 * 10^{-14}$, the convergence threshold used by the CMA-ES. This can be attributed to the fact that by taking the euclidean distance between the parameter vector describing the desired parameter vector and the actual parameter vector, all parameters equally contribute to the distance between the two vectors. Since D_A is an important parameter that is measured in smaller scale than the other parameters, its contribution to the search direction is overpowered by the others and the generated signal is blurred. And in fact, emphasizing the diffusion relation improved the convergence.

Coping with this problem brings us to our last approach. Here a desired minimal relation of $\frac{D_A}{D_B} = 0.1$ is used as a constraint (for the CMA-ES the parameter settings in Tab. 1 were used). Compared to all other approaches, it was only outperformed by the approach searching in a small already known region. Schematic pictures of the resulting patterns are given in Fig. 2. When again looking at the number of evaluated settings having a suitable diffusion constant relation, it turned out that for this last setup more than 50 percent were sufficiently small. The results for all approaches are shown in Fig. 4.

5   Conclusions

In this paper, we have studied the problem of parameter estimation for ODE models of genetic regulatory networks in order to generate spatial gene expression patterns over a population of cells. We have tested variants from two types of objective functions, one abandoning all domain knowledge and three objective functions integrating domain knowledge in different ways.

Already for small systems like the considered Brusselator with six parameters, the first approach failed to identify suitable parameter settings. A naive variation of this method drastically restricting the search space to a region known to be promising in principle failed as well. Only for very small parts of the decision space it was possible to identify good solutions, indicating that no real optimization took place but mere sampling.

The last two variants produced promising results. Both have in common that the search process is guided towards a region of in principle good solutions. Following this direction both approaches succeeded in identifying good parameter sets. The two variants are (1) using a single point which is known to be good as an attractor for the search process and (2) using knowledge about dependencies between parameters to guide the search process, with variant (2) producing the better results and therefore beeing the method of choice for the given problem.

Additionally, it might be interesting to combine the used approaches and thereby further improve the method.

As key results of this study it can be concluded that on the one hand side optimization on the given problem domain without having additional domain knowledge seems to be intractable. If domain knowledge becomes available on the other hand there are strategies allowing to identify good solutions to the problem.

This study only represents preliminary work. The focus of our work is on dealing with more complex networks both when considering the number of involved species and the number of cells in the simulated system. Additionally it is planned to expand the model to three dimensional setups. For these systems we are not only interested in the mere pattern formation but in the formation of specific patterns visible in our real world target *Arabidopsis thaliana*.

Acknowledgment

Tim Hohm has been supported by the European Commission under the Marie Curie Research Training Network SY-STEM, Project 5336.

References

1. H. de Jong. Modeling and simulation of genetic regulatory systems: A literature review. *Journal of Computational Biology*, 9(1):67–103, 2002.
2. A. Gierer et al. Regeneration of hydra from reaggregated cells. *Nature New Biology*, 239:98–101, 1972.
3. N. Hansen and A. Ostermeier. Adapting arbitrary normal mutation distributions in evolution strategies: the covariance matrix adaptation. In *Conf. on Evolutionary Computation*, pages 312–317. IEEE, 1996.
4. N. Hansen and A. Ostermeier. Completely Derandomized Self-Adaptation in Evolution Strategies. *Evolutionary Computation*, 9(2):159–195, 2001.
5. A. Hunding, S. Kauffman, and B. Goodwin. Drosophila Segmentation: Supercomputer Simulation of Prepattern Hierarchy. *J. theor. Biol.*, 145:369–384, 1990.
6. H. Jönsson et al. Modeling the organization of the WUSCHEL expression domain in the shoot apical meristem. *Bioinformatics*, 21:i232–i240, 2005.
7. H. Meinhardt. *The algorithmic beauty of sea shells*. Springer Verlag, 1995.
8. P. Mendes and D. B. Kell. Non-linear optimization of biochemical pathways: applications to metabolic engeneering and parameter estimation. *Bioinformatics*, 14(10):869–883, 1998.
9. C. G. Moles, P. Mendes, and J. R. Banga. Parameter estimation in biochemical pathways: A comparison of global optimization methods. *Genome Research*, 13(11):2467–2474, 2003.
10. I. Prigogine and R. Lefever. Symmetry Breaking Instabilities in Dissipative Systems. *J. chem. Phys.*, 48:1695–1700, 1968.
11. K.-Y. Tsai and F.-S. Wang. Evolutionary optimization with data collocation for reverse engeneering of biological networks. *Bioinformatics*, 21(7):1180–1188, 2005.
12. A. Turing. The chemical basis for morphogenesis. *Philos. Trans. R. Soc. Lond.*, B, 237:37–72, 1952.
13. E. O. Voit. *Computational Analysis of Biochemical Systems*. Cambridge University Press, Cambridge, UK, 2000.

Evolutionary Search for Improved Path Diagrams

Kim Laurio[1], Thomas Svensson[2], Mats Jirstrand[3], Patric Nilsson[1],
Jonas Gamalielsson[1], and Björn Olsson[1]

[1] Systems Biology Research Group, University of Skövde, Sweden
[2] Biovitrum AB, Göteborg, Sweden
[3] Fraunhofer-Chalmers Research Center for Industrial Mathematics, Göteborg, Sweden

Abstract. A path diagram relates observed, pairwise, variable correlations to a functional structure which describes the hypothesized causal relations between the variables. Here we combine path diagrams, heuristics and evolutionary search into a system which seeks to improve existing gene regulatory models. Our evaluation shows that once a correct model has been identified it receives a lower prediction error compared to incorrect models, indicating the overall feasibility of this approach. However, with smaller samples the observed correlations gradually become more misleading, and the evolutionary search increasingly converges on suboptimal models. Future work will incorporate publicly available sources of experimentally verified biological facts to computationally suggest model modifications which might improve the model's fitness.

1 Introduction

Several algorithms have been proposed and evaluated for the problem of inferring gene regulatory networks from observations. For an overview, see (Wessels, van Someren et al. 2001). However, the focus has often been on the reconstruction of the entire network from scratch. This work investigates a particular combination of path analysis (Wright 1934) and evolutionary search for comparison and improvement of existing pathway models of gene regulatory networks. Instead of trying to develop algorithms which start from scratch in every new model building effort, we are more interested in developing methods which help us analyze existing models in the light of new data. Our aim is to develop methods which can automatically check if new data rejects a model, or some model features. Such methods could be used for continuously monitoring and updating a database of pathways, and alerting users when new data arrives which contradicts the models they rely on. Another envisioned application is a system that accepts as input a rough model of a set of variables and their relations, and proceeds to refine it based on both existing knowledge and new observations. Current research on biological ontologies and semantic web technology indicates that this is receiving increased attention within the biological sciences (Bodenreider and Stevens 2006).

Regulatory interactions among genes can (to some extent) be represented as models in the form of path diagrams, as exemplified in figure 1. A path diagram is a directed graph which may contain cycles. A path diagram is also a graphical model

E. Marchiori, J.H. Moore, and J.C. Rajapakse (Eds.): EvoBIO 2007, LNCS 4447, pp. 114–121, 2007.

describing a theory about causal relationships among measured and unmeasured variables (Loehlin 1998). Double-headed arrows represent residual correlations between variables and arrows are drawn between variables where one is considered to be a function of the other (Wright 1934). Each arrow (single- or double-headed) has an attached floating-point value - called a path coefficient - which represents the strength of the interaction.

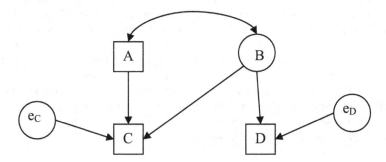

Fig. 1. A path diagram representing the relations between the measured variables A, C and D and the unmeasured variables B, e_C and e_D. Here A and B are correlated due to some unspecified interaction. Variable A influences C and variable B influences both C and D. Variables C and D are further influenced by the unspecified and unmeasured, error factors e_C and e_D.

The coefficients define a set of model-implied correlation values between any pair of variables in the diagram, according to a set of rules defined by (Wright 1934). The correlation between any two variables is the sum of the values of all compound paths between them. The value of a compound path is the product of the involved coefficients. A compound path, according to Wright's rules, consists of a sequence of distinct variables, is formed by doing at most one traversal along a double-headed arrow, and includes no backwards traversal after the first forward traversal. In figure 1 the correlation between C and D is the sum of the compound paths $C{\leftarrow}A{\leftrightarrow}B{\rightarrow}D$ and $C{\leftarrow}B{\rightarrow}D$. The coefficient for $B{\rightarrow}D$ is used in both these compound paths, hence it must be optimized to fit both. For details on path diagram creation, their usage and constraints on their usage, the reader is referred to (Loehlin 1998).

Using path diagrams, alternative theories can be formulated as alternative models, and these can easily be compared to each other as well as to observations on the basis of the model-implied correlations. Given sufficient training data the path coefficients can be optimized to minimize the differences between the observed correlations and the model-implied ones. Our search is using this as a fitness function.

In contrast, (Bay, Shrager et al. 2003) compare a set of model-implied vanishing *partial* correlations to observations. They do it by trying to improve a regulatory network structure by identifying which variables are directly linked, and which are indirectly linked to each other. They further restrict their networks to acyclic graphs. Once their network structure has stabilized they do a second pass to estimate signs for the links between variables. In our approach we do not separate the structure learning and path coefficient estimation into separate stages. The result of the path coefficient estimation is instead used as a hint for better fitting structures during the construction of the next population. Neither is the approach we use restricted to acyclic graphs.

Obviously, these comparisons between correlations are quite dependent on the quality and quantity of the observations. To further complicate matters, even with a model that has a perfect match there is still the possibility of some other model matching equally well, despite not representing the same theory (Lee and Hershberger 1990). That is, there could be an entire set of models which fit the data equally well, while representing incompatible causal claims about the system. In general, the amount of statistical error we can calculate does not necessarily correlate very well to the amount of causal error that our path diagram models contain (Hayduk 1996). A significant statistical error is a very useful and clear sign of model rejection, something which needs to be dealt with. However, the converse is not quite as useful. Even a small, often insignificant, statistical error does not prove that the model captures the functional relations which would be confirmed in a controlled experiment. This is evident from the fact that pairwise statistical correlation does not, by itself, imply anything about which of the concerned variables is the effect and which is the cause. Thus, it is easy to predict that there will be functional arrows in the path diagrams which point in the wrong direction.

2   Combining Path Analysis with Evolutionary Search

To test how well the combination of path diagrams and evolutionary search performs, we created artificial "gene expression" data sets based on previously published and a (partially) experimentally verified yeast gene regulatory pathway (Yeang, Mak et al. 2005), shown in figure 2. To generate the actual data we used the tool Tetrad IV – a Java and web-enabled version of Tetrad II (Scheines 1994) – to build a structural equation model (Hayduk 1996) corresponding to the yeast gene regulatory pathways. We used randomized weights for the path coefficients.

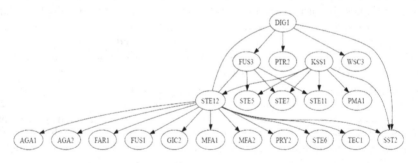

Fig. 2. The model of yeast regulatory pathway adapted from (Yeang, Mak et al. 2005) which was used for generating test data for the evaluation of the method

One difference from the yeast pathway of (Yeang, Mak et al. 2005) is that Tetrad IV does not support the specification of mutual influences which form a cycle between variables which have mutual inhibition relations in the pathway. These were therefore replaced with correlation edges (e.g. between STE12 and DIG1 in figure 2). We created sample sets of sizes 6, 12, 24, 48, 100 and 1000 to simulate the whole spectrum of realistic to unrealistic amounts of data. Each element of each sample is simply a "snapshot image" of the values of all the variables of the model. Since the model contains 21 variables, each element of each sample contains 21 values, and it is from these values that the pairwise, observed, correlations are estimated.

For all the experiments reported below, the initial models were created according to three simple heuristics: 1) For each variable in the path diagram, add one causal arrow to the variable with strongest correlation. 2) Make two passes over all variables, and for each one that has a higher number of incoming edges than outgoing, change all edges into outgoing ones. 3) Set the initial path coefficients to the observed pairwise correlations.

The path coefficient optimization is a simple steepest-ascent hill-climber. For each path coefficient a change from the old value to the *average* of the old value and a pseudorandom number with approximately uniform distribution between -1.0 and 1.0 is evaluated, one after another. Path coefficient changes which improve the model error E (1) are kept. This iterates for a user-selectable time-period. We used 100ms for the optimization each model. The measure optimised is the sum of the squared differences of the model-implied correlations and the observed correlations of all variable-pairs X and Y in the path diagram:

$$E = \Sigma \ (\text{model\_implied\_r}(X,Y) - \text{observed\_r}(X,Y))^2 \qquad (1)$$

No cross-over operations were used in the evolutionary search. The code developed for performing mutations favour adding (approx. a 50% chance) a functional arrow between variables with strong correlations, over removal of the weakest arrow or correlation (40%), over adding a totally random functional arrow (9%), over adding a totally random correlation edge (1%). We used a population size of 25, and included single-individual elitism and tournament selection (tournament size 2). All runs were allowed to last for 300 seconds, since initial testing revealed that all models converged in that amount of time.

3 Results

The models' sensitivity and specificity values (Attia 2003) were calculated with the original model (figure 2) as reference and ignoring the edge directions. Sensitivity is the proportion of edges in the original model that exist in the compared model, while specificity is the proportion of absent edges in the original model that are also absent in the compared model:

$$\text{TruePositiveCount} \ / \ (\text{TruePositiveCount} + \text{FalseNegativeCount}) \qquad (2)$$

$$\text{TrueNegativeCount} \ / \ (\text{TrueNegativeCount} + \text{FalsePositiveCount}) \qquad (3)$$

3.1 Model Initialised on a Sample of Size 1000

The initialisation assumes that a variable generally has a larger number of arrows starting from it than pointing to it. It also assumes that directly interacting variables generally receive higher correlation values than indirectly interacting ones. To investigate the effect of these heuristics we compared the sensitivity and specificity of the model immediately after initialisation (figure 3) with those of the original model (figure 2). Visual inspection of the initialised model reveals that many edges from the initial model are missing, and the model consists of three islands of variables. However, since there are so few extra edges the specificity remains high. These effects are reflected in the sensitivity and specificity values of 0.65 and 0.99, respectively. Many variable pairs are not linked by a direct edge in the original model, and these contribute to the excellent specificity for the initialized model since the heuristics favour initial models with few arrows. Thus, correctly absent edges contribute strongly to the high specificity values.

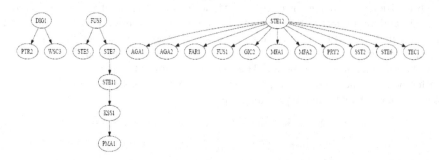

Fig. 3. Model initialised from a sample of size 1000

3.2   Model Initialised and Optimised on a Sample of Size 1000

The evolutionary optimization which starts from the initialised model was successful in correcting many of the remaining errors. The only guidance that the evolutionary search receives during the optimization, is in the form of the error value E. The optimised model, shown in figure 4, has increased the sensitivity to 0.88 and the specificity was reduced to 0.94. This indicates that mainly correct direct edges have been added by the evolutionary search, resulting only a slight reduction of the specificity.

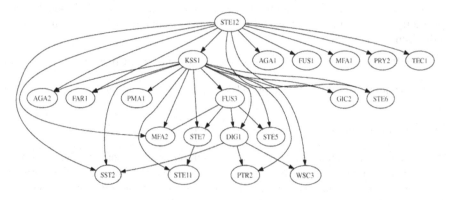

Fig. 4. Model after initialization and evolutionary optimization on a sample of size 1000

3.3 Summary of Results

Table 1 presents the results of our experiments. For different sample sizes we report typical values for the error E, sensitivity and specificity. As expected, problems accumulate when models are modified according to pairwise correlations that are created from smaller samples. The error E gets worse with smaller samples as well, and this predominantly affects the sensitivity. Edges are missing from the model, but the evolutionary search cannot really decide which edges to add to the model since the signal to noise ratio gets worse with small samples. It is clear from the values shown in the table that a sample size of six is too small for this approach. A sample size of 12 is a clear improvement, even if the usefulness of a model with a sensitivity of .42 is doubtful. Somewhere above 24 samples the gains in sensitivity from increasing the sample size start decreasing. Finally, it might be difficult to argue for the added expense of additional experiments after reaching a sample size of about 100.

Table 1. Final error E, sensitivity and specificity values for various models and sample sizes. "Initialized model" shows results from applying heuristics and path coefficient optimization. "Evolutionarily optimized model" shows the results of heuristics, evolutionary search for topological changes and path coefficient optimization. "Original model" shows values obtained when using the known topology of the path diagram and optimization of path coefficients.

Sample size	Initialized model			Evolutionarily optimized model			Original model		
	Error	Sen	Spe	Error	Sen	Spe	Error	Sen	Spe
6	24.8	.23	.94	9.65	.38	.86	16.34	1.0	1.0
12	8.92	.42	.95	3.12	.42	.89	4.67	1.0	1.0
24	9.92	.42	.96	3.02	.61	.90	2.22	1.0	1.0
48	7.96	.62	.98	1.27	.73	.94	1.55	1.0	1.0
100	12.11	.65	.99	.43	.84	.94	.47	1.0	1.0
1000	8.95	.65	.99	.10	.88	.94	.08	1.0	1.0

4   Discussion and Conclusions

We have presented a combination of path diagrams and evolutionary search for modelling gene regulatory networks and tested its performance on data generated from a simulated yeast gene regulatory pathway. Similarly to relevance networks (Butte, Tamayo et al. 2000), path diagrams allow multiple connections for each variable and can include disparate data sources in one model. In contrast to relevance networks, our approach takes into account the indirect influences between variables in the network and the arrow directions are significant. In comparison to Bayesian networks (Pearl 1988; Friedman, Linial et al. 2000) path diagrams are not restricted to acyclic graphs and do not require discrete variables.

Our evaluation shows that once a correct model has been identified, it consistently and significantly receives a lower error E compared to incorrect models, thus indicating the overall feasibility of this approach. We have further done experiments (data not shown) on performing random topological changes to the original model and it clearly showed that any changes to the structure of the original model leads to a larger error, confirming these observations. However, finding the best fitting model with an uninformed search process is extremely unlikely for any nontrivial problem size and there is often a large number of suboptimal models which fit with approximately similar prediction errors to a given set of observations.

We observe, not very surprisingly, that the search process gets trapped in local minima with increasing frequency when the signal-to-noise ratio gets worse with insufficient sample sizes. Even after successful optimization the models can encode very different, and often conflicting, causal theories, but in a realistic situation it is impossible to decide solely from a given set of observations which theory is more correct than the other. Only with additional information is it possible to discriminate among alternative models with similar errors. This additional information could come from expert knowledge or databases not used during construction of the models, or from new experiments which are specifically targeted towards discrimination between the remaining model alternatives.

5   Future Directions

It is very clear from these results that access to sufficient data of high quality is a prerequisite for good models. Biased or too small samples will generate a lot of spurious hits, which can perhaps eventually be computationally filtered out by having access to raw data from many independent experiments. Large amounts of data can suppress the spurious hits if proper, statistical, caution (Ioannidis 2005) is used in the fusion of data. However, for realistic situations with less than optimal data, we are currently examining different path diagram mutation strategies in which we try to incorporate existing knowledge about experimentally verified biological facts. For example, finding (Wernicke 2006) and using sets of network motifs frequently observed in *S. cerevisiae* (Lee, Rinaldi et al. 2002), could perhaps be used to guide the evolutionary search, either for the initialization of the search or to suggest model revisions. To accomplish this, to use known facts from one organism in order to bootstrap the learning for a new one, we are looking at using GO-annotation data.

We are also investigating the possible benefits of using semantic alignments of biological pathways (Gamalielsson and Olsson 2005) to generate candidate solutions for the initial population. Our plan is to use a measure of semantic similarity between Gene Ontology annotation terms to score pathway alignments, in order to suggest changes to existing subgraphs in models during search and optimization.

Finally, we are investigating methods based on the results of (Pearl 2001). Even if basic assumptions of the methods presented by Pearl are violated, perhaps there are still some useful hints available that the evolutionary search can benefit from.

References

Attia, J. (2003). "Moving beyond sensitivity and specificity: using likelihood ratios to help interpret diagnostic tests." Australian Prescriber **26**: 111-113.

Bay, S. D., J. Shrager, et al. (2003). "Revising Regulatory Networks: From Expression Data to Linear Causal Models." Journal of Biomedical Informatics **35**: 289-297.

Bodenreider, O. and R. Stevens (2006). "Bio-ontologies: current trends and future directions." Briefings in Bioinformatics **7**(3): 256-274.

Butte, A. J., P. Tamayo, et al. (2000). "Discovering functional relationships between RNA expression and chemotherapeutic susceptibility using relevance networks." Proceedings of the National Academy of Sciences of the United States of America **97**(22): 12182-12186.

Friedman, N., M. Linial, et al. (2000). "Using Bayesian networks to analyze expression data." Journal of Computational Biology **7**(3-4): 601-620.

Gamalielsson, J. and B. Olsson (2005). GOSAP: Gene Ontology based Semantic Alignment of Biological Pathways. Technical report HS-IKI-TR-05-005. Skövde, University of Skövde, Sweden.

Hayduk, L. A. (1996). LISREL issues, debates and strategies, Johns Hopkins Press.

Ioannidis, J. P. A. (2005). "Why Most Published Research Findings Are False." PLoS Medicine **2**(8): e124.

Lee, S. and S. Hershberger (1990). "A simple rule for generating equivalent models in structural equation modeling." Multivariate Behavioural Research **25**: 313-334.

Lee, T. I., N. J. Rinaldi, et al. (2002). "Transcriptional regulatory networks in S. cerevisiae." Science **298**: 799-804.

Loehlin, J. C. (1998). Latent Variable Models - an introduction to factor, path and structural analysis, Lawrence Erlbaum Associates.

Pearl, J. (1988). Probabilistic reasoning in intelligent systems: Networks of plausible inference. San Francisco, Morgan Kauffmann Publishers Inc.

Pearl, J. (2001). Causality – Models, Reasoning and Inference, Cambridge University Press.

Scheines, R., Spirtes, P., Glymour, C., Meek, C. (1994). TETRAD II: Tools for Discovery, Lawrence Erlbaum Associates.

Wernicke, S. (2006). "Efficient Detection of Network Motifs." IEEE/ACM Transactions of Computational Biology and Bioinformatics **3**(4): 347-359.

Wessels, L. F. A., E. P. van Someren, et al. (2001). "A comparison of genetic network models." Pacific Symposium on Biocomputing **6**: 508-519.

Wright, S. (1934). "The Method of Path Coefficients." Annals of Mathematical Statistics **5**(3): 161-215.

Yeang, C.-H., H. C. Mak, et al. (2005). "Validation and refinement of gene-regulatory pathways on a network of physical interactions." Genome Biology **6**(R62).

Simplifying Amino Acid Alphabets Using a Genetic Algorithm and Sequence Alignment*

Jacek Lenckowski and Krzysztof Walczak

Institute of Computer Science, Warsaw University of Technology
ul. Nowowiejska 15/19, 00-665 Warszawa, Poland

Abstract. In some areas of bioinformatics (like protein folding or sequence alignment) the full alphabet of amino acid symbols is not necessary. Often, better results are received with simplified alphabets. In general, simplified alphabets are as universal as possible. In this paper we show that this concept may not be optimal. We present a genetic algorithm for alphabet simplifying and we use it in a method based on global sequence alignment. We demonstrate that our algorithm is much faster and produces better results than the previously presented genetic algorithm. We also compare alphabets constructed on the base of universal substitution matrices like BLOSUM with our alphabets built through sequence alignment and propose a new coefficient describing the value of alphabets in the sequence alignment context. Finally we show that our simplified alphabets give better results in a sequence classification (using k-NN classifier), than most previously presented simplified alphabets and better than full 20-letter alphabet.

Keywords: amino acid alphabet, sequence alignment, substitution matrices, protein classification.

1 Introduction

Proteins are represented by 20 symbols, usually each symbol is a letter of the English alphabet and corresponds to one amino acid. However, in some cases, a simplified alphabet can be more convenient. In a simplified alphabet there are less than 20 symbols - each symbol from the simplified alphabet replaces a set of symbols from the original alphabet. In general, a simplified alphabet is built by grouping sets of symbols from a starting alphabet together - each group is represented by a new symbol. In this paper we consider if a simplified alphabet, which is as universal as possible, is the best choice for sequence alignment and for classifying a new protein. We propose a fast genetic algorithm for finding simplified alphabets and compare it to previously presented methods, where standard substitution matrices like BLOSUM are used. Next, we present a method of creating simplified alphabets with no substitution matrices based on the presented

---

* The research has been partially supported by grant No 3 T11C 002 29 received from Polish Ministry of Education and Science.

E. Marchiori, J.H. Moore, and J.C. Rajapakse (Eds.): EvoBIO 2007, LNCS 4447, pp. 122–131, 2007.

genetic algorithm and sequence alignment. We compare simplified alphabets using a new proper alignment coefficient. We also show that simplified alphabets can improve the correctness of amino acid sequence classification with respect to the full alphabet.

2   Previous Work

Simplified alphabets are used in several fields of bioinformatics. In sequence alignment (for protein classification), using simplified alphabets can provide more reliable results (see [1]) because it is often possible to change one amino acid to another in a protein sequence without any changes of the protein's properties. There are also some works ([3], [6], [11]) pertaining to how many amino acids are needed to fold a protein properly and whether all 20 amino acids are necessary to build all proteins that occur in Nature. There is a previously presented genetic algorithm proposed in [10] and a branch and bound algorithm designed in [2]. We compare produced alphabets to those based on the residue pair counts for the MJ and BLOSUM matrices ([7]), and to those based on the correlation coefficient ([9]).

3   Methods

It has been shown (see [2]) that alphabet simplifying can be related to the problem of set partitioning. The purpose is to divide an original set into disjoint subsets which completely cover the initial set. For example we can join any two symbols together and replace them by one new symbol. Starting from an original 20-letter amino acid alphabet we can repeat this operation as long as the alphabet has at least two symbols. It has also been shown that the problem of set partitioning is a hard computational problem because of the number of possible ways to create k subsets from a set of n elements can be calculated recursively following this expression:

$$S(n,k) = k * S(n-1,k) + S(n-1,k-1), 2 <= k <= n-1 \qquad (1)$$

where S(n,1) = S(n,n) = 1.

For a 20-letter alphabet and any count of subsets, there are more than 51×10^{12} possible simplified alphabets. When we consider, for example, only 8-letter simplified alphabets, there are more than 15×10^{12} such alphabets. The question is, which alphabet is the best and whether such an optimal alphabet exists.

3.1   Simple Rating Schema

One of the simplest ways to rate a reduced alphabet is to use one of the popular substitution matrices, like BLOSUM. In [2] a method has been proposed based on counting the total score for an alphabet. When we choose a substitution matrix, the score of a full 20-symbol alphabet is the sum of 20 exact matches, which is equal to the alignment of two identical sequences. When we create a new symbol, for example {CD}, from two symbols C and D, and when we align this new symbol {CD} with the same symbol, we should consider four possibilities: C:C, C:D, D:C and D:D because the new symbol can replace either the symbol C or D from the original alphabet. For simplicity it has been proposed to assume that each possibility has a 25% chance to be correct, so the calculated score of {CD} is the average of corresponding scores: C:C, D:D, C:D and D:C. In general, when we create a new symbol from n amino acid symbols, the score of the simplified symbol is counted as the average of positions in the substitution matrix corresponding to all pairs of starting symbols which are in the set replaced by the new symbol.

3.2   Previous Algorithms

In [2] an algorithm is presented, based on a branch and bound technique, which finds the best (partly described earlier) alphabet without checking all possibilities. A genetic algorithm for finding the best simplified alphabets has been presented in [10]. In that concept, the problem of a crossover has been resolved in the following way: Two parent solutions are randomly selected and then the sets of starting symbols, which correspond to one simplified symbol, from each parent solution (where the solution means a simplified alphabet) are divided in half with a k-means algorithm. This operation leads to separate the closely-related items in each cluster. Such kind of clusters (not the parent's original clusters) are used for a crossover. Clusters in the child solution are created by selecting a random amino acid and combining the clusters containing the selected acid from the parent solutions. No mutation is used.

3.3   Genetic Algorithm

We present another genetic algorithm in which we use three genetic operators - a natural selection, a crossover and a mutation. Each individual is represented with a sequence of 20 numbers. One position in the sequence corresponds to one amino acid, and the number in each position denotes the simplified symbol to which the given amino acid belongs.

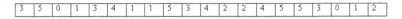

Fig. 1. The example of an individual. The simplified alphabet consists of six symbols: CR (0), DGHS (1), LMT (2), AEJQ (3), FKN (4), BIOP (5), from the starting alphabet: A, B, C, D, E, F, G, H, I, J, K, L, M, N, O, P, Q, R, S, T.

In the natural selection we use the same alphabet rating method which has been also used in [2] and [10]. This rating method is used here to compare results of algorithms, but in the later part of this paper we propose other rating methods based on sequence alignment. The natural selection is a tournament selection with the tournament group of size two. The amount of those tournaments is equal to the number of individuals in a population. After the natural selection the number of individuals in considered population is the same as before, but there are some individuals which appear more than once. The crossover is made in a way similar to that presented in [12]. With a certain probability for each individual, the second parent is randomly chosen from the rest of the population. Then, one number (simplified symbol) from the first parent is randomly chosen and the positions which contain this number in the first parent are rewritten to the second parent. The result of this operation is a new generation individual which joins the new generation population. Because the crossover probability may not be 1 (not all individuals are used in the crossover), the new generation can be less numerous than the parent generation. In such situations, the proper number of individuals from parent generation is added to the new generation.

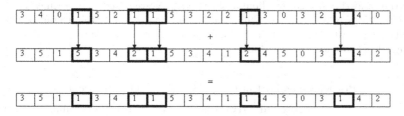

Fig. 2. The example of two individuals' crossover. The chosen simplified symbol is "1". The next generation individual is similar to the second parent, but it has "1" on those positions on which the first parent has "1".

The mutation operator changes each position in each individual (sequence) with a certain probability, which means that some amino acids in some individuals are moved to other subsets (simplified symbols). The starting configurations of individuals are created randomly using the uniform distribution. It is possible to create individuals with different sets of numbers (which corresponds to alphabets of different size). For example, one individual may have numbers from 0 to 4 while another could have numbers from 0 to 6.

After each crossover or mutation process, it is necessary to check the correctness of all individuals as, even if we use the same set of numbers in the individual representation, it is possible that in some individuals there will be no positions containing a given number, which would mean that the number of simplified symbols is too small. For example, when we search for a 5-letter alphabet, it is possible that in the representation of some individuals, not all symbols from the set 0,1,2,3,4 are present. If there is such a situation, the value on a randomly chosen position is changed to the number which is missed. According to the method used for rating the population, a situation, where we search for an alphabet from some range of size will promote, of course, individuals with the larger alphabet size, so when we use a considerably large number of generations (for example 100), we get as a result only the biggest possible alphabets.

3.4 Speed Comparison

Our algorithm outperforms the genetic algorithm presented by [10], what is shown in table 1. In our tests we use 100 individuals and 100 generations. The probability of crossover was set to 0.8 and the probability of mutation to 0.01. When we use 10 rounds of the algorithm in almost all cases, we get the optimal solution (according to the results of [2]). As we can see in table 1, our algorithm gives better solutions than the compared genetic algorithm and it is also roughly two orders of magnitude faster. When we compare computational times to the results of the branch and bound algorithm we can see that our algorithm is about four-five orders of magnitude faster. Of course, the presented times are not equally comparable because of the differences in systems and implementations used but they give a good enough idea of how much faster the new method is.

Table 1. Benchmark results of the presented algorithm compared to other genetic algorithm and the branch and bound algorithms. Average scores and times are taken from 10 rounds of algorithms. The alphabet's average score is the average quality of alphabets from 10 rounds of the algorithm and the best score is the best result received during all test rounds of algorithms.

Size	Previous genetic algorithm			Our genetic algorithm		
	Avg. score	Best found	Avg. time*	Avg. score	Best found	Avg. time
4	135.317	129.600	00:00:32	122.242	120.917	151 ms
5	118.173	109.362	00:00:31	107.123	105.705	149 ms
6	105.997	98.743	00:00:30	94.379	92.967	148 ms
7	90.848	83.133	00:00:31	82.352	81.000	155 ms
8	79.208	73.095	00:00:31	70.462	69.167	149 ms
9	65.300	62.833	00:00:34	60.546	58.500	151 ms

Size	Branch and bound	
	Best	Time*
4	120.917	01:21:00
5	105.705	04:23:03
6	92.967	06:52:26
7	81.000	05:58:45
8	69.167	02:51:23
9	58.500	01:01:20

* Computational times have been estimated (original times have been divided by 2), because the results presented by algorithms' authors were received on Pentium III processor clocked at 600MHz, which is about two magnitudes slower than a machine used in our tests (AMD Athlon 1,2GHz).

4   The New Method of Creating a Simplified Alphabet

Some works have been done about how to generate the best alphabet, but most of it was based on using the universal matrices. In [9], a method of reducing the alphabet size based on correlations indicated by the BLOSUM50 matrix has been presented. First correlation coefficients for all pairs of matrix elements are counted and then the pair with the highest value is grouped together. The process is repeated until all the amino acids are divided into the desired number of sets. In [7] a method has been described for generating simplified alphabets based on deviation of conditional probability from random background. Authors have presented the alphabets based on the residue pair counts for the MJ matrix (for more details see [8]) and for BLOSUM50 matrix (see [4]).

4.1   Alphabet Generating with No Substitution Matrices

In this paper we present the method of generating simplified amino acid alphabets without using any substitution or other matrices. This technique will ensure that a generated simplified alphabet is built without any previous assumptions about the probability of emerging any amino acids and without any knowledge about the possibilities of replacing one amino acid by another in protein sequences. We use our genetic algorithm in this method, but we change the way in which the value of a simplified alphabet is counted. The presented genetic algorithm is fast enough to use the score of sequence alignment as the value function. In general, to determine how good a simplified alphabet is, we align two groups of sequences rewritten using a currently tested simplified alphabet. To get a reliable measure, we propose an alignment coefficient describing how good the alignment of a sequence with other sequences from the same protein family is and how bad the alignment with sequences from other families is. We assume that the simplified alphabet is better if it gives a higher score of alignment in the same family and a low score when we align sequences from different protein families. Any biological knowledge we use in our work is only to know some protein examples from a selection of chosen protein families.

4.2   The Alignment Coefficient

The alignment coefficient is counted this way: we assume that we have a function $A(x,y)$ whose value is the alignment score of sequences x and y (it is possible that x or y are sets of sequences, in which case the value of such a function is the sum of alignment scores between all pairs of sequences in which the first sequence is from set x and the second from y). We mark T as a set of test sequences - in our experiments one from each protein family. These sequences are aligned to all other sequences from their families and from other families. The set of all families is F and a single family from F is marked as f. Simple sequence from f is marked as f_i, while simple sequence from T is t and the real family of sequence t is f_t. The alignment coefficient is the proportion of the mean alignment score of sequences from T with sequences from their own families to the mean alignment score of sequences from T with sequences which belong to other families.

$$q = (\sum_{t \in T} \frac{(\sum_{f_i \in f_t} A(t, f_i))/|f_t|}{(\sum_{f \in F - f_t}(\sum_{f_i \in f} A(t, f_i))/|f|)/|F - 1|})/|T| \tag{2}$$

In every step of the genetic algorithm, each individual is rated according to the value of its alignment coefficient. The crossover and mutation operators work in the same way as we described previously. The starting configuration of each individual is randomly chosen again.

4.3   Test Settings

In our tests we use sequences from four protein families of the Immuglobulin superfamily (according to [5]). Four sequences (one from each family) belong to the test set and are aligned with other sequences from the following families: C1 set domains, C2 set domains, I set domains and V set domains. To rate a simplified alphabet, which is equivalent to counting the value of the alignment coefficient, we rewrite all sequences from the experiment using this alphabet changing all symbols from each subset to the same symbol chosen from this subset. We then align the sequences so that each test sequence is first aligned with sequences from its family while noting the average score. Next, each test sequence is aligned with sequences from all other families noting the average score again. The results of alignments are used to count the alignment coefficient. In our tests we have used a global (Smith-Waterman) alignment algorithm implemented in the package neobio.alignment from neobio.sourceforge.net. We have chosen the following values of parameters: match reward = 5, mismatch penalty = -3, gap penalty = -4. For the classification tests we use 240 sequences (from the same four protein families of the Immuglobulin superfamily) - 50 sequences are randomly chosen to a test set and the remaining is a training set. The classifier is a 10-NN classifier, where the distance is counted as the alignment score between sequences (the higher the alignment score, the smaller the distance). For each test sequence, 10 other sequences with highest alignment scores to it are found in the training set, and the class of the test sequence is the one, which contains the most sequences matching the selected 10. The error of classification is the percentage of incorrectly classified test sequences, an average from 50 experiments.

5   Experimental Results

We have made a series of experiments with various sizes of alphabets to compare the results of our method to the most popular simplified alphabets presented in the previous work ([2], [9] and [7]). We observe that there are certain similarities with alphabets presented previously, especially when the size of an alphabet is small, which can be seen, for example, for 5-letter alphabets - ACKP, DHMFS, GIV, NTWY, RQEL (our alphabet), IMVLFY, PGAST, HNQEDRK, W, C (the alphabet based on the scoring schema from [2]), MFILV, ACW, YQHPGTSN, RK, DE (the alphabet based on the residue pair counts for the MJ matrix [7]), IMVL, FWY, G, PCAST, NHQEDRK (the alphabet based on the residue pair

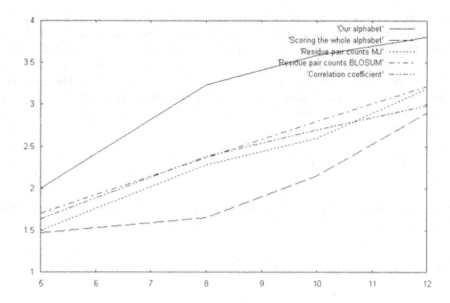

Fig. 3. Values of the alignment coefficient for different alphabet generating methods and alphabet size range from 5 to 12

Table 2. The error classification rate for simplified alphabets. The result for the 20-letter alphabet is 33.2%. We present the results of our alphabets (a), alphabets based on a scoring of the whole alphabet [2] (b), alphabets presented in [7] for MJ (c) and BLOSUM matrix (d), alphabets presented in [9] (e).

Size	Classification error in %				
	(a)	(b)	(c)	(d)	(e)
12	32	27.8	34.3	32.6	30.1
10	26.9	29.7	31.5	30.4	31.1
8	30.2	29.2	31	30.2	31

counts for the BLOSUM matrix [7]), ILMVC, ASGPT, FYW, EDQN, KRH (the alphabet based on the correlation coefficient [9]). When the alphabet is larger (less simplified), the discrepancies become more evident. These observations are also reflected in values of the alignment coefficient. On fig. 3 we present the values of the alignment coefficient for simplified alphabets generated by the described method. We can see that when alphabets are quite similar, the values of the coefficients are similar too. When alphabets differ more, what can be seen, for example, in 10-letter alphabets, the value of the coefficient of our alphabet is roughly 50 percent greater than for other alphabets. We also observe that when an alphabet has more than 10 symbols the value of the coefficient grows slowly and the differences between alphabets generated by presented methods become smaller. This corresponds to the previous results, according to which, 9-10 letter simplified alphabets allow storing almost as much information as full 20 amino acid alphabets. Therefore, for these sizes of alphabets, choosing the correct one is especially important. According to fig. 3, we can conclude that the results of alphabets generated using the deviation of conditional probability from random background are better than results from those generated using a correlation coefficient, and much better than results generated using a simple score schema proposed in [2].

In table 2 we present the classification error rate for considered simplified alphabets. We observe, that a 10-letter alphabet generated by our method yields the best result in this test. Almost all simplified alphabets provide better results than the full 20-letter alphabet, which confirms our hypothesis that simplified alphabets allow to better classify protein sequences.

6   Conclusions

In all sizes of alphabets, the alphabets generated using our method have had higher values of alignment coefficients. The optimal alphabet for sequence classifying is also the one produced by the new method. This confirms our hypothesis that universal simplified alphabets may not always be suitable for sequence alignment in classifying a new sequence and that good simplified alphabets can be better than the full 20-letter amino acid alphabet. When we can narrow the region of researches to certain protein families, we can then find and use alphabets which yield better classification, meaning greater differences in alignments with sequences from the correct protein family and sequences from other families. Furthermore, our method is flexible as it allows us to use as many sequences and protein families as desired. In view of the quantity of sequences used, the algorithm is linear scalable, however, we should remember that the global alignment, which is used to rate individuals, has the computational time complexity of $O(n^2)$. Our results therefore show that it is possible to generate useful simplified alphabets without any knowledge of the problem space. For example, we can simplify the amino acid alphabet even when we have no knowledge of the physical or chemical properties of amino acids. The presented method allows us to simplify alphabets of all kinds for many purposes, the one requirement being to have an objective value function to rate the alphabet. In this work, this kind of function is the alignment coefficient.

References

1. Andorf, C. M., Dobbs, D. L., Honavar1, V. G. Discovering protein function classification rules from reduced alphabet representation of protein sequences, Proceedings of the Conference on Computational Biology and Genome Informatics. Durham, North Carolina., (2002)

2. Cannata, N., Toppo, S., Romualdi, C., Valle, G. Simplifying amino acid alphabets by means of a branch and bound algorithm and substitution matrices, Bioinformatics vol.18 no.8, (2002) 1102–1108.
3. Fan, K., Wang, W. What is the Minimum Number of Letters Required to Fold a Protein?, J. Mol. Biol. 328, (2003) 921–926.
4. Henikoff, S., Henikoff, J.G. Amino acid substitution matrices from protein blocks, Proc. Natl Acad. Sci. USA, 89, (1992) 10915–10919.
5. Jaakkolay, T., Diekhansz, M., Hausslerz, D. A discriminative framework for detecting remote protein homologies, http://www.cse.ucsc.edu/research/compbio/research.html, (1999).
6. Li, T., Wang, J., Fan, K., Wang, W. How simple can the proteins be: from the prediction of the classes of protein structures, Modern Physics Letters B, vol.17, no. 5, (2003) 1–8.
7. Liu, X. , Liu, D., Qi, J., Zheng, W. Simplified amino acid alphabets based on deviation of conditional probability from random background, Physical Review E 66, (2002) 021906.
8. Miyazawa, S., Jernigan, R.L. Residue-Residue Potentials with a Favorable Contact Pair Term and an Unfavorable High Packing Density Term for Simulation and Threading. J. Mol. Biol. 256, (1996) 623–644.
9. Murphy, L. R., Wallqvist, A., Levy, R. M., Simplified amino acid alphabets for protein fold recognition and implications for folding, Protein Engineering vol.13 no.3, (2000) 149–152.
10. Palensky, M., Hesham, A. A Genetic Algorithm for Simplifying The Amino Acid Alphabet, Computer Society Bioinformatics Conference (CSB2003), (2003).
11. Romero, P., Obradovic, Z., Dunker, A.K. Folding minimal sequences: the lower bound for sequence complexity of globular proteins, FEBS Letters 462, (1999) 363–367.
12. Sakakibara, Y. Learning context-free grammars using tabular representations, Pattern Recognition 38, (2005) 1372–1383.

Towards Evolutionary Network Reconstruction Tools for Systems Biology

Thorsten Lenser, Thomas Hinze, Bashar Ibrahim, and Peter Dittrich

Friedrich Schiller University Jena
Bio Systems Analysis Group
Ernst-Abbe-Platz 1–4, D-07743 Jena, Germany
{thlenser,hinze,ibrahim,dittrich}@minet.uni-jena.de

Abstract. Systems biology is the ever-growing field of integrating molecular knowledge about biological organisms into an understanding at the systems level. For this endeavour, automatic network reconstruction tools are urgently needed. In the present contribution, we show how the applicability of evolutionary algorithms to systems biology can be improved by a domain-specific representation and algorithmic extensions, especially a separation of network structure evolution from evolution of kinetic parameters. In a case study, our presented tool is applied to a model of the mitotic spindle checkpoint in the human cell cycle.

1 Introduction

Reverse engineering of biochemical networks, making sense of rapidly growing molecular proteomics data, is a promising and important field at the crossroads of optimisation and model selection. Supplementing human-curated models with automatically generated, data-based models will enhance our understanding of the function of cells as a whole, which is at the core of systems biology.

Evolutionary algorithms (EA) and especially genetic programming (GP) have a long-established history as heuristic optimisation techniques [2,13,15]. Recently, methodologies adapted from this field have been used to evolve artificial biochemical networks, capable of performing arithmetic calculations [7] or specific behaviours such as oscillations and switching [10,17]. Others have used similar techniques to reconstruct metabolic pathways from time series data of chemical species [14]. While these attempts were successful for small networks, they also highlighted the complexity of evolving larger networks.

To expand our capability of evolving networks, improvements on these algorithms have to be investigated. In this contribution, we propose a separation of structural evolution of the network from kinetic parameter evolution, which yields a pronounced increase in the algorithm's fitness performance. Our studies show that this separation helps to prevent premature convergence when evolving networks performing arithmetic calculations. We suppose this happens because parameter fitting after each structural mutation smoothes their effect, which is usually rather strong. In this way, network parameters can adapt to a new topology before this topology is evaluated.

E. Marchiori, J.H. Moore, and J.C. Rajapakse (Eds.): EvoBIO 2007, LNCS 4447, pp. 132–142, 2007.

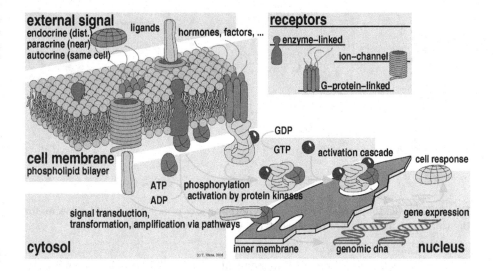

Fig. 1. Biological principle of signalling in eukaryotic cells: from arriving stimuli to specific cell response

Two other ideas are also investigated: the biologically-inspired mutation operator of species duplication, and the use of Akaike's Information Criterion (AIC) as a fitness function to evolve parsimonious models. By using the markup language SBML, the tool described here can work directly on systems biological problems, aiming at applications throughout this growing community.

In a case study, we apply our algorithm to a model of the human mitotic spindle checkpoint. By allowing the algorithm to introduce additional reactions, the performance of the model can be increased in comparison to a mere optimisation of parameters. Although biological plausibility is not considered, the example serves as a proof of concept for further investigations.

2   Modelling and Evolving Biochemical Networks

Biochemical reaction networks found in pro- and eukaryotic cells represent important components of life. Despite their high degree of complexity, they are hierarchically arranged in modular structures of astonishing order. The function of a cell emerges from the interplay of connected reaction processes. Three essential types of biochemical networks can be distinguished: metabolic, cell signalling (CSN), and gene regulatory (GRN) networks [1]. While metabolism consists of coupled enzymatically catalysed reactions supplying energy, CSNs and GRNs perform information processing of external and internal signals [6]. Malfunctions or perturbations within these networks are the cause of many diseases.

We have built a software tool implementing an evolutionary algorithm that evolves artificial biochemical networks performing pre-specified tasks. As a representation format, the systems biology standard SBML [9] is used, the most

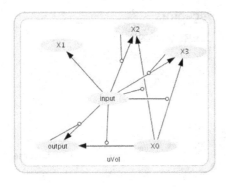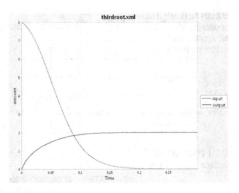

Fig. 2. Example solution and corresponding time series of *input* and *output* species for the third root network, produced using the CellDesigner [11] tool

common interchange format for biochemical models. This provides us with the opportunity to profit from an immense variety of tools developed for the analysis and interpretation of such models. The evolutionary algorithm used here employs eight different mutations:

- Addition / deletion of a species
- Addition / deletion of a reaction
- Connection / removal of an existing species to / from a reaction
- Duplication of a species with all its reactions
- Mutation of a kinetic parameter

While the first six and the last mutation have been used before, we are not aware of work that has used species duplication for the kind of network evolution discussed here. Crossover between networks is possible, but its effects are not part of this work and it has been disabled for all presented experiments.

Fitness evaluation in the algorithm is done by integrating the ODE system resulting from an individual model using the SBML ODE Solver Library [16], a tool designed precisely for that task. The resulting multidimensional time series is then compared to a target, and the weighted quadratic difference

$$f = 1/C \sum_{c=1}^{C} 1/S \sum_{i=1}^{S} 1/m_{c,i} \sum_{j=1}^{N} (x_{c,i}(t_j) - y_{c,i}(t_j))^2,$$

$$\text{with } m_{c,i} = \sum_{j=1}^{T} (x_{c,i}(t_j) + y_{c,i}(t_j))/2, \ i = 1, \ldots, S,$$

between the resulting time series x and the target time series y defines the fitness. Here, $i = 1, \ldots, S$ runs over the set of evaluated species, and $c = 1, \ldots, C$ runs over the fitness cases. Thus, fitness values are minimised, 0 being the absolute lower bound. If a steady state value is regarded as the result of the computation,

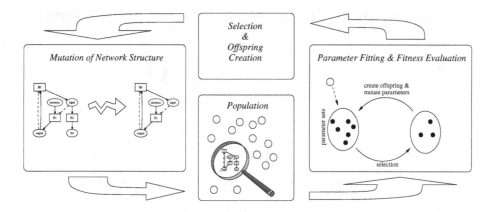

Fig. 3. Outline of the two-level evolutionary algorithm

a constant time series is the target and the first few timesteps are discarded. When Akaike's Information Criterion (see Section 4) is applied, the number of free parameters in the model k (kinetic parameters plus free initial conditions) and the number of data points $n = CTS$ are incorporated and the fitness is modified in the following way:

$$f_{\text{AIC}} = 2k + (n\log(f)) + 2k(k+1)/(n-k-1)$$

In this case, fitness is still minimised but can be negative without lower bound.

Selection is elitistic, with a certain percentage of the population surviving to the next generation, which is filled by mutants of survivors. It is possible to fit kinetic parameter before evaluating the model structure, a technique described in detail in the next paragraph. The software and all data shown in this paper is available from the authors upon request.

3   Separating Structural from Parameter Evolution

The evolution of an artificial network model can be separated into two parts: On the one hand, a set of species and reactions adequate for the task has to be found. On the other hand, the parameters of this model structure have to be optimised. The problem is analogue to model inference, where a dataset is used not only to fit the parameters of a model, but rather to choose a model structure together with a set of parameters. For nonlinear problems, this is still a largely unresolved challenge. Here, we show that a separation of model-structure evolution from parameter-fitting helps to prevent premature structural convergence.

In traditional GP approaches, parameters are usually evolved together with programme structure. In our approach, we use the opportunity to differentiate mutation and selection on the model structure from parameter fitting. To this end, a two-level evolutionary algorithm was implemented (Fig. 3), where the

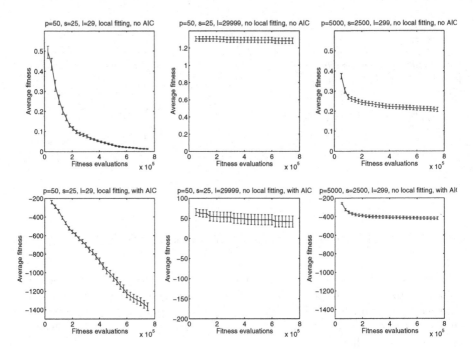

Fig. 4. Average best-fitness with standard error over ten runs of the evolution of logarithm networks. Left column: two-level EA, middle column: one-level for many generations (note the different scale), right column: one-level with a large population. Upper row is without AIC, lower row uses AIC. Headers: p = population size, s = number of survivors, I = number of generations.

upper level evolves a model structure in analogy to GP, while the lower level takes care of the parameters with an evolution strategy (ES) [3].

In order to test the effect of this separation on the performance, we evolved networks supposed to perform two tasks: calculating the third root and logarithm of a positive real number. Here, "calculating" means that the input is set as initial concentration of species *input*, while the output is read from the steady state concentration of species *output*. Therefore, the target time series for the *output* species is simply the desired output value, constant over a period of time, where the first few timepoints are excluded from the fitness evaluation. An example solution for a third root network is shown in Fig. 2. While the third root has been observed to be solvable but substantially more difficult than a square root network [7], no precise solution to the logarithmic problem is known yet. Therefore, the best possible approximation to the logarithm is sought. In this work, the main focus is not put on the evolved networks, but rather on the evolutionary process.

Three different strategies were used, each with and without AIC:

1. Two-level evolution using ES for local fitting (upper level: (25+25)-elitist selection, 29 generations, only structural mutations; lower level: (5,15)-ES, 99 generations, only parameter mutations)

2. One-level evolution running for more generations ((25+25)-elitist selection, 29999 generations, structural and parameter mutations)
3. One-level evolution employing a larger population ((2500+2500)-elitist selection, 299 generations, structural and parameter mutations)

The parameter settings were chosen such that the number of fitness evaluations and the ratio of structural vs. parameter mutations are identical, enabling an objective comparison. The one-level strategies invested the saved fitness evaluations into more generations (2) or more individuals (3). In the ES, adaptive stepsizes were disable to make the results comparable. Computations were carried out as single-processor runs on a cluster of workstations equipped with two Dual Core AMD Opteron(tm) 270 processors running Rocks Linux.

Results of the evolution of a logarithm-network (Fig. 4) show that the two-level structure of the algorithm improves fitness development drastically in comparison to a larger number of generations, while it prevents the premature convergence seen with a larger population. A large population seems to enable the algorithm to guess a good initial network, but it is unable to improve upon this. In contrast, the two-level approach improves the network continuously, yielding significantly better results in the end.

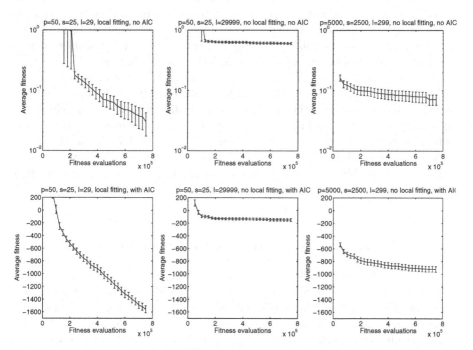

Fig. 5. Average best-fitness with standard error over 100 runs of the evolution of third root networks. Left column: two-level EA, middle column: one-level for many generations, right column: one-level with a large population. Upper row (log scale) is without AIC, lower row uses AIC. Headers: see Fig. 4.

Figure 5 shows the average fitness development for the third root task. Results are similar to Fig. 4, although not as pronounced. Again, the two-level strategy drastically outperforms the setting with more generations, while its initial progress is slower than for a large population. However, the large-population approach converges too early, while the two-level setting continuous to improve in a smooth fashion. In this task, the networks were also required to be mass-conserving, i.e. it was demanded that a feasible configuration of molecular masses for the different species exists. This constraint might explain the slower rate of convergence in comparison to the logarithm-trials.

As a control test, we performed a random search with the same parameters, replacing mutations in the evolutionary algorithm with creation of new random individuals. The EA drastically outperformed random search, resulting in fitness values one order smaller after 750000 fitness evaluations (data not shown). Even though random search finds good initial solutions, it cannot narrow its search and thus lacks the ability to fine-tune the network for the desired calculation.

4   Using AIC to Evolve Parsimonious Models

Another focus of our investigations was the effect of using Akaike's Information Criterion (AIC) as a fitness measure. This measure weights the goodness-of-fit of a model against the number of its free parameters. Given that more parameters will lead to a better fit, but not always to a better explanation of a dataset, AIC formalises a compromise between free parameters and data-fitting. For an overview of information-theory model selection tools, including AIC, see [4].

To investigate AIC, we compared fitness values and free parameters after runs with AIC with those without. Our results are mixed: while AIC has a tendency to reduce model size (not shown), it can drastically affect fitness development, especially for the one-level approaches (Fig. 4 and 5). It seems that AIC either causes premature convergence to small models with bad function, or models achieve the desired function while size increases. This effect is explained by considering that AIC assumes stochastic data, which target time-courses here are not. When models can be fitted perfectly to desired values, the goodness-of-fit dominates the number of free parameters.

5   Species Duplication - A Soft Mutation Operator

A major problem with evolving biochemical networks seems to be the often deleterious effect of structural mutations on network behaviour. Additions and deletions of species and reactions usually change the resulting time series drastically, especially for smaller networks. Therefore, we are looking for "softer" mutation operators. Inspired by biology, one such operator is the duplication of a species and all the reactions it participates in. When the rate constants of all reactions producing the species are halved, this operator does not affect the concentrations of non-mutated species. Later on, deletions and rate mutations can exploit the additional freedom gained by duplication.

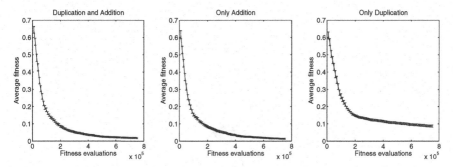

Fig. 6. Average fitness development with standard error for logarithm networks, results from 100 independent runs. Left: species addition and duplication together, middle: only addition, right: only duplication. Also shown are the best 5 runs per setting (gray). Global selection is (50+100)-elitistic, local fitting is a (1+10)-ES, and 50 generations were calculated.

Our results (Fig. 6) show that even though species duplication alone is inferior to addition of new species with random reactions, combining both operators does not yield an inferior result. However, it is still open under which conditions the combined approach improves the random addition of new species.

6 Case Study: The Human Cell Cycle Spindle Checkpoint

Segregation of newly duplicated sister chromatids into daughter cells during anaphase is a critical event in each cell division cycle. Any mishap in this process gives rise to aneuploidy that is common in human cancers and some forms of genetic disorders [5]. Eukaryotic cells have evolved a surveillance mechanism for this challenging process known as the spindle checkpoint. The spindle checkpoint monitors the attachment of kinetochores to the mitotic spindle and the tension exerted on kinetochores by microtubules and delays the onset of anaphase until all the chromosomes are aligned at the metaphase plate [8].

To demonstrate the usefulness of our approach in systems biology, we applied combined structural- and parameter-optimisation to a recent model by Ibrahim et al. [12] of the mitotic spindle checkpoint. This model, which is originally crafted by hand according to literature and laboratory data, describes in details the concentration dynamics of 17 species, namely Mad2, Mad1, BubR1, Bub3, Mad2*, Mad1*, BubR1*, BubR1:Bub3, APC, Cdc20, MCC, MCC:APC, Cdc20:Mad2, and APC:Cdc20, CENPE, Mps1 together with Bub1, and the kinetochore as a pseudospecies. Different kinetochores are represented by three compartments coupled by diffusion, each with the same 11 reaction rules. The last four species represent input signals to the model, reflected in the rate constants of certain reactions. The model corresponds to biological experimental results, which characterise the main components of the mitotic checkpoint.

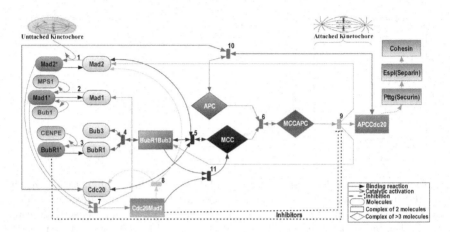

Fig. 7. Schematic network model of mitotic spindle checkpoint. Figure taken from [12].

Table 1. Steady-state concentrations of APC:Cdc20 for four settings of the cell: All kinetochores unattached (1), one attached (2), two attached (3), all three attached (4). The unoptimised model has all parameters set to 0.1, the parameter optimised one has the values from [12], and the structurally optimised model is the result of the procedure described here. Note that the fitness function used is different to the one in [12].

	Level 1	Level 2	Level 3	Level 4	Fitness
Desired value	0	0	0	0.3	0
Without optimisation	0.086154	0.0865285	0.0869323	0.0872706	27.0122
Parameter optimised	0.010924	0.011051	0.011359	0.298768	0.470562
Structure optimised	0.000106	0.000051	0.000037	0.29972	0.000077

As an optimisation target, the concentration of the central species APC:Cdc20 is supposed to be low as long as not all kinetochores are attached, but to rise to a higher value when they all are. In [12], this target has been combined with behaviour from knockout-experiments (which we do not consider here) to fit the rate constants. Here, we test which results can be achieved when the algorithm is allowed to add additional reactions. Any reaction given in the original model cannot be deleted. In future, it will be interesting to loosen this, which could lead to an evolutionary model reducer isolating only the important parts of a given model.

Our results, summarised in Table 1, show that performance of the model has improved compared to the optimisation of rate constants alone. In principle, this could also help to uncover additional structure in the data and to propose additional features of the system which can then be verified experimentally. To achieve biological plausibility, these additional features have to be constrained, which has not been observed in this proof of concept. The best model evolved contains only two additional reactions compared to the original,

$$BubR1_Z \rightarrow Mad1_X^* + BubR1_Y^*$$
$$BubR1_X^* + Cdc20_Y \rightarrow Mad2_X + Cdc20_Y,$$

but these require species from different compartments to interact, which is not intended in the model. We are currently extending our work to include plausibility constraints.

7   Conclusions

As our results for the third root and logarithm tasks indicate consistently, separating structural network evolution from parameter evolution in a two-level algorithm improves the fitness performance significantly. It can be expected that the inclusion of an adaptive stepsize - which has been excluded here to focus on the separation effect - will deepen this advantage. This result is especially interesting as it is in contrast to traditional GP approaches, where parameters are usually evolved simultaneously to the programme structure.

Results for species duplication show that this operator indeed has a beneficial effect on the algorithm, but cannot be used alone, i.e. without the addition of species with random connections. The right balance between creative potential and soft adaptations in different stages of the run seems to be crucial here. For Akaike's Information Criterion, results were unexpected: instead of facilitating the evolution of parsimonious models with a good fitness, evolved solutions were either stuck to small size (usually in one-level approaches), or were of the same size as models evolved without AIC. While the first aspect results from the general tendency of one-level approaches to premature convergence, the second aspect can be explained by the noise-free target data that was used, allowing an almost perfect fit in which the size term in AIC is dominated by the logarithm of goodness-of-fit.

In Section 6, we show that the demonstrated approach can be used to automatically improve realistic models. The next steps are clearly visible now: plausibility constraints have to be included in order to restrict the evolution to solutions that are biologically meaningful. With this in mind, interesting results from this field can be expected in the near future.

Acknowledgements. We thank Stephan Diekmann and Eberhard Schmitt at the FLI Jena for fruitful collaboration on the spindle checkpoint model, and Anthony Liekens at the Technical University of Eindhoven for support with the computations. Funding from the EU (ESIGNET, project no. 12789), Federal Ministry of Education and Research (BMBF, grant 0312704A) and German Academic Exchange Service (DAAD, grant A/04/31166) is gratefully acknowledged. The ESIGNET project also provided funding for the cluster computing facility utilised for this work.

References

1. B. Alberts, A. Johnson, J. Lewis. *Essential Cell Biology.* Garland Publishing, 2003
2. W. Banzhaf, P. Nordin, R.E. Keller, F.D. Francone. *Genetic Programming, An Introduction: On The Automatic Evolution of Computer Programs And Its Applications.* Morgan Kaufmann, 1998
3. H. Beyer and H. Schwefel. *Evolution strategies.* Natural Computing 1:3-52, 2002
4. K.P. Burnham, D.R. Anderson. *Model selection and inference : a practical information-theoretic approach.* Springer, 1998
5. E. Chung, R.-H. Chen. *Spindle Checkpoint Requires Mad1-bound and Mad1- free Mad2.* Molecular Biology of the Cell 13, pp. 1501-1511, 2002
6. B.L. Cooper, N. Schonbrunner, G. Krauss. *Biochemistry of signal transduction and regulation.* Wiley-VCH, 2001
7. A. Deckard and H.M. Sauro. *Preliminary Studies on the In Silico Evolution of Biochemical Networks.* ChemBioChem 5:1423-1431, 2004
8. G. Fang. *Checkpoint Protein BubR1 Acts Synergistically with Mad2 to Inhibit Anaphase-promoting Complex.* Molecular Biology of the Cell 13, pp. 755-766, 2002
9. A. Finney, M. Hucka. *Systems biology markup language: Level 2 and beyond.* Biochem Soc Trans, 31(Pt 6):1472–1473, 2003.
10. P. François, V. Hakim. *Design of Genetic Networks With Specified Functions by Evolution in silico.* PNAS 101:580-585, 2004
11. A. Funahashi, N. Tanimura, M. Morohashi, H. Kitano. *CellDesigner: a process diagram editor for gene-regulatory and biochemical networks.* BIOSILICO 1:159-162, 2003
12. B. Ibrahim, S. Diekmann, E. Schmitt, P. Dittrich. *Compartmental Model of Mitotic Spindle Checkpoint Control Mechanism.* BMCBioinformatic, Submitted Paper, 2006
13. J.R. Koza. *Genetic Programming: On the Programming of Computers by Means of Natural Selection.* Cambridge, MA: MIT Press, 1992
14. J.R. Koza, W. Mydlowec, G. Lanza, J. Yu, M.A. Keane. *Automatic Synthesis of Both the Topology and Sizing of Metabolic Pathways using Genetic Programming.* Proceedings of the Genetic and Evolutionary Computation Conference (GECCO-2001), pp. 57–65, Morgan Kaufmann, 2001
15. W.B. Langdon, R. Poli. *Foundations of Genetic Programming.* Springer, 2002
16. R. Machne, A. Finney, S. Muller, J. Lu, S. Widder, C. Flamm. *The SBML ODE Solver Library: a native API for symbolic and fast numerical analysis of reaction networks.* Bioinformatics 22(11), pp. 1406-7, 2006
17. S.R. Paladugu, V. Chickarmane, A. Deckard, J.P. Frumkin, M. McCormack, H.M. Sauro. *In Silico Evolution of Functional Modules in Biochemical Networks.* IEE Proceedings-Systems Biology 153(4), 2006

A Gaussian Evolutionary Method
for Predicting Protein-Protein Interaction Sites

Kang-Ping Liu[1] and Jinn-Moon Yang[1,2,3,*]

[1] Institute of Bioinformatics, National Chiao Tung University, Hsinchu, Taiwan
[2] Department of Biological Science and Technology, National Chiao Tung University, Hsinchu, Taiwan
[3] Core Facility for Structural Bioinformatics, National Chiao Tung University, Hsinchu, Taiwan
moon@faculty.nctu.edu.tw

Abstract. Protein-protein interactions play a pivotal role in modern molecular biology. Identifying the protein-protein interaction sites is great scientific and practical interest for predicting protein-protein interactions. In this study, we proposed a Gaussian Evolutionary Method (GEM) to optimize 18 features, including ten atomic solvent and eight protein 2nd structure features, for predicting protein-protein interaction sites. The training set consists of 104 unbound proteins selected from PDB and the predicted successful rate is 65.4% (68/104) proteins in the training dataset. These 18 parameters were then applied to a test set with 50 unbound proteins. Based on the threshold obtained from the training set, our method is able to predict the binding sites for 98% (49/50) proteins and yield 46% successful prediction and 42.3% average specificity. Here, a binding-site prediction is considered successful if 50% predicted area is indeed located in protein-protein interface (i.e. the specificity is more than 0.5). We believe that the optimized parameters of our method are useful for analyzing protein-protein interfaces and for interfaces prediction methods and protein-protein docking methods.

Keywords: Atomic solvation parameter, Gaussian evolutionary method, protein-protein interactions, protein-protein binding site.

1 Introduction

Protein-protein interactions play a pivotal role in modern molecular biology. Study of the energetics and mechanism of protein-protein association is a matter of great scientific and practical interest. Identifying the interface between two interacting proteins can reduce the search space required by docking algorithms to predict the structures of complexes and provides important information to identify the function of a protein.

It is widely accepted that protein structural knowledge on a residue and atom level share common properties that can be used to distinguish a protein-protein interacting

---

* Corresponding author.

E. Marchiori, J.H. Moore, and J.C. Rajapakse (Eds.): EvoBIO 2007, LNCS 4447, pp. 143–154, 2007.

interface from the rest of a protein [1-4]. The hydrophobic interaction is one of the major contributors to the affinity of the association [5, 6]. Fernandez-Recio *et al.* [7, 8] successfully extracted the desolvation properties of protein surface to construct atomic solvation parameters for predicting protein-protein interaction sites. However, no single attribute absolutely identifies interface from the rest of a protein [3]. Combination of more physical–chemical properties [2, 9-12] and computational methods are needed for predicting protein-protein interaction sites.

In this study, according to the concept of atomic ASA-based model, we examine the difference in parameters of hydrophobic and structure between the protein interface and the rest of the protein surface for a set of 104 protein–protein interfaces. Briefly, we combine secondary structure information with atomic solvation parameters [7, 8] and optimize these parameters using the GEM [13-18] method for developing an interface prediction. These 18 parameters composed of 10 of the atom properties derived form Fernandez-Recio *et al.* [7, 8] and 8 of the secondary structure properties from DSSP [19]. Based on these visualized optimizing parameters, we are able to predict and analyze interface residues of proteins which are not included in the training set, without any prior knowledge of the binding partner.

2   Materials and Methods

Figure 1 shows the scheme of our method for predicting protein-protein interaction binding site. First, we prepared a training data set which consisted of 52 complexes (104 chains) from a widely used benchmark [20] . For each chain in this data set, the protein surface and interacting interface are derived from the protein structures collected in Protein Data Bank (PDB). Based on these characteristics, we calculated all surface residues scores and trained the GEM parameters to distinguish interacting residues from non-interacting residues by ranking scores of surface residues. The surface residue with the score lower than a given threshold was predicted as an interface residue. For the predicted area of a protein, the specificity and sensitivity [21] were applied to measure performance. Specificity was defined as number of interface residues in predicted area/number of predicted area residues. Sensitivity was defined as number of interface residues in predicted area/number of interface residues. A prediction was deemed a success if predicted area with over 50% specificity [11]. Based on these measure factors as the scoring function, the GEM method optimized atomic and structure parameters to find best solution of average specificity and success rate from 104 training proteins. These optimized parameters were used to predict the protein-protein binding sites for 50 testing proteins if the score of a interface was lower than a given threshold obtained from the training set.

2.1   Data Sets

The aim of this work is to study energetics and mechanism of protein-protein association and improve protein-protein docking algorithm by identifying protein-protein interfaces.

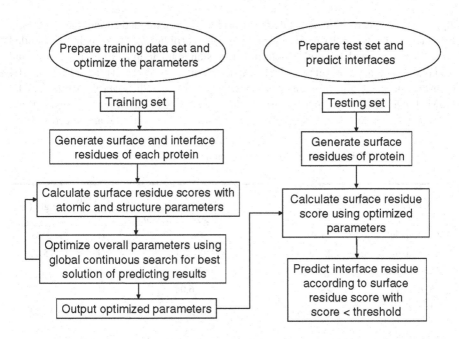

Fig. 1. Flowchart of the GEM for predicting protein-protein interaction sites

Thus, we used 104 unbound protein chains according to 52 protein complexes selected from the protein-protein docking benchmark [20] which was a non-redundant data set contains all test cases used in previous docking studies. The benchmark complements the CAPRI effort, and has the advantage of using a much larger set of test cases. We discarded 7 difficult test cases, which have significant conformational change for more than half of the interface backbone residues between unbound and bound structures, in order to avoid instable predicting system caused by more flexible protein structure. The data set consisted of 22 (44 chains) enzyme-inhibitor complexes, 19 (38 chains) antibody-antigen complexes, and 11 other complexes (22 chains).

2.2 Parameters for Protein Interface Prediction

In order to describe interface properties, we used atomic and structure parameters which combined attributes of hydrophobic properties and structure properties. The atomic parameters consisted of 10 hydrophobic parameters (δ) for protein-protein binding [7, 8] and 8 secondary structure parameters (σ) based on DSSP [19] (Table 1). We referred to the atomic parameters of Fernandez-Recio et al. [8] as the initial values of our parameters. For example, a parameter of C aliphatic is selected from the random positive value (0, 300), since value of C aliphatic from atomic parameters is positive. The strategy ensures the optimized parameters conformed to the behavior of biology. The 2nd structure types were defined as DSSP [19] program, that the definition of 'H' is alpha helix, 'B' is residue in isolated beta-bridge, 'E' is extended strand, participates in beta ladder, 'G' is 3-helix (3/10 helix), 'I' is 5 helix (pi helix), 'T' is hydrogen bonded turn, 'S' is bend and 'others' is determination stands for loop or irregular. Loops and irregular elements are often called "random coil" or "coil". Training set of 104 high-resolution crystal structures was then used to train the overall parameters by GEM. Each value of parameters was used to calculate score of surface residue and optimized by GEM directly according to the specificity, sensitivity, and success rate.

Table 1. Parameters for protein-protein interaction sites prediction

Atom type	Optimized hydrophobic value (δ) (GEM)	Atomic Solvation Parameters (Fernandez-Recio *et al.*)	2nd structure type	weight ($\square$)
C aliphatic	8.607	19.18	H	3.06
C aromatic	257.134	110.80	G	15.12
N uncharged	-11.500	-39.10	I	-5.14
Nζ in Lys+	-295.247	-126.04	B	-13.74
Nη1,Nη2 in Arg+	-26.958	-62.56	E	12.22
O hydroxyl	-6.091	-42.55	T	15.22
O carbonyl	-1.619	-31.28	S	18.18
O- in Glu, Asp	-41.291	-68.77	others	13.84
S in SH	281.919	25.76		
S in Met	298.668	5.06		

2.3 Surface, Interface, and Surface Patch

Surface and interface residues of the proteins were identified according to the atom coordinates in the PDB. Solvent accessible surface areas (ASA) for each residue are calculated using DSSP [19]. A residue is defined to be a surface residue if its ASA is at least 25% of its nominal maximum area [22] as defined by Rost and Sander [23] . The distance-based definition of interface used here was that a surface residue is defined as an interface residue if its side-chain center is within 4.5Å of the side-chain center of a residue belonging to another chain in the complex. There is no indication that this criterion is better or worse than others [24]. Ofran et al. [25], for example, used a cut-off distance that if any of residue atoms < 6Å from any atom of the other protein. Sternberg and coworkers [26] set a cut-off distance of 8 Å between the C$_\beta$ atoms.

A patch was defined as a surface area decided by a residue at a protein surface as a centre with different radius sizes. For example, a protein with 100 surface residues would have 100 surface patches. To ensure that we did not measure through the protein, the following procedure was followed. A C$_\beta$ of every surface residue on the protein (a C$_\alpha$ for glycine) was used to calculate distances between all surface residues. The patch was grown from a single starting (seed) residue and subsequently used to generate series of surface patch. Different-sized surface patches were generated by selecting all surface residues at different radius (here, $r = 1, 2, \cdots, 20\text{Å}$) from a given seed [8]. The final patch size of the seed was according to the lowest score of a surface patch. Scores of different-sized surface patches were calculated based on the parameters of the atomic ASA of their component residues and secondary structures. This process was iterated using the newly acquired residues, until the total number of residues in the patch was equal to the total number of residues in the surface.

2.4 Scoring Function

The desolvation properties of a given surface residue were calculated using an atomic ASA-based model [27, 28], and we combined new attributes of secondary structure properties to consider interface structure information.

The surface residue score S of each residue k in a protein surface is calculated as

$$S_k = \min(-\sum_{i \in shpere_r}(\delta_{SA(i)} \times ASA_i + \sigma_{SS(i)})),$$

where the atom i within the sphere of a defined radius r and the C_β of residue k as the center where r ranges from 1 to 20 Å; $\delta_{SA(i)}$ is the weight of the atom type $SA(i)$; ASA_i is the solvent accessible of the surface area of the atom i; $\sigma_{SS(i)}$ is the weight for the 2nd structure type $SS(i)$. The lowest score of the surface patch is selected for the final score of the residue k. To assess the surface residue score S of a residue k is with respect to the whole protein, the S_k is transformed into Z_k and is calculated as

$$Z_k = \frac{S_k - \overline{S}}{\gamma},$$

where $\overline{S}$ and γ are the average surface patch score and the standard deviation, respectively, of all of the residues in a protein surface. Residues with Z_k lower than a given threshold (Z_{cut}) are taken as the predicted interfaces. In other words, we calculate score of each surface residue and identify interface residues according to their scores. If scores of residues are lower than Z_{cut}, we mark these residues to be protein-protein interaction site. In this paper, the Z_{cut} is set to -1.5.

The specificity, sensitivity, and success rate were used to evaluate the performance of a computational method. For the predicted interfaces of a protein (p), the specificity (ψ_p) and sensitivity [21] are define as the number of interface residues in predicted area/ total number of residues in the predicted area and as the number of interface residues in predicted area/ total number of the interface residues, respectively. A prediction was deemed a success (ω_p) for a protein if specificity is more than 0.5 [11].

2.5 GEM Algorithm

GEM is a multi-operator approach that combines three mutation operators: decreasing-based Gaussian mutation, self-adaptive Gaussian mutation, and self-adaptive Cauchy mutation. It incorporates family competition and adaptive rules for controlling step sizes to construct the relationship among these three operators. To balance the search power of exploration and exploitation, each of operators is designed to compensate for the disadvantages of the other. The details of GEM described in previous works [16, 18], have been successfully applied to some specific problems, such as protein-ligand docking, drug screening, and protein side-chain prediction [13-15, 17, 18].

Here, we provided an outline of the GEM for predicting protein-protein interaction sites, which can be represented by adjustable variables of atomic and structure parameters (Table 1) as

$$(\delta_1, \delta_2, ..., \delta_{10}, \sigma_1, \sigma_2, ..., \sigma_8),$$

where δ is the atomic parameter and σ is the structure parameter of surface residue scoring function. The values of parameters are then used in the surface residue scoring function and predicting results are presented as specificity and success rate. In order to determine that the performance of adjustable parameters, we use a fitness function which combines specificity and success rate for GEM training. In this work, the GEM to optimize atomic and structure parameters for identifying protein interfaces by minimizing the fitness function which is given as

$$Fitness = -\sum_{p=1}^{N}(\psi_p + \omega_p),$$

where N is the total number of training proteins (here N is 104) ; ψ_p is the specificity of predicting results of the protein p based on the training values of atomic parameters. In order to raise success rate (ω_p) of prediction, the ω_p was added into the fitness function. The value of ω_p is depend on ψ_p. When ψ_p is more than 0.5, ω_p is set to 1. Conversely, ω_p is set to 0 if ψ_p is less than 0.5,

The core idea of our GEM was to design multiple operators (decreasing-based Gaussian mutation and self-adaptive Cauchy mutation) that cooperate using the family competition model, which is similar to a local search procedure [13-15, 17, 18]. Here, the steps for protein-protein binding site prediction are briefly described as follows:

1. Initialize the atomic and structure parameters of surface residue scoring function. The initial values for the parameters are randomly generated in the feasible region (-300, 300). We used a strategy, which is according to the value of atomic solvent parameters proposed by Fernandez-Recio et al.[8], to ensure optimized value conformed to the behavior of biology and increase the search speed. For example, a parameter of C aliphatic is selected from the random positive value (0, 300), since value of C aliphatic from atomic solvent parameters is positive. Evaluate the objective value of each solution, which consists of 18 feature values, in the population (with N solutions) based on the fitness function.
2. Generate a new quasi-population with N offspring by applying genetic operators. Evaluate the objective values of these offspring.
3. Use selection operators to select N solutions from both parent and offspring solutions.
4. Repeat steps 2 and 3 until one of the terminal criteria is satisfied.

The GEM parameters used in this paper are listed in Table 2 such as population size, initial step sizes of Gaussian mutations, recombination probability, and family competition length in this work. The GEM optimization stops when either the convergence is below certain threshold value or the iterations exceed a maximal preset value which was set to 200. These parameters were selected after many attempts to predict interaction sites for test proteins with various initial values.

Table 2. Gaussian Evolutionary Method parameters

Parameter	Value
Population size (N)	200
Step size of Gaussian mutations	$v = 0.2$ and $\lambda = 0.8$ (in radius)
Recombination probability	0.2
Family competition length	$L = 3$
Number of maximum generations	200

3 Results and Discussions

The results of the GEM method for the training set are summarized in Table 3. GEM is able to predict the location of the interface on 65.4% (68/104) proteins in the training dataset. The average specificity and sensitivity are 57.8% and 35.5%, respectively. If we only train atomic parameters by GEM and use optimized atomic parameters to predict training set, the success of prediction is decreasing to 51.0% and the average specificity and sensitivity are 44.9% and 27.6%, respectively. In addition, if we used atomic parameters based on Fernandez-Recio et al. [8] without any optimization, the

Table 3. Summary of training results from 104 unbound proteins

	Atomic and 2nd parameters (GEM [a])			Atomic parameters (GEM [a])			Atomic parameters (Fernandez-Recio et al. [b])		
	Success	Specificity	Sensitivity	Success	Specificity	Sensitivity	Success	Specificity	Sensitivity
Enzyme Inhibitor	79.5% (35/44)	67.1%	27.9%	65.9% (29/44)	56.0%	28.7%	45.5% (20/44)	45.9%	25.3%
Antibody antigen	60.5% (23/38)	54.3%	33.2%	39.5% (15/38)	37.0%	30.1%	44.7% (17/38)	37.2%	25.1%
Others	45.5% (10/22)	45.2%	18.1%	40.9% (9/22)	36.2%	21.2%	22.7% (5/22)	32.3%	19.1%
Average	65.4% (68/104)	57.8%	35.5%	51.0% (53/104)	44.9%	27.6%	37.5% (39/104)	39.8%	23.9%

[a] The parameters which are optimized by GEM.
[b] The parameters which are based on Fernandez-Recio et al. [8].

performance of prediction is the worse. Combination of physical–chemical properties [2, 9-12] and using computational methods to assist the finding of best parameters are useful for predicting protein-protein interaction sites.

Table 4. Prediction specificities of 50 unbound proteins

PDB	GEM	Fernandez Recio *et al.*	PDB	GEM	Fernandez Recio *et al.*	PDB	GEM	Fernandez Recio *et al.*
1a19A	1.00	0.89	1eztA	0.16	0	1pco_	0.00	0.18
1a2pA	0.66	—[a]	1f00I	0.00	—	1pne_	0.33	—
1a5e_	0.60	0.2	1f5wA	0.60	1	1poh_	0.66	0.8
1acl_	0.22	0	1fkl_	0.33	—	1ppp_	0.50	0.3
1ag6_	1.00	—	1flzA	0.33	0.9	1rgp_	0.00	1
1aje_	0.50	0.4	1fvhA	0.52	0.71	1selA	1.00	0.2
1ajw_	0.50	—	1g4kA	1.00	1	1vin_	1.00	1
1aueA	0.00	0.8	1gc7A	0.21	—	1wer_	0.30	—
1avu_	0.33	0.7	1gnc_	1.00	0.1	1xpb_	0.42	—
1b1eA	1.00	0.7	1hh8A	0.00	—	2bnh_	0.45	—
1bip_	1.00	1	1hplA	0.00	—	2cpl_	0.66	—
1ctm_	0.31	0.63	1hu8A	0.00	0	2f3gA	1.00	—
1cye_	0.00	0.29	1iob_	0.25	—	2nef_	0.00	0.9
1d2bA	0.50	1	1j6zA	0.54	1	2rgf_	—	1
1ekxA	0.00	0.22	1jae_	0.52	1	3ssi_	1.00	—
1ex3A	0.60	1	1lba_	0.00	—	6ccp_	0.00	—
1eza_	0.08	—	1nobA	0.00	—	average	0.42	0.38

[a] — means that there are no results of prediction.

The overall accuracy of GEM in predicting the protein-protein interaction sites of 50 test proteins is shown in Table 4. In order to test our performance of predicting protein-protein interaction sites, we tested our parameters against a accompanying paper of Fernandez-Recio *et al.* [8] and found that our atomic parameters performed at least as well as their atomic solvation parameters. The results of this test are summarized in Table 4, in which it can be seen that our method can predict 98% (49/50) proteins among the testing set and have 42.3% average specificity, better than Fernandez-Recio's results which can predict 60% protein among whole testing set and have 37.8% average specificity.

Figure 2 shows six examples of the predicted results of using GEM method for the training set (Figures 2a, 2b and 2c) and testing set (Figures 2d, 2e and 2f). Even if the predicted binding-site areas (red) do not totally match the interface area, GEM can predict satisfied results for these cases. For example, the specificity of 2cpl (Figure 2d) 0.66 for GEM is better than the result from the Fernandez-Recio *et al.* [8] which can't predict the interaction site for this protein. Although our prediction result of 1ctm (Figure 2e) is worse than the result from the Fernandez-Recio *et al.* [8] (the specificity of results are 0.31 and 0.63, respectively), our predicting area fits the interaction site

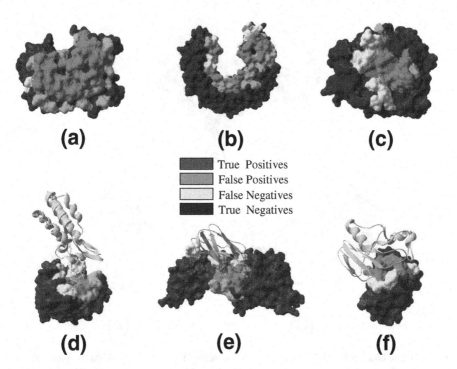

Fig. 2. Six predicted examples of using GEM method: (a) hyhel-63 Fab (1dqj_r); (b) ribonuclease inhibitor (1dfj_r); (c) β-Trypsin (2ptc_r); (d) Cyclophilin a (2cpl), (e) Cytochrome f (1ctm), (f) barstar (1a19A). The partner molecule(s) in the bound conformation after superimposition of the corresponding molecule in the complex is represented in ribbon in (d), (e), and (f). Predicted interface and non-interface residues, identified by the GEM, are colored as follows: red is the true positives (TP), actual interface residues that are predicted as such; blue is the true negatives (TN), non-interface residues that are predicted as such; yellow is the false negatives (FN), interface residues that are misclassified as non-interface residues; green is the false positives (FP), non-interface residues that are misclassified as interface residues.

well. The GEM method (specificity is 1.00) outperform Fernandez-Recio *et al.* [8] (specificity is 0.89) on barstar (Figure 2f).

The factors causing GEM method to predict the wrong protein-protein binding sites can be divided into four categories. The first factor is that a protein structure consists of symmetrical hydrophobic cores (Figure 3a) and our method can not discriminate them. As shown in Figure 3a, this protein has two fibronectin type III modules whose hydrophobic cores merge in the domain-domain interface and our prediction is almost invariably symmetrical. In the secondary category, the protein has two large binding sites (Figure 3b): a large ligand (e.g. heme) and a protein-protein binding sites. The predicted area of our method is located nearby heme propionate, this result may due to the residues nearby the heme are more hydrophobic than protein-protein interaction site. In the secondary category, the binding site of a protein is hydrophilic. The surface

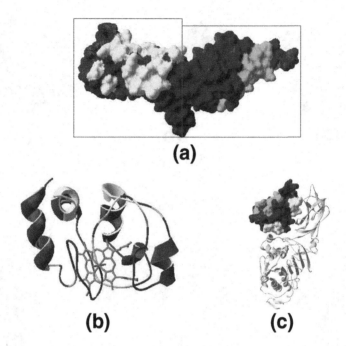

Fig. 3. Three bad predicted examples. (a) The protein (PDB code 1ahw_l) has a symmetric core that is the fibronectin type (b) The protein (PDB code 1wej_l) has two binding sites that are ligand heme (purple) binding site and protein-protein binding site. (c) The binding site of the protein (PDB code 1pco) is hydrophilic. The predicted and interface areas are colored with green and yellow, respectively, the other areas are colored with blue.

of colipase can be divided into a rather hydrophilic part, interacting with 1pco (lipase), and a more hydrophobic part, formed by the tips of the fingers [29]. This suggests that interface of 1pco is more hydrophilic than the surface, and our method do not prove to be very useful in this case. The final factor is that the protein-protein binding interface of a protein is significantly changed from the unbinding state to the binding state.

4   Conclusion

We have developed a method for predicting protein-protein binding sites using the GEM method. Our method successfully predicted the location of the binding site on 65.4% of the 104 proteins in training set. In addition, we applied the GEM to predict 50 unbound proteins and obtained 46% successful prediction.

This method can be further improved on several aspects. First, we notice that hydrophilic effect is the main force of protein-protein interaction in some cases. This is due to the fact that most interfaces of training set are hydrophobic and our parameters perform this characteristic faithfully. Therefore, it may be useful to classify interfaces of training set according to hydrophobic or hydrophilic, and each protein has two predicting areas which are hydrophobic patch and hydrophilic patch. Second, sequence conservation tends to be important attribute to identify protein-protein interface [30]. Finally, we will apply our approach to other data set and to study the behavior.

Acknowledgement

We are grateful to both the hardware and software supports of the Structural Bioinformatics Core Facility at National Chiao Tung University. J.-M. Yang was supported by National Science Council and the University System at Taiwan-Veteran General Hospital Grant.

References

1. Jones S, Thornton JM: Principles of protein-protein interactions. *Proceedings of the National Academy of Sciences of the United States of America* 1996, 93(1):13-20.
2. Jones S, Thornton JM: Prediction of protein-protein interaction sites using patch analysis. *Journal of Molecular Biology* 1997, 272(1):133-143.
3. Jones S, Thornton JM: Analysis of protein-protein interaction sites using surface patches. *Journal of Molecular Biology* 1997, 272(1):121-132.
4. Nooren IM, Thornton JM: Diversity of protein-protein interactions. *EMBO Journal* 2003, 22(14):3486-3492.
5. Vakser IA, Aflalo C: Hydrophobic docking: a proposed enhancement to molecular recognition techniques. *Proteins: Structure, Function and Genetics* 1994, 20(4):320-329.
6. Young L, Jernigan RL, Covell DG: A role for surface hydrophobicity in protein-protein recognition. *Protein Science* 1994, 3(5):717-729.
7. Fernandez-Recio J, Totrov M, Abagyan R: Identification of protein-protein interaction sites from docking energy landscapes. *Journal of Molecular Biology* 2004, 335(3):843-865.
8. Fernandez-Recio J, Totrov M, Skorodumov C, Abagyan R: Optimal docking area: a new method for predicting protein-protein interaction sites. *Proteins: Structure, Function, and Bioinformatics* 2005, 58(1):134-143.
9. Fariselli P, Pazos F, Valencia A, Casadio R: Prediction of protein--protein interaction sites in heterocomplexes with neural networks. *European Journal of Biochemistry* 2002, 269(5):1356-1361.
10. Keil M, Exner TE, Brickmann J: Pattern recognition strategies for molecular surfaces: III. Binding site prediction with a neural network. *Journal of Computational Chemistry* 2004, 25(6):779-789.
11. Neuvirth H, Raz R, Schreiber G: ProMate: a structure based prediction program to identify the location of protein-protein binding sites. *Journal of Molecular Biology* 2004, 338(1):181-199.
12. Zhou HX, Shan Y: Prediction of protein interaction sites from sequence profile and residue neighbor list. *Proteins: Structure, Function and Genetics* 2001, 44(3):336-343.
13. Yang JM: Development and evaluation of a generic evolutionary method for protein-ligand docking. *Journal of Computational Chemistry* 2004, 25(6):843-857.
14. Yang JM, Chen CC: GEMDOCK: a generic evolutionary method for molecular docking. *Proteins: Structure, Function, and Bioinformatics* 2004, 55(2):288-304.
15. Yang JM, Horng JT, Kao CY: A genetic algorithm with adaptive mutations and family competition for training neural networks. *International Journal of Neural Systems* 2000, 10(5):333-352.

16. Yang JM, Kao CY: A family competition evolutionary algorithm for automated docking of flexible ligands to proteins. *IEEE Transactions on Information Technology in Biomedicine* 2000, 4(3):225-237.

17. Yang JM, Shen TW: A pharmacophore-based evolutionary approach for screening selective estrogen receptor modulators. *Proteins: Structure, Function, and Bioinformatics* 2005, 59(2):205-220.

18. Yang JM, Tsai CH, Hwang MJ, Tsai HK, Hwang JK, Kao CY: GEM: a Gaussian Evolutionary Method for predicting protein side-chain conformations. *Protein Science* 2002, 11(8):1897-1907.

19. Kabsch W, Sander C: Dictionary of protein secondary structure: pattern recognition of hydrogen-bonded and geometrical features. *Biopolymers* 1983, 22(12):2577-2637.

20. Chen R, Mintseris J, Janin J, Weng Z: A protein-protein docking benchmark. *Proteins: Structure, Function and Genetics* 2003, 52(1):88-91.

21. Bradford JR, Westhead DR: Improved prediction of protein-protein binding sites using a support vector machines approach. *Bioinformatics* 2005, 21(8):1487-1494.

22. Yan C, Dobbs D, Honavar V: A two-stage classifier for identification of protein-protein interface residues. *Bioinformatics* 2004, 20 Suppl 1:I371-I378.

23. Rost B, Sander C: Conservation and prediction of solvent accessibility in protein families. *Proteins: Structure, Function, and Genetics* 1994, 20(3):216-226.

24. Ansari S, Helms V: Statistical analysis of predominantly transient protein-protein interfaces. *Proteins: Structure, Function, and Bioinformatics* 2005, 61(2):344-355.

25. Ofran Y, Rost B: Predicted protein-protein interaction sites from local sequence information. *FEBS Letters* 2003, 544(1-3):236-239.

26. Sternberg MJ, Gabb HA, Jackson RM: Predictive docking of protein-protein and protein-DNA complexes. *Current Opinion in Structural Biology* 1998, 8(2):250-256.

27. Eisenberg D, McLachlan AD: Solvation energy in protein folding and binding. *Nature* 1986, 319(6050):199-203.

28. Wesson L, Eisenberg D: Atomic solvation parameters applied to molecular dynamics of proteins in solution. *Protein Science* 1992, 1(2):227-235.

29. Egloff MP, Sarda L, Verger R, Cambillau C, van Tilbeurgh H: Crystallographic study of the structure of colipase and of the interaction with pancreatic lipase. *Protein Science* 1995, 4(1):44-57.

30. Caffrey DR, Somaroo S, Hughes JD, Mintseris J, Huang ES: Are protein-protein interfaces more conserved in sequence than the rest of the protein surface? *Protein Science* 2004, 13(1):190-202.

Bio-mimetic Evolutionary Reverse Engineering of Genetic Regulatory Networks

Daniel Marbach, Claudio Mattiussi, and Dario Floreano

Ecole Polytechnique Fédérale de Lausanne (EPFL),
Laboratory of Intelligent Systems,
CH-1015 Lausanne, Switzerland
{Daniel.Marbach,Claudio.Mattiussi,Dario.Floreano}@epfl.ch
http://lis.epfl.ch

Abstract. The effective reverse engineering of biochemical networks is one of the great challenges of systems biology. The contribution of this paper is two-fold: 1) We introduce a new method for reverse engineering genetic regulatory networks from gene expression data; 2) We demonstrate how nonlinear gene networks can be inferred from steady-state data alone. The reverse engineering method is based on an evolutionary algorithm that employs a novel representation called Analog Genetic Encoding (AGE), which is inspired from the natural encoding of genetic regulatory networks. AGE can be used with biologically plausible, nonlinear gene models where analytical approaches or local gradient based optimisation methods often fail. Recently there has been increasing interest in reverse engineering *linear* gene networks from steady-state data. Here we demonstrate how more accurate *nonlinear dynamical models* can also be inferred from steady-state data alone.

Keywords: Systems Biology, Gene Networks, Reverse Engineering, Steady-State Data, Genetic Algorithm, Analog Genetic Encoding (AGE).

1 Introduction

Genetic regulatory networks perform fundamental information processing and control mechanisms in the cell. Regulatory genes code for proteins that enhance or inhibit the expression of other regulatory and/or non-regulatory genes, thereby forming a complex web of interactions (Fig. 1a). Inference and simulation of gene networks may contribute substantially to our biological knowledge in the post-genomic era. Practical applications may have a strong impact on biotech and pharmaceutical industries, potentially setting the stage for rational redesign of living systems and predictive, model-based drug design [1]. Technologies to assay gene expression levels in terms of mRNA concentrations are advancing at a fast pace. Using oligonucleotide chips or quantitative PCR for instance, it is possible to probe a set of genes of interest that are part of an uncharacterized gene network (henceforth known as *target network*) under different conditions. The goal of reverse engineering is inferring the target gene regulatory network from this experimental data.

The choice of a suitable reverse engineering method depends on the type of model used to describe the target network. Here we focus on models that represent

E. Marchiori, J.H. Moore, and J.C. Rajapakse (Eds.): EvoBIO 2007, LNCS 4447, pp. 155–165, 2007.
© Springer-Verlag Berlin Heidelberg 2007

a genetic regulatory network as a dynamical system described by a system of ordinary differential equations. The linear model [1,2,3,4,5], which is based on a first-order approximation of gene expression dynamics, is by far the most widely used gene network model. Its main advantage is that reverse engineering can be tackled analytically using standard techniques of system identification [1,2,3,4,5,6]. However, gene regulation is known to be strongly nonlinear. Hence, the linearization is generally only valid in a small regime, i.e., close to a specific steady-state [1,3,5]. This implies that valuable data from perturbation experiments with a strong effect on the network (e.g., gene knockouts) cannot be used because they fall outside the valid regime of the first-order approximation [1,5]. Furthermore, the inferred linear model is unlikely to correctly predict network response under strong perturbations [1,5], as can be expected in disease for instance.

As both quantity and quality of experimental data improve, we can aim at a more biologically plausible, faithful reconstruction of the target network. This requires the conception of adequate inference methods that can handle complex, nonlinear gene models, where analytical approaches and local gradient based optimisation often fail. In this paper we propose a bio-mimetic approach based on artificial evolution [7] using Analog Genetic Encoding (AGE) [8,9], an artificial genetic representation that has already proven its merits in benchmark problems in the fields of analog electronic circuits [8,9] and artificial neural networks [10]. Unlike other reverse engineering algorithms based on global optimisation techniques such as simulated annealing [11,12] or conventional evolutionary algorithms [4,13,14,15], AGE allows simultaneous inference of model structure and numerical parameter values. Furthermore, AGE mimics the evolutionary process of incremental complexification that natural gene regulatory networks are subjected to.

In the past, most reverse engineering studies used time-series gene expression data. However, time-series data are more difficult to obtain experimentally than steady-state data and their information content is lower (samples in a time-series are not independent). Indeed, there has been a recent trend towards approaches based on steady-state perturbation data [1,3,16,17]. These studies use analytical approaches based on first-order approximations. Here we demonstrate for the first time – to the best of our knowledge – how *nonlinear* models (network structure and parameters) can be inferred from steady-state data alone.

2   Evolutionary Reverse Engineering with AGE

The first step in the reverse engineering process generally consists in the choice of a gene network model type (e.g. the sigmoid model introduced in Sect. 3). AGE is not constrained to a specific model type, but can be used with a large class of nonlinear models termed *analog networks* [9]. An analog network is composed of a collection of devices connected by links of different strengths. Here, devices are genes and links correspond to regulatory interactions[1]. Without limiting

---

[1] In this paper we consider the simplest case where all devices are of the same type, but AGE can also handle heterogeneous networks [9]. We plan to use several device types in the future for more complex models of gene-protein networks, for instance.

ourselves to a specific model type, we assume that genes are characterized by a vector of internal parameters $\mathbf{p}$ (e.g. decay rate, maximum transcription rate, etc.) and regulatory links have a single parameter called weight w. Within this framework, reverse engineering requires the specification of the network structure (size, topology) and the specification of the numerical values of all gene parameter vectors $\mathbf{p}$ and connection weights w. Using AGE, we encode these elements in a bio-inspired artificial genome. The reverse engineering process then amounts to the artificial evolution of gene networks that best match the experimental gene expression data (see Sect. 3.2). Apart from the bio-inspired genotype and mutation operators, the evolutionary algorithm is similar to a standard genetic algorithm [7]. It acts on a population of gene networks, which are encoded as described below. At each generation, fitter individuals are selected with higher probability for reproduction. Offspring are produced from the selected genomes by applying crossover and mutation operators as described in Sect. 2.2.

2.1 Genetic Encoding

We stress that AGE is not supposed to be a detailed model of the workings of gene networks, but a bio-inspired genotype that abstracts key features distinguishing the biological encoding from traditional artificial encodings used in genetic algorithms. Nature has chosen a digital encoding of genomes based on sequences of characters. Similarly, the AGE genome is constituted by one or more chromosomes, which are sequences of characters drawn from a genetic alphabet. The genetic alphabet used here has 26 nucleotides, which we designate with letters 'A'-'Z'. In nature, the beginning and the end of genes are marked by signals encoded in the DNA (promoters and terminators). Analogously, we use special nucleotide patterns ('GN' and 'TE') termed 'tokens' to delimit genes in the artificial genome as illustrated in Fig. 1. Consequently, genes may be located anywhere in the genome. Sequences that are not part of a gene are non-coding.

In a cell, the potential regulatory interaction between two genes A and B is not encoded *explicitly* in the genome, but follows *implicitly* from a biochemical process which depends among other things on: i) The coding region of gene A, which encodes the characteristics of protein A; ii) The regulatory region of gene B, which contains the potential binding sites for the regulatory protein (Fig. 1a; Note that there are other mechanisms of gene regulation not discussed here). Thus, the strength of the interaction is implicitly encoded by the respective coding and regulatory sequence. One of the consequences of the implicit encoding is that a *single* mutation in a coding or regulatory sequence may affect zero, one or several regulatory interactions simultaneously. In contrast, in an explicit (direct) encoding a single mutation affects only one characteristic of the network.

Analogously, artificial genes in AGE have a regulatory and a coding region as shown in Fig. 1b. The regulatory influence w_{ij} of gene j on gene i is implicitly encoded in the coding region $s_{\mathrm{cod},j}$ and the regulatory region $s_{\mathrm{reg},i}$ via an interaction map I_{w} that abstracts the complex biochemical process of transcriptional regulation: $w_{ij} = I_{\mathrm{w}}(s_{\mathrm{cod},j}, s_{\mathrm{reg},i})$. The interaction map is based on local alignment

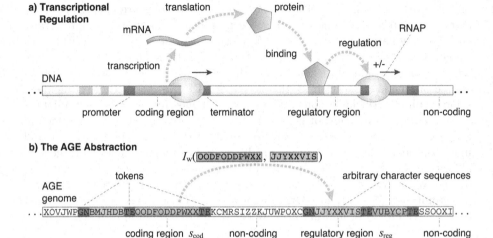

Fig. 1. a) Simplified representation of transcriptional regulation. A gene is transcribed by RNA Polyermase (RNAP). Proteins are synthesized from mRNA (translation). Gene regulatory proteins bind to specific sites, enhancing or repressing the transcription rate of the following gene. Genetic regulatory networks are complex webs of such regulatory interactions. b) AGE chromosome encoding two genes. Analogous to the natural encoding, the beginning and the end of genes are marked by special nucleotide patterns in the artificial genome (tokens 'GN' and 'TE'). Genes have a coding region s_{cod} and a regulatory region s_{reg}, which may interact via an interaction map $I_w(s_{cod}, s_{reg})$ that abstracts the complex biochemical process illustrated in a).

of the two sequences [8,9]. The closer the match between two subsequences of s_{cod} and s_{reg}, the stronger the interaction. Details are given in the Appendix.

In summary, decoding of the AGE genome involves the identification of valid genes (which must be correctly delimited by the corresponding tokens) and the subsequent application of the sequence interaction map to all pairs of coding and regulatory sequences. The interaction strength between two sequences may be zero, in which case there is no regulatory link between the two genes. Hence, the size of the decoded network is given by the number of genes in the genome and the topology and weights w follow from the computed interaction strengths.

For the gene parameters **p**, it is desirable to use the same encoding as for the interactions [8]. Consider first the case where genes have a single parameter p. The value of p is decoded analogously to the weights by a sequence interaction map: $p = I_p(s_{p,1}, s_{p,2})$, where $s_{p,1}$ and $s_{p,2}$ are two additional sequences appended to the coding region of genes as shown in Fig. 2. Further gene parameters can be encoded by appending an additional pair of sequences for every additional parameter. Implementation details of I_p are also given in the Appendix.

2.2   Genetic Operators

One of the key features of AGE is the possibility to apply a wide range of biologically inspired genetic operators that are believed to play important roles

Fig. 2. In order to encode a gene parameter p, genes must have two additional sequences $s_{p,1}$ and $s_{p,2}$. Thus, a valid gene with one gene parameter has four sequences separated by tokens 'TE'. The token 'GN' to the left of the valid gene is not followed by the necessary four tokens 'TE', thus it is not coding.

in the evolution and complexification of natural genetic regulatory networks [8]. From the point of view of the genetic operators, the tokens that delimit the genes have no special meaning – there is no distinction between tokens, coding and non-coding genome fragments. The operators described below are applied probabilistically to randomly chosen parts of the genome (see Sect. 3.2).

- *Nucleotide deletion, insertion, and substitution*: A character is removed, inserted, or substituted in the genome. Random characters from the genetic alphabet are used for insertions and substitutions.
- *Chromosome fragment deletion, transposition, and duplication*: Two points are chosen in a chromosome and the intervening genome fragment is deleted, transferred or copied to another point of the genome.
- *Chromosome deletion/duplication*: A chromosome is deleted/duplicated.
- *Crossover*: Chromosomes of parents are recombined using homologous crossover, which is based on the search of a homologous crossover point [8].

The application of the genetic operators can invalidate genes (e.g., through invalidation of a token) and transform the corresponding fragments into non-coding genome, which may play the role of an evolutionarily useful repository of genetic fragments. On the other hand, new genes can be created, for example through the appearance of new tokens or the duplication of a genome fragment.

3   Experiments

When applying a novel reverse engineering technique directly to biological data, performance evaluation is difficult because the target network is unknown. Thus, we first test AGE using synthetic expression data generated in simulation from an *in silico* target gene network. Subsequently the inferred networks are compared with the target network in order to validate the results. This is a standard approach to assess the performance of reverse engineering methods [5].

3.1   The Test Case

Gene Network Model. We demonstrate the application of AGE using a standard sigmoid model [11,12,13,18] defined by the system of state equations:

$$dx_i/dt = m_i \cdot \sigma\Big(\sum_{j \in R_i} w_{ij} x_j \Big) - \delta_i x_i, \qquad (1)$$

where x_i is the expression level of gene i, m_i is the maximum transcription rate, and δ_i is the degradation rate. R_i is the set of regulators of gene i and w_{ij} represents the regulatory influence of gene j on gene i (positive for enhancers, negative for repressors). The activation function is a sigmoid $\sigma(z) = 1/(1+e^{-z})$.

In the experiments reported in this paper we use steady-state expression levels as input data for the inverse problem, though AGE could as well be applied to time-series data. At steady-state, the state equations become a set of algebraic equations:

$$0 = p_i \cdot \sigma\Big(\sum_{j \in R_i} w_{ij} x_j\Big) - x_i, \quad \text{with } p_i = m_i/\delta_i. \tag{2}$$

Synthetic Target Network and Expression Data. We employ the topology of a nine-gene subnetwork of the *E.coli* SOS pathway as described in [1] as test case (see Fig. 4). We refer to this topology as *SOS network*. There is no quantitative model of the SOS network available in the literature. Hence, numerical parameter values for the weights w and the parameter p of the steady-state equation introduced above are sampled randomly (see Appendix). The sign of the weights is set according to the SOS network topology (positive for enhancers and negative for repressors). The resulting *in silico* target gene networks are *random targets* with a realistic topology, which is a more biologically plausible approach than random generation of both topology and parameters [5,13].

Synthetic gene expression data is obtained by applying a perturbation to the *in silico* target network and computing the steady-state expression levels of all genes[2]. This process is repeated for different perturbations to gather the necessary experimental data for reverse engineering. In our experiments we use two different types of perturbations that are commonly used for gene network inference: Gene knockouts, i.e. silencing of a particular gene, and gene over-expression, which consists in artificially boosting the transcription rate of a gene. A gene knockout can be simulated by setting the rate parameter m_i to zero. Consequently, the expression level x_i of this gene at steady-state will be zero. Over-expression is simulated by doubling the m_i value of the affected gene.

For the experiments reported below we generate expression data from the *in silico* SOS network for the wild type (unperturbed network) and for 9 knockout and 9 over-expression experiments (knockout and over-expression of each gene).

3.2   The Evolutionary Algorithm

The weights w and the single gene parameter p of the steady-state sigmoid model are encoded in the artificial genome as explained in Sect. 2. Details of the sequence interaction maps are given in the Appendix.

Since we do not yet model noise, the least squares optimisation criterion is suitable for the definition of the fitness. The goal of the evolutionary algorithm

---

[2] We compute the steady-states numerically using Powell's method of the GNU Scientific Library (GSL, http://www.gnu.org/software/gsl).

is to minimize the fitness function: $f(\hat{\mathbf{X}}) = \sum_{i=1}^{M} \sum_{j=1}^{N} (x_{ij} - \hat{x}_{ij})^2$, where $\mathbf{X}$ is the synthetic gene expression data generated from the target network (element x_{ij} corresponds to the expression level of gene j in experiment i) and $\hat{\mathbf{X}}$ are the corresponding expression levels in the inferred network. M denotes the number of different perturbation experiments and N is the number of genes. Thus, figuratively speaking, the reverse engineering process amounts to finding the gene network that best fits the target expression data.

We use the following parameters for the evolutionary algorithm. The population size is 200. We use elitism, i.e., the best individual is protected from replacement. At each generation, 40 parents are chosen using tournament selection [7]. From the 40 parents, 200 new individuals are created and genetic operators are applied with:

- Probability of homologous crossover (per individual)   0.1
- Prob. of nucleotide deletion, insertion and substitution (*per nucleotide*) 0.001
- Prob. of chromosome fragment deletion, transposition and duplication (per chromosome)   0.01
- Prob. of chromosome deletion and duplication (per chromosome)   0.001

The choice of the parameters listed above is not critical. They were chosen heuristically based on a series of test runs and the experiences reported in [8,9]. We observe no significant difference in the quality of results obtained with different standard selection and replacement strategies. Note that the mutation rates were chosen such that the more disruptive whole chromosome and chromosome fragment mutations occur less frequently than single nucleotide mutations.

3.3   Results

The results of a batch of ten reverse engineering runs using synthetic data from a randomly initialized SOS gene network as explained above are shown in Fig. 3. For each run, we record the fitness $f(\hat{\mathbf{X}})$ of the best individual at every generation of the evolutionary algorithm. Of course, in addition to a good fit of the expression data, the structure of the inferred networks should match the target gene network. In a real biological application the structure of the target network is unknown, but in the synthetic test case employed here the accuracy of the inferred networks can be quantified. To this end we use the mean square error $E(\hat{\theta})$ of all parameters of the reverse engineered network $\hat{\theta}$ (including all weights w_{ij} and gene parameters p_i) compared to the true parameter values θ of the *in silico* target gene network: $E(\hat{\theta}) = 1/K \cdot \sum_{l=1}^{K} (\theta_l - \hat{\theta}_l)^2$, where θ_l denotes the l-th element of parameter vector θ and K is the total number of parameters. We refer to $E(\hat{\theta})$ as *estimation error* of the inferred network. In addition we also count the number of *false positives* and *false negatives*[3].

The ten runs shown in Fig. 3 were executed for 100'000 generations of the evolutionary algorithm. Average computation time was roughly two days for

---

[3] We count as *false positive* when a target weight $w_{ij} = 0$ and the absolute value of the inferred weight $|\hat{w}_{ij}| > 0.1$; A *false negative* occurs when $w_{ij} \neq 0$ and $|\hat{w}_{ij}| < 0.1$.

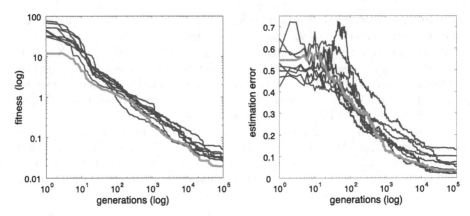

Fig. 3. Reverse engineering of the SOS network – ten runs. The fitness $f(\hat{\mathbf{X}})$ (left) and the estimation error $E(\hat{\theta})$ (right) are plotted for the best individual of each run. As the fitness is optimized (i.e., minimized), the inferred networks match the structure and parameters of the target network with increasing accuracy (the estimation error goes down). The run with the best final fitness is highlighted. For details, see main text.

one run on a standard desktop PC. All ten runs achieve a fitness below 0.1. Since fitness is a *sum* of square errors, the individual expression levels are fitted extremely accurately with a relative error in the order of 1%. As the reverse engineering algorithm optimizes the fitness, the estimation error of the inferred networks goes down. Four out of ten runs inferred the SOS network with high precision[4] (final estimation error between 0.02 and 0.03). In other words, these runs closely matched the structure, weights and gene parameters of the target network. The other runs converged at estimation errors of about 0.1.

In a set of reverse engineering runs, one would like to choose the inferred network with the lowest estimation error $E(\hat{\theta})$. However, since $E(\hat{\theta})$ is unknown in a real application, this is not possible. Hence, we choose the inferred network with the best fitness as the most plausible reconstruction of the target network[5]. Here, the best run (see Fig. 3) achieves a fitness of 0.02 and the corresponding network has a very low estimation error of 0.03 with only one false positive and one false negative out of a total of 81 possible connections (see Fig. 4b).

In other experiments with target SOS networks with different random initializations of the parameter values we have obtained the same quality of results. AGE infers networks with very good fitness in every run. Roughly 40% of the runs also achieve very low estimation errors (i.e., they correctly infer the target, having only very few false positives and false negatives). Simpler networks, e.g.,

---

[4] Further analysis indicates that the accuracy achieved by the best runs corresponds to a lower bound given by the discretization of the search space due to the quantization of the parameters and weights.

[5] In a real application, one should not just consider the inferred network with the best fitness – which merely corresponds to the network with the highest *a posteriori* probability – but analyze all well-scoring (i.e., probable) inferred networks [13].

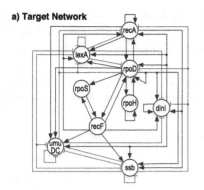
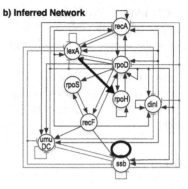

Fig. 4. a) Topology of the *E. coli* SOS network [1]. Arrows are enhancing, Tee ends denote inhibitory interactions. b) The topology inferred by the run with the best final fitness is identical except for one false positive (bold) and one false negative (encircled).

cascades of size six, are inferred correctly in every run. In addition, we have also tested gradient descent and observed that it is not successful at inferring the sigmoid model – even when restarted many times – because it prematurely converges to local optima with very bad fitness and high estimation error.

4   Conclusion

We have introduced a new approach for reverse engineering genetic regulatory networks. The method is based on artificial evolution with a bio-inspired genetic encoding (AGE), which allows simultaneous inference of numerical parameter values and model structure (network size, topology and – in heterogeneous networks – the type of the nodes). AGE is not constrained to a specific gene network model type, but can be used with a large class of nonlinear models.

Using a standard sigmoid model as test case, we have successfully reverse engineered the *E. coli* SOS network from synthetic steady-state gene expression data. The SOS network arguably has a more complex and densely connected topology than typical target networks used for inference methods based on global optimisation techniques [11,12,13,14,15]. Thus, our results demonstrate the competitiveness of AGE for inference of complex nonlinear gene regulatory networks.

There has been a recent trend towards reverse engineering methods using steady-state perturbation data [1,3,16,17]. However, those approaches are based on first-order approximations. Here we propose for the first time – to the best of our knowledge – the use of steady-state perturbation data for reverse engineering of *nonlinear* models. Considering the advantages of steady-state with respect to time-series data and based on the encouraging results of our test case, we believe this to be an extremely promising approach.

Experiments reported here were conducted with synthetic, noise-free expression data. We are currently working both on an application to real expression data and on a more realistic *in silico* test case based on a mechanistic model of a well-characterized gene-protein network. In addition, we also intend to take advantage of the flexibility of AGE in order to explore novel, more biologically plausible gene network models than the sigmoid model employed here.

Acknowledgments. We thank Peter Dürr, Sara Mitri, Tim Stirling and Fanny Riedo for discussions and comments on the manuscript. This work was supported by the Swiss National Science Foundation, grants no. 620-58049 and 200021-112060.

References

1. Gardner, T.S., di Bernardo, D., Lorenz, D., Collins, J.J.: Inferring genetic networks and identifying compound mode of action via expression profiling. Science **301**(5629) (2003) 102–5
2. D'Haeseleer, P., Wen, X., Fuhrman, S., Somogyi, R.: Linear modeling of mRNA expression levels during CNS development and injury. Pac Symp Biocomput (1999) 41–52
3. Brazhnik, P.: Inferring gene networks from steady-state response to single-gene perturbations. J Theor Biol **237**(4) (2005) 427–440
4. Corne, D., Pridgeon, C.: Investigating issues in the reconstructability of genetic regulatory networks. In: Congress on Evolutionary Computation. (2004)
5. Yeung, M.K.S., Tegnér, J., Collins, J.J.: Reverse engineering gene networks using singular value decomposition and robust regression. PNAS **99**(9) (2002) 6163–8
6. Ljung, L.: System identification: Theory for the user. Prentice Hall, Upper Saddle River, NJ (1999)
7. Bäck, T., Fogel, D., Michalewicz, Z., eds.: Evolutionary Computation 1: Basic Algorithms and Operators. Institute of Physics, Bristol (2000)
8. Mattiussi, C.: Evolutionary synthesis of analog networks. PhD thesis, Ecole Polytechnique Fédérale de Lausanne, Lausanne (2005)
9. Mattiussi, C., Floreano, D.: Analog Genetic Encoding for the Evolution of Circuits and Networks. IEEE Transactions on Evolutionary Computation (2006, To appear)
10. Dürr, P., Mattiussi, C., Floreano, D.: Neuroevolution with Analog Genetic Encoding. In: Parallel Problem Solving from Nature - PPSN IX. Volume 4193 of LNCS. (2006) 671–680
11. Reinitz, J., Sharp, D.H.: Mechanism of eve stripe formation. Mech Dev **49**(1-2) (1995) 133–58
12. Jaeger, J., Surkova, S., Blagov, M., Janssens, H., Kosman, D., Kozlov, K.N., Manu, Myasnikova, E., Vanario-Alonso, C.E., Samsonova, M., Sharp, D.H., Reinitz, J.: Dynamic control of positional information in the early drosophila embryo. Nature **430**(6997) (Jul 2004) 368–371
13. Wahde, M., Hertz, J., Andersson, M.: Reverse engineering of sparsely connected genetic regulatory networks. In: Proceedings of the 2nd Workshop of Biochemical Pathways and Genetic Networks. (2001)
14. Noman, N., Iba, H.: Inference of gene regulatory networks using S-system and differential evolution. In: GECCO'05. (2005)
15. Kimura, S., Ide, K., Kashihara, A., Kano, M., Hatakeyama, M., Masui, R., Nakagawa, N., Yokoyama, S., Kuramitsu, S., Konagaya, A.: Inference of S-system models of genetic networks using a cooperative coevolutionary algorithm. Bioinformatics **21**(7) (2005) 1154–63

16. Kholodenko, B.N., Kiyatkin, A., Bruggeman, F.J., Sontag, E., Westerhoff, H.V., Hoek, J.B.: Untangling the wires: a strategy to trace functional interactions in signaling and gene networks. PNAS **99**(20) (2002) 12841–12846
17. de la Fuente, A., Brazhnik, P., Mendes, P.: Linking the genes: inferring quantitative gene networks from microarray data. Trends Genet **18**(8) (Aug 2002) 395–398
18. Weaver, D.: Modeling regulatory networks with weight matrices. In: Pacific Symposium on Biocomputing. (1999)

Appendix

We give here only a short description of the sequence interaction maps I_w and I_k introduced in Sect. 2.1 due to space limitation. For details, refer to Refs. [8,9].

The sequence interaction map $I_w(s_{cod}, s_{reg})$ that decodes the weights w_{ij} from the respective coding and regulatory regions of genes i and j is defined as a composed map $N_w(L(s_{cod}, s_{reg}))$, which is formed by a generic interaction map $L(s_{cod}, s_{reg})$ and a network-specific map $N_w(i)$. The generic interaction map $L(s_{cod}, s_{reg})$ is defined as the local alignment score of the two sequences [8], using the same alignment parameters as Ref. [9]. Figuratively speaking, the closer the match between two subsequences of s_{cod} and s_{reg}, the stronger the interaction. Simpler techniques of sequence comparison such as exact matching or Hamming distance would compromise evolvability [8]. The network-specific map $N_w : [i_{min}, i_{max}] \mapsto [0, w_{max}]$ transforms integer local alignment scores i into floating-point weights. Alignment scores smaller than the threshold i_{min} are mapped to zero (no interaction), scores greater than i_{max} are truncated to w_{max}, and scores in between are mapped linearly onto the positive interval $[0, w_{max}]$. In the experiments reported in this paper, we used $i_{min} = 11$, $i_{max} = 31$ and $w_{max} = 2$.

The alignment score given by the interaction map is always positive. In order to represent negative weights, genes actually have two sequences s_{cod+} and s_{cod-} corresponding to enhancing and repressing regulatory activity (for clarity, only one sequence s_{cod} was mentioned in Sect. 2.1). The weight is defined by the stronger interaction: $w = +I_w(s_{cod+}, s_{reg})$ if $I_w(s_{cod+}, s_{reg}) \geq I_w(s_{cod-}, s_{reg})$ and $w = -I_w(s_{cod-}, s_{reg})$ otherwise.

Analogously to I_w, the sequence interaction map $I_p(s_{p,1}, s_{p,2})$ used for decoding gene parameter p from the respective sequences $s_{p,1}$ and $s_{p,2}$ is implemented as a composed map $N_p(L(s_{p,1}, s_{p,2}))$, using the same generic interaction map L defined above. $N_p : [i_{min}, i_{max}] \mapsto [p_{min}, p_{max}]$ maps the integer local alignment scores onto the interval of parameter p analogously to N_w described above.

Finally, we briefly discuss the random sampling of numerical parameter values for the *in silico* target network (Sect. 3.1). Weights w_{ij} were initialized uniformly in the range $[0.15, 1.5]$ and parameters p_i in the range $[1/2, 2]$. These ranges were selected empirically with the goal to obtain rich nonlinear dynamics in the target networks, i.e., so that on average the total regulatory input for the sigmoid activation functions was neither completely saturated nor constrained to a very small, almost linear regime.

Tuning ReliefF for Genome-Wide Genetic Analysis

Jason H. Moore and Bill C. White

Dartmouth College, One Medical Center Drive, Lebanon, NH 03756
Jason.H.Moore@dartmouth.edu, Bill.C.White@dartmouth.edu

Abstract. An important goal of human genetics is the identification of DNA sequence variations that are predictive of who is at risk for various common diseases. The focus of the present study is on the challenge of detecting and characterizing nonlinear attribute interactions or dependencies in the context of a genome-wide genetic study. The first question we address is whether the ReliefF algorithm is suitable for attribute selection in this domain. The second question we address is whether we can improve ReliefF for selecting important genetic attributes. Using simulated genetic datasets, we show that ReliefF is significantly better than a naïve chi-square test of independence for selecting two interacting attributes out of 10^3 candidates. In addition, we show that ReliefF can be improved in this domain by systematically removing the worst attributes and re-estimating ReliefF weights. Our simulation studies demonstrate that this new Tuned ReliefF (TuRF) algorithm is significantly better than ReliefF.

1 Introduction

1.1 The Problem Domain: Human Genetics

Biological and biomedical sciences are undergoing an information explosion and an understanding implosion. That is, our ability to generate data is far outpacing our ability to interpret it. This is especially true in the domain of human genetics where it is now technically feasible to measure thousands of DNA sequence variations from across the human genome. For the purposes of this paper we will focus exclusively on the single nucleotide polymorphism or SNP which is a single nucleotide or point in the DNA sequence that differs among people. It is anticipated that at least one SNP occurs approximately every 100 nucleotides across the $3*10^9$ nucleotide human genome. An important goal in human genetics is to determine which of the many thousands of SNPs are useful for predicting who is at risk for common diseases such as prostate cancer, cardiovascular disease, or bipolar depression. This "genome-wide" approach is expected to revolutionize the genetic analysis of common human diseases [1], [2].

The charge for computer science and bioinformatics is to develop algorithms for the detection and characterization of those SNPs that are predictive of human health and disease. Success in this genome-wide endeavor will be difficult

E. Marchiori, J.H. Moore, and J.C. Rajapakse (Eds.): EvoBIO 2007, LNCS 4447, pp. 166–175, 2007.
© Springer-Verlag Berlin Heidelberg 2007

due to nonlinearity in the genotype-to-phenotype mapping relationship that is due, in part, to epistasis or nonadditive gene-gene interactions. Epistasis was recognized by Bateson [3] nearly 100 years ago as playing an important role in the mapping between genotype and phenotype. Today, this idea prevails and epistasis is believed to be a ubiquitous component of the genetic architecture of common human diseases [4]. As a result, the identification of genes with genotypes that confer an increased susceptibility to a common disease will require a research strategy that embraces, rather than ignores, this complexity [4], [5], [6]. The implication of epistasis from a data mining point of view is that SNPs need to be considered jointly in learning algorithms rather than individually. Because the mapping between the attributes and class is nonlinear, the concept difficulty is high. The challenge of modeling attribute interactions has been previously described [7]. Due to the combinatorial magnitude of this problem, intelligent feature selection strategies are needed. The goal of this paper is to evaluate the ReliefF algorithm as a statistical filter in this domain.

1.2   A Simple Example of the Concept Difficulty

Epistasis can be defined as biological or statistical [5]. Biological epistasis occurs at the cellular level when two or more biomolecules physically interact. In contrast, statistical epistasis occurs at the population level and is characterized by deviation from additivity in a linear mathematical model. Consider the following simple example of statistical epistasis in the form of a penetrance function. Penetrance is simply the probability (P) of disease (D) given a particular combination of genotypes (G) that was inherited (i.e. $P[D|G]$). A single genotype is determined by one allele (i.e. a specific DNA sequence state) inherited from the mother and one allele inherited from the father. For most single nucleotide polymorphisms or SNPs, only two alleles (encoded by A or a) exist in the biological population. Therefore, because the order of the alleles is unimportant, a genotype can have one of three values: AA, Aa or aa. The model illustrated in Table 1 is an extreme example of epistasis. Let's assume that genotypes AA, aa, BB, and bb have population frequencies of 0.25 while genotypes Aa and Bb have frequencies of 0.5 (values in parentheses in Table 1). What makes this model interesting is that disease risk is dependent on the particular combination of genotypes inherited. Individuals have a very high risk of disease if they inherit Aa or Bb but not both (i.e. the exclusive OR function). The penetrance for each individual genotype in this model is 0.5 and is computed by summing the products of the genotype frequencies and penetrance values. Thus, in this model there is no difference in disease risk for each single genotype as specified by the single-genotype penetrance values. This model was first described by Li and Reich [8]. Heritability or the size of the genetic effect is a function of these penetrance values. In this model, the heritability is maximal at 1.0 because the probability of disease is completely determined by the genotypes at these two DNA sequence variations. This is a special case where all of the heritability is due to epistasis. As Freitas reviews [7], this general class of problems has high concept difficulty.

Table 1. Penetrance values for genotypes from two SNPs

	AA (0.25)	Aa (0.50)	aa (0.25)
BB (0.25)	0	1	0
Bb (0.50)	1	0	1
bb (0.25)	0	1	0

1.3   Three Data Mining Challenges

Moore and Ritchie [9] have outlined three significant challenges that must be overcome if we are to successfully identify genetic predictors of health and disease. First, powerful data mining and machine learning methods will need to be developed to statistically model the relationship between combinations of DNA sequence variations and disease susceptibility. Traditional methods such as logistic regression have limited power for modeling high-order nonlinear interactions [10]. A second challenge is the selection of genetic variables or attributes that should be included for analysis. If interactions between genes explain most of the heritability of common diseases, then combinations of DNA sequence variations will need to be evaluated from a list of thousands of candidates. Filter and wrapper methods will play an important role here because there are more combinations than can be exhaustively evaluated. A third challenge is the interpretation of gene-gene interaction models. Although a statistical model can be used to identify DNA sequence variations that confer risk for disease, this approach cannot be translated into specific prevention and treatment strategies without interpreting the results in the context of human biology. Making etiological inferences from computational models may be the most important and the most difficult challenge of all [5].

1.4   Research Questions Addressed

The goal of the present study was to evaluate the ReliefF algorithm as a filter method for selecting interesting SNPs prior to combinatorial modeling. Is filtering a good approach for attribute selection and pre-processing? How does ReliefF perform in this problem domain? Is ReliefF better than a naïve chi-square test of independence that is commonly employed in genetic analysis? An additional goal of the study was to improve upon the ReliefF algorithm for this problem domain. Is it possible to improve the performance of ReliefF? The manuscript is organized in the following manner. Section 2 introduces the ReliefF algorithm. Section 3 describes our modification of ReliefF called Tuned ReliefF or TuRF. Section 4 describes our methods for generating and analyzing artificial genetic datasets for evaluating the ReliefF and TuRF algorithms. Section 5 summarizes the experimental results while Section 6 provides a discussion of the results and the study conclusions.

2   The ReliefF Filter

There are many different statistical and computational methods for determining the quality of attributes. Our goal is to identify those methods that are capable of identifying attributes that predict class primarily through dependencies or interactions with other attributes. Kira and Rendell [11] developed an algorithm called Relief that is capable of detecting attribute dependencies. Relief estimates the quality of attributes through a type of nearest neighbor algorithm that selects neighbors (instances) from the same class and from the different class based on the vector of values across attributes. Weights (W) or quality estimates for each attribute (A) are estimated based on whether the nearest neighbor (nearest hit, H) of a randomly selected instance (R) from the same class and the nearest neighbor from the other class (nearest miss, M) have the same or different values. This process of adjusting weights is repeated for m instances. The algorithm produces weights for each attribute ranging from -1 (worst) to +1 (best). The Relief pseudocode is outlined below:

set all weights $W[A] = 0$
for $i = 1$ to m **do**
   randomly select an instance R_i
   find nearest hit H and nearest miss M
   for all A **do**
     $W[A] = W[A] - diff(A, R_i, H)/m + diff(A, R_i, M)/m$
   end for
end for

The function $diff(A, I_1, I_2)$ calculates the difference between the values of the attribute A for two instances I_1 and I_2. For nominal attributes such as SNPs it is defined as:

$$diff(A, I_1, I_2) = \begin{cases} 0 & : & genotype(A, I_1) = genotype(A, I_2) \\ 1 & : & otherwise \end{cases}$$

The time complexity of Relief is $O(m*n*a)$ where m is the number of instances randomly sampled from a dataset with n total instances and a attributes.

Kononenko [12] improved upon Relief by choosing n nearest neighbors instead of just one. This new ReliefF algorithm has been shown to be more robust to noisy attributes [12], [13], [14] and is widely used in data mining applications. Benchmarking of a C++ program for ReliefF on a 2.0 Ghz Opteron processor with 2 GB RAM demonstrated that a dataset with 200 instances and 10^6 attributes could be processed in approximately four minutes. Thus, this algorithm is very fast and certainly practical for high-dimensional genetic datasets. In the present study we evaluated the ReliefF algorithm using a neighborhood of 10 instances. We also set m to be the whole sample size.

3   Tuned ReliefF (TuRF)

ReliefF is able to capture attribute interactions because it selects nearest neighbors using the entire vector of values across all attributes. However, this advantage is also a disadvantage because the presence of many noisy attributes can reduce the signal the algorithm is trying to capture. We propose a "tuned" ReliefF algorithm (TuRF) that systematically removes attributes that have low quality estimates so that the ReliefF values if the remaining attributes can be re-estimated. The pseudocode for TuRF is outlined below:

```
let a be the number of attributes
for i = 1 to n do
    estimate ReliefF
    sort attributes
    remove worst n/a attributes
end for
return last ReliefF estimate for each attribute.
```

The motivation behind this algorithm is that the ReliefF estimates of the true functional attributes will improve as the noisy attributes are removed from the dataset. In the present study we evaluated the TuRF algorithm using a neighborhood of 10 instances with n in the above algorithm set to 10 for an a of 1000. Thus, the 10% of worst attributes were removed at each of the $i = 1$ to n iterations (TuRF 10%). We also evaluated setting n to 100 so that the worst 1% of attributes were removed each iteration (TuRF 1%).

4   Data Simulation and Analysis

The goal of the simulation study is to generate artificial datasets with high concept difficulty to evaluate the power of ReliefF in the domain of human genetics. We first developed thirty-five different penetrance functions (see Section 1.2) that define a probabilistic relationship between genotype and phenotype where susceptibility to disease is dependent on genotypes from two SNPs in the absence of any independent effects. These thirty-five models all had heritabilities ranging from 0.01 (small effect) to 0.4 (large effect). Each functional SNP had two alleles with frequencies of 0.4 and 0.6. All models with full precision are available upon request. Each of the models was used to generate 100 replicate datasets with sample sizes of 200, 400, 800, 1600 and 3200. This range of sample sizes represents a spectrum that is consistent with small to medium size genetic studies. Each dataset consisted of an equal number of cases (sick) and controls (healthy). ReliefF, TuRF 10%, TuRF 1% and chi-square were applied to each of the datasets. The 1000 SNPs were sorted and the top 50, 100, 150, 200, 250, 300, 350, 400, 450, and 500 SNPs out of 1000 were selected. From each subset we counted the number of times the two functional SNPs were selected out of each set of 100 replicates. This proportion is an estimate of the power or how likely we are to find the true SNPs if they exist in the dataset. The number of times each method found the correct two SNPs was statistically compared. A difference in counts (i.e. power) was considered statistically significant at a type I error rate of 0.05.

5   Experimental Results

We find that the power of ReliefF to pick (filter) the correct two functional SNPs or attributes was consistently better ($P \leq 0.05$) than a naïve chi-square test of independence across subset sizes and models when the sample size was 800 or larger. These results suggest that ReliefF is capable of identifying interacting SNPs with a small to moderate genetic effect size in moderate sample sizes. Figure 1 summarizes the average power of chi-square (circles) and ReliefF (triangles). Each point in the plot is an average across the five models used in the simulation. For all sample sizes, the power of ReliefF improves with increasing SNP subset size much more rapidly than the power of chi-square, and this gap between the two methods widens significantly as the sample size is increased.

Next we compared the power of TuRF 10 % and TuRF 1% to the power of ReliefF. We find that the TuRF algorithm was consistently better ($P \leq 0.05$) than ReliefF across small SNP subset sizes (50, 100, and 150) Figure 1 summarizes the average power of ReliefF (triangles), TuRF 10% (+s) and TuRF 1% (Xs). Each point in the plot is an average across the five models used in the simulation. The power of the two methods asymptotically approach one another for larger SNP subset size. As the sample size increases, the advantage of TuRF becomes most apparent at 1600 subjects, a sample size achieved in many genetic epidemiological studies. At these larger sample sizes, TuRF is better able to filter the functional, interacting SNPs for a moderate subset size, making it well suited as a filter. Note that the TuRF algorithm that removed 1% of worst attributes each iteration consistently performed better than the TuRF algorithm that removed 10% each time.

6   Discussion

Human genetics is transitioning away from the traditional approach of studying one gene at a time to a genome-wide approach that uses emerging chip-based technologies to measure thousands of DNA sequence variations across the human genome. The rationale for this latter approach is that it is unbiased and no assumptions need to be made about which genes might be involved in the disease process. Given we have incomplete knowledge about gene function, this seems like a reasonable approach. Regardless, the technology is now available and will result in an information explosion over the next few years. The advantage of the genome-wide approach is that we now have much of the necessary information in hand to identify disease susceptibility genes. The disadvantage is that the

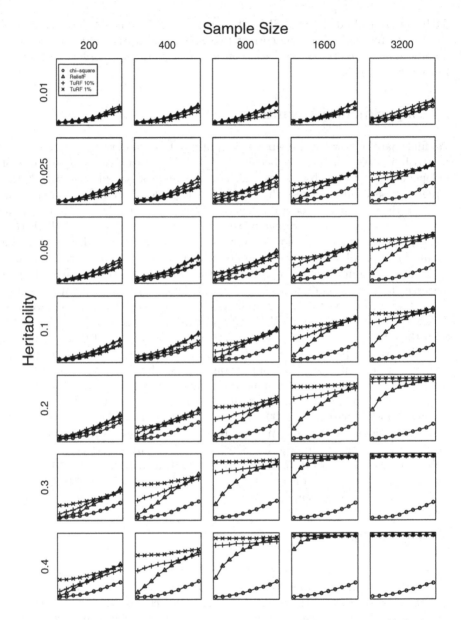

Fig. 1. Comparison of chi-sqaure (circles), ReliefF (triangles) TuRF 10% (+s) and TuRF 1% (Xs) for filtering SNPs. The x-axis of each plot indicates the number of SNPs (50 to 500) filtered. The y-axis of each plot indicates the power (0 to 100).

statistical and computational methods are not yet ready to find those genes that impact disease susceptibility primarily though complex interactions with other genes and with environmental exposures.

There are two important data mining issues that need to be addressed if we are to be successful in our endeavor to identify and characterize complex genetic effects from among 1000 or more attributes. First, we need classifiers that are able to model the nonlinearity in the relationship between the attributes and disease class. Some progress has been made in this domain using feature construction and classification methods such as multifactor dimensionality reduction (MDR) [15], [16], [17] [18], [19], [20]. Second, we need both filter and wrapper strategies that can be combined with methods such as MDR in the context of genome-wide datasets [21], [22], [23]. The goal of the present study was to evaluate and improve upon ReliefF for selecting a subset of attributes that can then be exhaustively modeled using a classifier.

The main conclusion of this study is that the ReliefF algorithm is able to identify nonlinear attribute dependencies in genome-wide datasets. We showed that ReliefF has better power than a naïve or myopic chi-square test of independence across a range of sample sizes. While this was encouraging, it was clear that ReliefF was not perfect. In fact, the power of ReliefF only approached a reasonable level (80%) in the largest datasets when nearly 500 attributes were being selected out of the 1000. It is well known that the power of ReliefF is significantly impacted by the number of noisy attributes and the number of instances in the dataset [13]. This is because ReliefF looks at the entire vector of attributes when estimating the quality of each individual attribute. As a result, the algorithm is very sensitive to the context that the functional attributes find themselves in.

Can ReliefF be improved? A number of different ReliefF algorithms have been proposed, each with a different motivation. For example, RReliefF [13] was developed specifically for regression problems. Selective Sampling ReliefF was developed to reduce the time complexity by using kd-trees to select the most informative set of instances [24]. Our goal here was to improve ReliefF by systematically removing attributes with the worst weights followed by re-estimating of the weights for the remaining attributes. The reasoning is that the signal should improve by removing those attributes that are most likely to be noise. The results of our simulation study demonstrate that our new Tuned ReliefF (TuRF) algorithm is significantly better than ReliefF. In fact, the TuRF algorithm had greater than 80% power to pick the correct two interacting SNPs in a subset of only 50 out of 1000 attributes. ReliefF had less than 50% power.

We conclude that the Relief family of algorithms is suitable for genome-wide genetic analysis. While these algorithms are not perfect, they are very capable of identifying nonlinear gene-gene interactions even in the presence of hundreds of noisy SNPs. The next step is to explore the power of these methods across a wide range of different genetic models, effect sizes, allele frequencies, and sample sizes. We anticipate ReliefF will play an important role in data mining strategies in the domain of human genetics. Indeed, ReliefF is already included in the open-source multifactor dimensionality reduction (MDR) software package (www.epistasis.org/mdr.html) that was designed specifically for detecting gene-gene interactions. The availability of Relief algorithms in the MDR software package and others such as Weka [25] opens the door to routine use of this powerful algorithm in the human genetics research community.

Acknowledgment

This work was supported by NIH grants LM009012, AI59694, HD047447 and RR018787.

References

1. Hirschhorn, J.N., Daly, M.J.: Genome-wide association studies for common diseases and complex traits. Nature Reviews Genetics **6** (2005) 95–108
2. Wang, W.Y., Barratt, B.J., Clayton, D.G., Todd, J.A.: Genome-wide association studies: Theoretical and practical concerns. Nature Reviews Genetics **6** (2005) 109–118
3. Bateson, W.: Mendel's Principles of Heredity. Cambridge University Press, Cambridge (1909)
4. Moore, J.H.: The ubiquitous nature of epistasis in determining susceptibility to common human diseases. Human Heredity **56** (2003) 73–82
5. Moore, J.H., Williams, S.W.: Traversing the conceptual divide between biological and statistical epistasis: Systems biology and a more modern synthesis. BioEssays **27** (2005) 637–46
6. Thornton-Wells, T.A., Moore, J.H., Haines, J.L.: Genetics, statistics and human disease: Analytical retooling for complexity. Trends in Genetics **20** (2004) 640–7
7. Freitas, A.: Understanding the crucial role of attribute interactions. Artificial Intelligence Review **16** (2001) 177–199
8. Li, W., Reich, J.: A complete enumeration and classification of two-locus disease models. Human Heredity **50** (2000) 334–49
9. Moore, J.H., Ritchie, M.D.: The challenges of whole-genome approaches to common diseases. JAMA **291** (2004) 1642–3
10. Moore, J.H., Williams, S.W.: New strategies for identifying gene-gene interactions in hypertension. Annals of Medicine **34** (2002) 88–95
11. Kira, K., Rendell, L.A.: A practical approach to feature selection. In: Machine Learning: Proceedings of the AAAI'92 (1992)
12. Kononenko, I.: Estimating attributes: Analysis and extension of relief. Machine Learning: ECML-94 (1994) 171–182
13. Robnik-Sikonja, M., Kononenko, I.: Theoretical and empirical analysis of relieff and rrelieff. Machine Learning **53** (2003) 23–69
14. Robnik-Sikonja, M., Kononenko, I.: Comprehensible interpretation of relief's estimates. In: Proceedings of the Eighteenth International Conference on Machine Learning. (2001) 433–440
15. Hahn, L.W., Moore, J.H.: Ideal discrimination of discrete clinical endpoints using multilocus genotypes. In Silico Biology **4** (2004) 183–94
16. Hahn, L.W., Ritchie, M.D., Moore, J.H.: Multifactor dimensionality reduction software for detecting gene-gene and gene-environment interactions. Bioinformatics **19** (2003) 376–82

17. Moore, J.H.: Computational analysis of gene-gene interactions in common human diseases using multifactor dimensionality reduction. Expert Review of Molecular Diagnostics **4** (2004) 795–803
18. Moore, J.H., Gilbert, J.C., Tsai, C., Chiang, F.T., Holden, W., Barney, N., White, B.C.: A flexible computational framework for detecting, characterizing, and interpreting statistical patterns of epistasis in genetic studies of human disease susceptibility. Journal of Theoretical Biology **241** (2006) 252–261
19. Ritchie, M.D., Hahn, L.W., Moore, J.H.: Power of multifactor dimensionality reduction for detecting gene-gene interactions in the presence of genotyping error, phenocopy, and genetic heterogeneity. Genetic Epidemiology **24** (2003) 150–157
20. Ritchie, M.D., Hahn, L.W., Roodi, N., Bailey, L.R., Dupont, W.D., Parl, F.F., Moore, J.H.: Multifactor dimensionality reduction reveals high-order interactions among estrogen metabolism genes in sporadic breast cancer. American Journal of Human Genetics **69** (2001) 138–147
21. Moore, J.H., White, B.C.: Exploiting expert knowledge in genetic programming for genome-wide genetic analysis. Lecture Notes in Computer Science **4193** (2006) 969–977
22. Moore, J.H.: Genome-wide analysis of epistasis using multifactor dimensionality reduction: feature selection and construction in the domain of human genetics. In: Knowledge Discovery and Data Mining: Challenges and Realities with Real World Data. IGI (2007)
23. Moore, J.H., White, B.C.: Genome-wide genetic analysis using genetic programming: The critical need for expert knowledge. In: Genetic Programming Theory and Practice IV. Springer (2007)
24. Lui, H., Motoda, H., Yu, L.: A selective sampling approach to active feature selection. Artificial Intelligence **159** (2004) 49–74
25. Frank, E., Hall, M., Trigg, L., Holmes, G., Witten, L.: Data mining in bioinformatics using weka. Bioinformatics **20** (2004) 2479–2481

Dinucleotide Step Parameterization of Pre-miRNAs Using Multi-objective Evolutionary Algorithms

Jin-Wu Nam[1,2], In-Hee Lee[3], Kyu-Baek Hwang[4],
Seong-Bae Park[5], and Byoung-Tak Zhang[1,2,3]

[1] Graduate Program in Bioinformatics
[2] Center for Bioinformation Technology
[3] Biointelligence Laboratory, School of Computer Science and Engineering
Seoul National University, Seoul 151-742, Korea
[4] School of Computing, Soongsil University, Seoul 151-746, Korea
[5] Department of Computer Engineering
Kyungpook National University, Daegu 702-701, Korea
btzhang@bi.snu.ac.kr

Abstract. MicroRNAs (miRNAs) form a large functional family of small noncoding RNAs and play an important role as posttranscriptional regulators, by repressing the translation of mRNAs. Recently, the processing mechanism of miRNAs has been reported to involve Drosha/DGCR8 complex and Dicer, however, the exact mechanism and molecular principle are still unknown. We thus have tried to understand the related phenomena in terms of the tertiary structure of pre-miRNA. Unfortunately, the tertiary structure of RNA double helix has not been studied sufficiently compared to that of DNA double helix. The tertiary structure of pre-miRNA double helix is determined by 15 types of dinucleotide step (d-step) parameters for three classes of angles, i.e., twist, roll, and tilt. In this study, we estimate the 45 d-step parameters (15 types by 3 classes) using an evolutionary algorithm, under several assumptions inferred from the literature. Considering the trade-off among the four objective functions in our study, we deployed a multi-objective evolutionary algorithm, NSGA-II, to the search for a nondominant set of parameters. The performance of our method was evaluated on a separate test dataset. Our study provides a novel approach to understanding the processing mechanism of pre-miRNAs with respect to their tertiary structure and would be helpful for developing a comprehensible prediction method for pre-miRNA and mature miRNA structures.

1 Introduction

The tertiary structure of RNAs is deeply related with their processing and functions, and knowing it helps to resolve their binding mechanisms with other molecules in cells. It can be described by several angle parameters, such as twist, roll, and tilt. However, the tertiary structure of RNA has not been studied much yet, compared to the secondary structure [1] or the tertiary structure of DNA [2]. Crystallography methods facilitated the elucidation of the structural parameters of DNA double helix [2, 3].

Recently, the known functional range of RNA has been expanded gradually since a new posttranscriptional regulator, (i.e., the microRNA (miRNA)), was found [10].

E. Marchiori, J.H. Moore, and J.C. Rajapakse (Eds.): EvoBIO 2007, LNCS 4447, pp. 176–186, 2007.

miRNAs are defined as single-stranded RNAs of ~22 nucleotides (nt) in length, generated from endogenous transcripts that can form local hairpin structures [11]. The local hairpin structures are processed by the nuclear RNase type III enzyme, Drosha, releasing the hairpin-shaped intermediates (pre-miRNAs) which are typically 60-70 nt [12]. After exported to the cytoplasm, the pre-miRNAs are cleaved by another RNase III type enzyme Dicer and then are processed into the miRNAs [13]. However, the structural mechanism associated to the recognition and processing of pre-miRNA still remains unknown. Elucidation of the structural mechanism is one of the most crucial problems towards the understanding of molecular basis of miRNA processing. In a recent report on the structural mechanism, it is shown that the cleavage site by Drosha is distant from a terminal loop by about two-turn helices[14]. This biological knowledge can be used for the parameterization of tertiary structure of pre-miRNA.

The parameterization of tertiary structure is a computationally intensive work involving a large number of parameters and several objectives. Conventional approaches to the parameterization include iterative algorithms and weighted linear sum methods. Iterative methods optimize each of the objectives one by one until a self-consistent state has been reached [4]. Weighted linear sum methods address the multiple objective problem by choosing suitable weighting parameters between objectives [5]. However, the choice of weighting parameters for different objectives becomes another challenging problem when the different objectives are dependent on each other. In this case, *a priori* knowledge about the dependency structure is needed for finding suitable weight values.

An alternative method is to adopt a multi-objective optimization algorithm where all objectives are simultaneously optimized. Multi-objective optimization algorithms have widely been applied to the problems where the trade-off relation (dependency) among the objectives exists [6-8]. They produce not a single parameter set but a variety of parameter sets with various trade-offs for the objective functions. Multi-objective optimization algorithms find the optimal solution by comparing the candidate solutions based on the dominance relationship. When comparing two solutions with respect to the dominance relationship, the fitness value of each objective is considered together. Therefore, there is no information distortion in the multi-objective optimization algorithm, whereas the weighted linear sum method inevitably distorts some information while summarizing the individual fitness values [9].

We introduce a novel approach for the parameterization of pre-miRNA structure using multi-objective evolutionary algorithms (MOEAs). In our knowledge, there has been no reported research on the parameterization of pre-miRNAs in terms of their tertiary structure. In specific, we focus on the dinucleotide step (d-step) parameters of double helix structure of pre-miRNA. Results of this study may help to understand the mechanism as well as can be used as an integral part for the prediction of pre-miRNAs. This paper largely consists of three parts; the first section describes the implementation of MOEAs for the parameterization of RNA tertiary structure; the second part demonstrates the results of a case study about pre-miRNAs; discussion and future work are given in the last part.

2   Materials and Methods

2.1   Tertiary Structure of RNA

The double helix structure of RNA is determined by three angle parameters; twist (W), roll (R), and tilt (T) as described in Figure 1. Twist is a main angle of the helix structure and decides whether it is left-handed or right-handed (Figure 1(b)). Roll is a rotation angle bending along the main groove and a minor groove, and it opens the grooves (Figure 1(c)). Tilt is a rotation angle where a plane of base pair moves up and down, and it destabilizes the stacking energy (Figure 1(d)). The angle parameters determine the position of each base and backbone in three dimensional spaces. They are basically

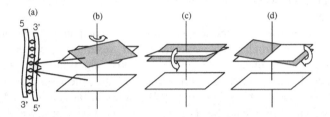

Fig. 1. The angle parameters of RNA tertiary structure. (a) The double helix structure of RNA; (b) Twist; (c) Roll; (d) Tilt. The rhombus denotes the plane of base pair.

defined with dinucleotide step (d-step) parameters, as described in the following. Here, we can define three net angles as described in Equations (1)~(3) [15]. Net twist ($\tilde{W}_j$) is the cumulative twist from the first to the last d-step (L d-steps) of the jth example. Net roll ($\tilde{R}_j$) and net tilt ($\tilde{T}_j$) are the cumulative angles from the first to the last d-step of the jth example.

$$\tilde{W}_j = \sum_{i=1}^{L} W_i \tag{1}$$

$$\tilde{R}_j = \sum_{i=1}^{L} \left[-T_i \cdot \sin\left(\sum_{k=1}^{i} W_k \right) + R_i \cdot \cos\left(\sum_{k=1}^{i} W_k \right) \right] \tag{2}$$

$$\tilde{T}_j = \sum_{i=1}^{L} \left[T_i \cdot \cos\left(\sum_{k=1}^{i} W_k \right) + R_i \cdot \sin\left(\sum_{k=1}^{i} W_k \right) \right], \tag{3}$$

where W_i, T_i, and R_i are d-step parameters for twist, tilt, and roll angles, respectively.

2.2   RNA Dinucleotide Step

The secondary structure of a single RNA sequence can be predicted using the nearest neighbor model [16]. It considers thermodynamic interaction between base pair i and $i+1$, called dinucleotide step (d-step). Here, each d-step has three tertiary structural parameters: twist, roll, and tilt angles. The d-steps can be classified into the three types according to bases' tendency or effect over geometric space. In this study, we use the following ten d-steps.

- Pyrimidine(Py) – Purin(Pr) steps: UA(UA), CA(UG), CG(CG)
- Pr – Py steps: AA(UU), AG(CU), GA(UC), GG(CC)
- Pr – Pr steps: AU(AU), GU(AC), GC(GC)

Here, the complement pair of a dinucleotide step is given in parentheses. Also, we need the following five additional d-steps for insertion or deletion. They represent an internal bulge in RNA structure.

- -A(U-), -G(C-), -U(A-), -C(G-), --

Here, '-' denotes deletion. To summarize, we have to consider 15 d-steps for each angle parameter (twist, roll, and tilt) and thus we should search for the 45 d-steps parameter values (15 d-steps by 3 angle classes).

2.3 Description of the Objective Functions

Here we present the four objective functions used to search for the optimal d-step parameters of double helix structure of pre-miRNA. For the optimization of the d-step parameters, we exploit some prior knowledge, which is specific to the pre-miRNAs. Since the same preprocessing mechanism is applied to all pre-miRNAs, their tertiary structures should be similar. The first function restricts the net twist should be similar across all miRNAs and the second implies the sense-antisense pair should have similar twist values. The third and fourth denotes that the net roll values and the net tilt values should be similar for all examples. All objective functions should be minimized here.

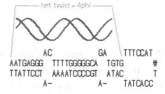

Fig. 2. The net twist of pre-miRNA

2.3.1 Twist: Mean of Difference and Standard Error

All net twists from a Drosha cleavage site to a Dicer cleavage site should be similar to be cropped by the same processing mechanism. A previous experiment reported that the net twist is about two-turn helices (4π) as depicted in Figure 2 [14]. Based on this, we can define the first objective function, which minimizes the difference from 4π and the standard errors of the net twists. If there are N pre-miRNAs, the mean of difference (MOD) and the standard error of the net twists are defined as:

$$MOD = \frac{\sum_{j=1}^{N}\left|4\pi - \tilde{W}_j^c\right|}{N} \quad (4)$$

$$SE_W = \sqrt{\frac{\sum_{j=1}^{N}\left(E(\tilde{W}_j^c) - \tilde{W}_j^c\right)^2}{N}}, \quad (5)$$

where $\tilde{W}_j^c$ is the net twist of the jth pre-miRNA calculated by a set of d-step parameters, C. From Equations 4 and 5, the first objective function can be defined as follow:

$$f_1 = MOD + SE_W \tag{6}$$

Here, SE_W was added because the net twist should be uniformly distributed around the two-turn helices.

2.3.2 Twist: Mean of Difference Between Both Strands
The stem structure of RNA is bi-directional: (1) from the stem-end of the mature miRNA to the loop-end of the mature miRNA and (2) vice versa. Thus, there are two net twists for a sense and an antisense strand, on a stem structure of pre-miRNAs. Here, a tuple $(S_j^c, \bar{S}_j^c)$ specifies the net twists of the sense and antisense strands of the jth pre-miRNA calculated by a set of d-step parameters, C. We can assume that there is no difference of the net twists between both strands. Hence, we can define the second objective function f_2 using the mean of difference between both strands (MDS).

$$f_2 = MDS = \frac{\sum_{j=1}^{N}\left(\left|S_j^c - \bar{S}_j^c\right|\right)}{N} \tag{7}$$

2.3.3 Roll: Standard Error of Net Rolls
The third objective function is designed to minimize the standard error of net rolls. The standard error of net rolls is defined as follow:

$$f_3 = SE_R = \sqrt{\frac{\sum_{j=1}^{N}(E(\tilde{R}_j^c) - \tilde{R}_j^c)^2}{N}} \tag{8}$$

where, $\tilde{R}_j^c$ is the net roll of jth pre-miRNA calculated by a set of d-step parameters, C.

2.3.4 Tilt: Standard Error of Net Tilts
The last objective function minimizes the standard error of net tilts. The standard error of net tilts is defined as follow:

$$f_4 = SE_T = \sqrt{\frac{\sum_{j=1}^{N}(E(\tilde{T}_j^c) - \tilde{T}_j^c)^2}{N}}, \tag{9}$$

where $\tilde{T}_j^c$ is the net roll of the jth pre-miRNA calculated by a set of d-step parameters, C.

2.4 Multi-objective Optimization

A multi-objective optimization is to find a set of decision vectors $\vec{x}^*$ which minimize (or maximize) m objective vectors (functions) $\vec{f}(\vec{x})$ at the same time.

$$\vec{x}^* = \arg\min_{\vec{x}} (\vec{f}(\vec{x})) \qquad (10)$$

$$\vec{f}(\vec{x}) = \left(f_1(\vec{x}), f_2(\vec{x}), ..., f_m(\vec{x}) \right), \qquad (11)$$

where $f_i(\vec{x})$s are objective functions. The decision vector (parameters) $\vec{x}$ consists of n real values $(x_1, x_2, ..., x_n)$ belong to the feasible region $S \subset \Re^n$ [8].

MOEAs find the optimal decision vectors, exploiting the dominance relationship for comparing the candidate solutions, not integrating the individual objective functions as a single function. The dominance relationship can be defined as follows [6].

$$\forall i \in \{1, 2, ..., m\}, f_i(\vec{x}_1) \leq f_i(\vec{x}_2) \qquad (12)$$

$$\exists i \in \{1, 2, ..., m\}, f_i(\vec{x}_1) < f_i(\vec{x}_2) \qquad (13)$$

If a decision vector $\vec{x}_1$ is not worse than $\vec{x}_2$ across all the objective functions, $\vec{x}_1$ is said to dominate $\vec{x}_2$. If there is no dominance between the decision vectors, they are said to be non-dominated each other. If a decision vector, $\vec{x}_1$ is not dominated by all other decision vectors in the whole search space, the vector is defined as Pareto-optimal solution [6]. The non-dominated set of the entire search space is the Pareto-optimal set. The Pareto-optimal set in the objective space is called the Pareto-optimal front.

2.5 Implementation of the MOEA

We used non-dominated sorting genetic algorithm (NSGA-II), one of the most popular MOEAs, for the optimization of our d-step parameters [6]. NSGA-II can handle a number of objectives through ranking by non-dominated sorting procedure (Figure 3(a-1)).

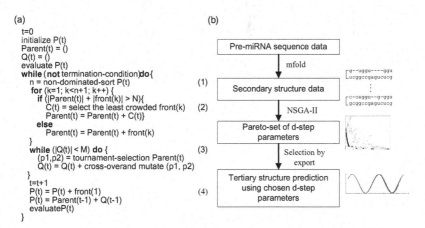

Fig. 3. (a) Pseudo-code of the implemented NSGA-II, (1) creates n fronts; (2) produces parents of size N; (3) reproduce daughters of size M; (4) elitism strategy. (b) Flow chart for tertiary structure prediction using d-step parameters optimized by evolutionary algorithm.

It has shown good convergence and diversity performance on various problem domains. The crowding distance measure in NSGA-II overcomes a drawback of deciding the sharing parameter in the previous NSGA (Figure 3(a-2)). To handle multiple objectives more efficiently, we used the NSGA-II with the elitism strategy (Figure 3(a-4)). The entire NSGA-II procedure is summarized in Figure 3(a).

3 Results

3.1 Case Study: Pre-miRNAs

For the d-step parameterization of pre-miRNA, we used 38 pre-miRNA sequence data where the cleavage site by Drosha/DGCR8 complex is experimentally validated by Northern blotting (http://microrna.sanger.ac.uk/). To extract d-steps of pre-miRNAs, their secondary structures should be known. Here, we applied the mfold package (http://ww.bioinfo.rpi.edu/applications/mfold) for the prediction of secondary structures, and converted the results as pairwise type (Figure 3(b)). We also prepared 20 independent test examples using the same method. The proposed d-step parameter learning procedure is summarized in Figure 3(b).

3.2 Experimental Setting

To find the optimal parameters of dinucleotide steps using NSGA-II, we set several running parameter values as follows. Population size and the number of maximum generation were set to 1000, respectively. As genetic operators, we used the uniform

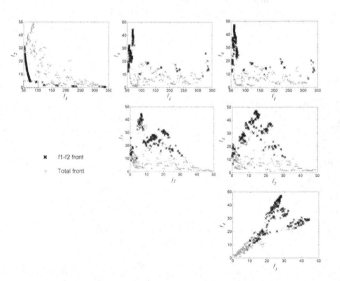

Fig. 4. The relation between the fitness functions. Each graph shows the distribution of total front set of all functions on the pair of functions. The black dots are the front set of function f_1 and function f_2. The gray dots are all front set for all functions.

crossover and 1-point mutation. The rate of crossover was set as 0.9 and the rate of mutation was one over 45, which is the number of total d-step parameters. A chromosome consists of a string with real values of the 45 d-step parameters corresponding to the 3 tertiary structure parameters for each 15 d-steps.

To reduce the search space, we imposed several constraints on the range of each angle (twist, roll, and tilt). For the range values, the angles discovered in DNA double helix were referred to. The twist of d-steps not including deletion is from 20° to 45° and the twist of d-steps including deletion is from 0° to 45°. Roll is from -10° to 10° and tilt is from -5° to 5°.

3.3 The Relation Among the Objectives

Given the running parameter set, we searched for the total front set for all given functions from f_1 to f_4. Figure 4 represents the total front set, a non-dominated set for four functions, which are plotted between pairs of functions. In the results, all pairs of functions excluding the pair of functions f_3 and f_4 show the trade-off relationship. However, the function pair f_3 and f_4 shows a liner relationship, which is reasonable because net roll and tilt are dependent on each other as in Equations 2 and 3.

Among the total front set, we wished to select several sets that are appropriate for our specific purpose. In this study, we focused the twist angle because it is a main contributor of the RNA tertiary structure and thus we first selected the front set determined by functions f_1 and f_2 (black dots in Figure 4). They included 227 parameter sets. Next, we chose eight parameter sets in a specific range where the fitness value of function f_1 is under 100 and the fitness value of function f_2 is under 5 (a small dotted rectangle in the upper left plot in Figure 4).

3.4 Learned Results

The fitness values of the eight selected parameter sets are given in Table 1. Table 1(a) describes the fitness values for the training dataset and Table 1(b) for the test dataset. We concluded that the front sets were optimized well with the given objective functions from the fact that the result on the test dataset is similar to that on the training dataset.

Table 1. The fitness values of each selected front members for training and test datasets

	f_1	f_2	f_3	f_4		f_1	f_2	f_3	f_4
1	72.4	4.7	38.3	25.5	1	82.5	7.1	28.6	21.4
2	72.1	4.9	39.7	24.7	2	83.7	7.7	32.6	18.5
3	93.4	4.2	10.2	13.0	3	82.3	4.1	17.5	14.3
4	76.2	4.3	32.1	24.0	4	90.9	6.0	30.8	20.9
5	**93.3**	**4.2**	**10.3**	**13.0**	**5**	**82.5**	**4.0**	**17.5**	**14.4**
6	94.2	4.1	10.2	12.8	6	82.2	4.2	17.2	14.2
7	95.0	3.8	10.1	13.4	7	80.0	4.2	17.5	14.7
8	73.0	4.4	35.0	27.2	8	82.4	6.8	30.4	20.0

(a) Fitness values for the training dataset   (b) Fitness values for the test dataset>

Table 2. The d-step parameter sets of DNA (*) and pre-miRNA. Row is the first nucleotide step and column is the second nucleotide step.

			Purine				Pyrimidine				Deletion
			A	A*	G	G*	U	T*	C	C*	-
Pyrimidine	T	W	34.8	40.0							
		R	7.2	2.6							
		T	4.8	0.0							
	C	W	32.7	36.9	35.0	31.1					
		R	6.5	1.1	6.1	6.6					
		T	-1.6	0.6	5.0	0.0					
Purine	A	W	35.9	35.8	32.9	30.5	34.0	33.4			
		R	6.8	0.5	0.6	2.9	10.0	-0.6			
		T	3.7	-0.4	3.3	-2.0	-0.1	0.0			
	G	W	34.9	39.3	34.9	33.4	34.3	35.8	37.6	38.3	
		R	9.6	-0.1	10.0	6.5	9.5	0.4	10.0	-0.7	
		T	5.0	-0.4	4.2	-1.1	4.6	-0.9	2.8	0.0	
Deletion	-	W	31.4		35.8		34.6		29.8		34.8
		R	9.4		10.0		6.1		-10.0		9.4
		T	5.0		4.7		4.7		3.9		3.5

3.5   The Representative Parameter Set

Among the eight front parameter sets, we selected a representative parameter set (the fifth parameter set in Table 1) for comparing with the parameter set of DNA, estimated by the crystallography experiment (Table 2). Table 2 demonstrates the 45 d-step parameters of pre-miRNA and the 30 d-step parameters of DNA. It should be noted that there are no deletions in the structure of DNA double-helices and their physical and chemical characteristics are different from those of pre-miRNAs. Nevertheless, several parameters of pre-miRNA have similar values to those of DNA.

3.6   Twist Change of Pre-miRNAs

We drew the twist of pre-miRNAs using the fifth parameter set of Table 2 in the secondary dimension using a sine function (Figure 5). Figure 5(a) displays the twist change for 38 training examples and Figure 5(b) shows the twist change of 20 test examples. The net twists of the training are about 4.1π around 2-turn helices. In both datasets, the variance of net twists increases as the bp-step increases. We guessed that the difference may result from the consecutive deletions after the 10th d-step. For

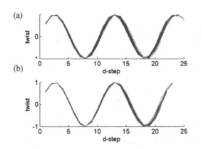

Fig. 5. Twist angles of the training and test datasets calculated using the fifth front parameter set. (a) Twist angles of the training examples (b) Twist angles of the test examples.

example, because '-A(A-)' or '-C(C-)' d-steps have relatively smaller angles than others, their consecutive d-steps will make more difference among twist angles of examples.

4 Discussion

In this study, we suggested an approach to the parameterization of pre-miRNAs using multi-objective evolutionary algorithms (MOEAs). In specific, we optimized the d-step parameters for twist, roll, and tilt angles, needed for predicting the tertiary structure of double helix of pre-miRNA, by introducing four fitness functions. The d-step parameter set of a representative front member had realistic values and shared similar patterns with that of DNA double helix. Also the results on the test dataset were consistent with those on the training dataset with respect to the fitness value and twist change. Combining these, our method has shown to be an effective tool for finding the Pareto-optimal front for this problem.

Multi-objective evolutionary algorithms search for a front, non-dominated optimal set, for given functions. Thus, we should determine the representative results among non-dominated optimal sets in the front. In this study, we mainly focused on the twist angle, a major component of tertiary structure, when choosing a representative front parameter set from the total front sets.

The tertiary structure of a double-stranded RNA is determined by coordinates in the three dimensional space originated form the twist, roll, and tilt angles. In the future work, we will analyze the characteristics of pre-miRNAs on the tertiary structure and investigate the association with the processing mechanism of Drosha/DGCR8 and Dicer RNase III type enzymes. As a matter of fact, binding of proteins or complexes to the DNA or RNA double helices causes conformational changes. The d-step parameters should also be changed by physical or chemical effects of the protein. Hence, dynamic conformational changes can cause difficulties in finding the optimal d-step parameter set. Mitigating the above problem would be another research direction.

Successful improvement of our study may help to understand the processing mechanism of pre-miRNAs as well as to be used as a crucial feature for the prediction of pre-miRNAs and mature miRNAs, which is one of the most challenging problems in miRNA study. These results can also be useful for designing an artificial short hairpin RNA as RNA interference system.

Acknowledgments

This work was supported by the Korea Science and Engineering Foundation(KOSEF) through the National Research Lab. Program funded by the Ministry of Science and Technology (No. M10400000349-06J0000-34910) and the BK21-IT program of the Korea Ministry of Education and Human Resources Development. The ICT at Seoul National University provided research facilities for this study. Kyu-Baek Hwang was supported by the Soongsil University Research Fund. Seong-Bae Park was supported by MIC&IITA through IT Leading R&D Support Project.

References

1. Zuker, M., Mfold web server for nucleic acid folding and hybridization prediction. *Nucleic Acids Res*, (2003). 31(13): p. 3406-15.
2. Olson, W.K., et al., DNA sequence-dependent deformability deduced from protein-DNA crystal complexes. *Proc Natl Acad Sci U S A*, (1998). 95(19): p. 11163-8.
3. Strahs, D. and T. Schlick, A-Tract bending: insights into experimental structures by computational models. *J Mol Biol*, (2000). 301(3): p. 643-63.
4. MacKerell, A.D. Empirical force fields: Overview and parameter optimization. in *In 43th Sanibel Symposium*. 2003.
5. Wang, J.a.K., P.A., Automatic parameterization of force field by systematic search and genetic algorithms. *Journal of Computational Chemistry*, (2001). 22: p. 1219-1228.
6. Deb, K.a.G., T. Controlled elitist non-dominated sorting genetic algorithm for better convergence. in *In Proceedings of the First International Conference on Evolutionary Multi-Criterion Optimization*. 2001.
7. Zitzler, E., K. Deb, and L. Thiele, Comparison of multiobjective evolutionary algorithms: empirical results. *Evol Comput*, (2000). 8(2): p. 173-95.
8. Mostaghim, S., Hoffman,M., Koenig,P.H., Frauenheim,T., and Teich,J. Molecular Force Field Parameterization using Multi-Objective Evolutionary Algorithms. in *In Proceedings of the Congress on Evolutionary Computation (CEC '04)*. 2004. Portland, U.S.A.
9. Shin, S.-Y., Lee,I.-H., Kim,D. and Zhang,B.-T., Multi-objective evolutionary optimization of DNA sequences for reliable DNA computing. *IEEE Transactions on Evolutionary Computation*, (2005). 9(2): p. 143-158.
10. Lagos-Quintana, M., et al., Identification of novel genes coding for small expressed RNAs. *Science*, (2001). 294(5543): p. 853-8.
11. Kim, V.N., Small RNAs: Classification, Biogenesis, and Function. *Mol Cells*, (2005). 19(1): p. 1-15.
12. Lee, Y., et al., The nuclear RNase III Drosha initiates microRNA processing. *Nature*, (2003). 425(6956): p. 415-9.
13. Kim, V.N., MicroRNA biogenesis: coordinated cropping and dicing. *Nat Rev Mol Cell Biol*, (2005). 6(5): p. 376-85.
14. Zeng, Y., R. Yi, and B.R. Cullen, Recognition and cleavage of primary microRNA precursors by the nuclear processing enzyme Drosha. *Embo J*, (2005). 24(1): p. 138-48.
15. Schlick, T., *Molecular modeling and simulation*. Interdisciplinary Applied mathematics, ed. S.S. Antman, et al. Vol. 21. 2002, New York: Springer-Verlag.
16. Mathews, D.H., Predicting a set of minimal free energy RNA secondary structures common to two sequences. *Bioinformatics*, (2005). 21(10): p. 2246-53.

Amino Acid Features for Prediction of Protein-Protein Interface Residues with Support Vector Machines

Minh N. Nguyen[1], Jagath C. Rajapakse[1,2], and Kai-Bo Duan[1]

[1] BioInformatics Research Centre, School of Computer Engineering,
Nanyang Technological University, Singapore
[2] Singapore-MIT Alliance, Singapore
{nmnguyen,asjagath,askbduan}@ntu.edu.sg

Abstract. Knowledge of protein-protein interaction sites is vital to determine proteins' function and involvement in different pathways. Support Vector Machines (SVM) have been proposed over the recent years to predict protein-protein interface residues, primarily based on single amino acid sequence inputs. We investigate the features of amino acids that can be best used with SVM for predicting residues at protein-protein interfaces. The optimal feature set was derived from investigation into features such as amino acid composition, hydrophobic characters of amino acids, secondary structure propensity of amino acids, accessible surface areas, and evolutionary information generated by PSI-BLAST profiles. Using a backward elimination procedure, amino acid composition, accessible surface areas, and evolutionary information generated by PSI-BLAST profiles gave the best performance. The present approach achieved overall prediction accuracy of 74.2% for 77 individulal proteins collected from the Protein Data Bank, which is better than the previously reported accuracies.

1 Introduction

The knowledge of protein-protein interaction is valuable for understanding mechanisms of diseases of living organisms and for facilitating discovery of new drugs. The identification of interface residues has many applications such as drug design, protein mimetic engineering, elucidation of molecular pathways [1,2], and understanding of disease mechanisms [3]. Proper identification of the residues at interfaces helps guiding of the processes of docking to build the structural models of protein-protein complexes [4].

Many computational techniques are available in the literature to predict protein-protein interface residues based on different characteristics of known protein-protein interaction sites [4,5,6,7,8,9]. Neural networks use residues in a local neighborhood or a window as inputs to predict protein-protein interface residues at a particular location of an amino acid sequence by finding an appropriate non-linear mapping. Protein-protein interaction sites are predicted from a neural network with sequence profiles of neighboring residues and solvent

E. Marchiori, J.H. Moore, and J.C. Rajapakse (Eds.): EvoBIO 2007, LNCS 4447, pp. 187–196, 2007.

exposures as inputs [6]. Fariselli *et al.* implemented a neural network method for predicting protein-protein interaction sites based on the information of evolutionary conservation and surface disposition [7]. Ofran and Rost proposed a neural network to predict interaction sites from local sequence information [8]. Chen and Zhou introduced a consensus neural network that combines predictions from multiple models with different levels of accuracy and coverage [4]. Input information derived from single sequences has been used by support vector machine (SVM) for predicting protein-protein interface residues [9]. Recently, Yan *et al.* proposed a two-stage classifier consisting of an SVM and a Bayesian network classifier that identifies interface residues primarily on the basis of sequence information [5].

Despite the existence of many approaches, the current success rates of existing approaches to protein-protein interface residue prediction are insufficient for practical applications; further improvement of the accuracy is necessary. Most of the existing techniques use conventional orthogonal encoding or information derived directly from amino acid sequences as inputs to predict protein-protein interaction residues. The biochemical properties of each amino acid residue have not exploited systematically in the prediction.

In this paper, we investigate into various features of amino acids previously used by various researchers and select the optimal combination to predict the residues at protein-protein interactions by using SVM which has a strong foundation in statistical learning theory [10]; its generalization capability is optimal compared to other statistical or machine learning methods in solving many biological problems [11]. SVMs are powerful and generally applicable tools in predicting protein secondary structure [12,13,14], relative solvent accessibility [15], accessible surface areas of amino acids [16], and cancer classification [17,18]. Apart from amino acid composition, we investigate hydrophobic characters of amino acids [19], secondary structure propensity of amino acids [20], accessible surface areas [16], and the evolutionary information generated by PSI-BLAST profiles into an encoding schema. We begin with all known features and reduce using a backward elimination procedure to find the optimal features.

The present approach achieves a substantial improvement of the prediction accuracy from 2.2% to 8.2% on a dataset of 77 individulal proteins collected from the Protein Data Bank compared to the previously reported best prediction accuracies with SVM methods [5,9]. Results of our experiments with the proposed encoding schema confirmed that the accessible surface areas and the evolutionary information of amino acids along the sequences can detect different sequence features to enhance the accuracy of protein-protein interaction residue prediction.

2   Feature Selection

The protein sequences are converted into feature vectors constructed from amino acid composition, hydrophobic characters of amino acids, secondary structure propensity of amino acids, accessible surface areas, and the evolutionary information generated by PSI-BLAST profiles. The backward stepwise selection is used for feature selection. We start with the full model of features and sequentially delete predictors by dropping of the features, which entrances the prediction accuracy.

Let us denote the given protein sequence by $\mathbf{r} = (r_1, r_2, \ldots, r_m)$ where $r_i \in \Omega_R = \{L, V, I, M, C, A, G, S, T, P, F, Y, W, E, D, N, Q, K, R, H\}$ is the set of 20 amino acid residues, and $\mathbf{a} = (a_1, a_2, \ldots, a_m)$ denotes the corresponding protein-protein interface sequence where $a_i \in \Gamma_A = \{N, I\}$; N is non-interface residue and I is interface residue; m is the length of the sequence. The prediction of the protein-protein interface sequence, $\mathbf{a}$, from an amino acid sequence, $\mathbf{r}$, is the problem of finding an optimal mapping from the space of Ω_R^m to the space of Γ_A^m.

2.1 Amino Acid Composition

Given the protein, the amino acid composition is the fraction of each amino acid. The fraction f_r of amino acid $r \in \Omega_R$ in the protein sequence $\mathbf{r}$ is given by

$$f_r = \frac{1}{m} \sum_{i=1}^{m} \delta(r_i = r) \tag{1}$$

where the Kronecker delta δ is one when the argument satisfies otherwise zero.

2.2 Hydrophobic Characters

In protein folding, polar residues prefer to stay outside the protein to prevent non-polar residues from exposing to polar solvent molecules, such as water molecules. The interactions between non-polar residue side chains are called to be hydrophobic interactions. The Kyte and Doolittles (K-D) method [19] is used to calculate the parameter of hydrophobic characters of the 20 amino acid residues. The hydrophobic character parameters of the 20 amino acid residues obtained by K-D method are [20]: (I, 4.5), (V, 4,2), (L, 3.8), (F, 2.8), (C, 2.5), (M, 1.9), (A, 1.8), (G, 0.4), (T, -0.7), (S, -0.8), (W, -0.9), (Y, -1.3), (P, -1.6), (H, -3.2), (E, -3.5), (N, -3.5), (Q, -3.5), (N, -3.5), (K, -3.9), (R, 4.0).

2.3 Secondary Structure Propensity

The secondary structure propensity characterizes the conformation propensity factor of α-helix, β-strand, and coil of the corresponding amino acid residue. The conformation parameters for each amino acid P_{rt}, where $r \in \Omega_R$ and $t \in \Omega_T = \{H, E, C\}$; H, E, and C denote the secondary structure elements α-helix, β-strand, and coil respectively, are calculated as follow:

$$P_{rt} = \frac{l_{rt}}{l_t} \tag{2}$$

where l_t is the score of all of the amino acid residues whose secondary structure has the conformation t and l_{rt} is the corresponding score of all amino acid residue r having conformation t.

2.4   Accessible Surface Areas

Information of accessible surface areas (ASA) directly reflects the degree to which the residues are in contact with the solvent molecules. The results of prediction of ASA have a significant impact in determining interacting residues in proteins and the prediction of protein-protein interactions [16]. The ASA values of the amino acids are derived from using a two-stage support vector regression [16].

2.5   Evolutionary Information Generated by PSI-BLAST Profiles

The PSI-BLAST profiles contain useful information to display correlations between hidden patterns of substitutions of amino acids in similar proteins and similar structures [21]. Firstly, the values of raw matrices of PSI-BLAST [22] are obtained from NR (Non-Redundant) or SWISS-PROT databases. The low-complexity regions, transmembrane regions, and coil-coil segments are then filtered from these databases by PFILT program [21]. Finally, the E-value threshold of 0.001, three iterations, BLOSUM62 matrix, a gap open penalty of 11, a gap extended penalty of 1 are used for searching the non-redundant sequence database to generate position specific scoring matrix (PSSM) profiles [21].

3   SVM Approach for Predicting Protein-Protein Interface Residues

Let $\mathbf{v}_i = (v_{i1}, v_{i2}, \ldots, v_{ik})$ be the vector representing coding of features of the residue r_i where k denotes the number of selected features. Let the input pattern to SVM at site i be $\mathbf{r}_i = (\mathbf{v}_{i-h}, \mathbf{v}_{i-h+1}, \ldots, \mathbf{v}_i, \ldots, \mathbf{v}_{i+h})$ where $\mathbf{v}_i$ denotes the center element, h denotes the width of window of input residues on one side; $w = 2h + 1$ is the width of the neighborhood around the element i. Fig. 1 represents the architecture of SVM approach for the prediction of protein-protein interface residues.

In our SVM approach, the classifier N/I maps the examples of class N to -1 and the examples of class I to +1. The input vectors, derived from a window of w amino acid residues, are transformed to a hidden-space and compared to the support vectors via a kernel function of each classifier [10]. The results are then linearly combined by using parameters α_i, $i = 1, 2, \ldots, n$ where n is the number of training exemplars, that are found by solving a quadratic optimization problem. The process of training of each classifier N/I is illustrated in the algorithm I:

Algorithm I: Classifier N/I
Inputs: Training set $\{(\mathbf{r}_i, q_i) : q_i \in \{+1, -1\}, i = 1, \ldots, n\}$
Maximize over α_i:
$Q = \sum_{i=1}^{n} \alpha_i - (1/2) \sum_{i=1}^{n} \sum_{j=1}^{n} \alpha_i \alpha_j q_i q_j \mathcal{K}(\mathbf{r}_i, \mathbf{r}_j)$
subject to
$0 \leq \alpha_i \leq \gamma$ and $\sum_{i=1}^{n} \alpha_i q_i = 0$
Outputs: Parameters α_i

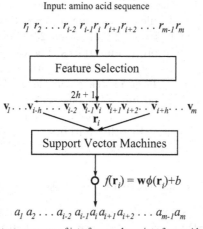

Input: amino acid sequence

$$r_1\ r_2 \cdots r_{i-2}\ r_{i-1}r_i\ r_{i+1}r_{i+2} \cdots r_{m-1}r_m$$

Feature Selection

$2h + 1$

$$\mathbf{v}_1 \cdots \mathbf{v}_{i-h} \cdots \mathbf{v}_{i-2}\ \mathbf{v}_{i-1}\mathbf{v}_i\ \mathbf{v}_{i+1}\mathbf{v}_{i+2} \cdots \mathbf{v}_{i+h} \cdots \mathbf{v}_m$$
$$\mathbf{r}_i$$

Support Vector Machines

$f(\mathbf{r}_i) = \mathbf{w}\phi(\mathbf{r}_i)+b$

$$a_1\ a_2 \cdots a_{i-2}\ a_{i-1}a_i a_{i+1}a_{i+2} \cdots a_{m-1}a_m$$

Output: sequence of interfaces and non-interface residues

Fig. 1. SVM approach for protein-protein interface residue prediction

When the cost function Q is optimized, algorithm I yields a classifier for N/I with maximum margin of separation [10]: the summations of the maximizing function Q run over all training patterns; $\mathcal{K}(\mathbf{r}_i, \mathbf{r}_j) = \phi(\mathbf{r}_i)\phi(\mathbf{r}_j)$ denotes the kernel function where ϕ is a mapping used to convert input vectors $\mathbf{r}_i$ into a high dimensional space; q_i encodes the protein-protein interface such that the output is a binary value; -1 if the protein-protein interface of the residue r_i is N or $+1$ if the protein-protein interface is I. γ is a positive constant used to decide the trade-off between the training error and the margin between the two classes. Once the parameters α_i are obtained from the above algorithm, the resulting discriminant function is known.

The resulting discriminant function f of a test vector $\mathbf{r}_j$ of the above classifier is given by

$$f(\mathbf{r}_j) = \sum_{i=1}^{n} q_i\alpha_i\mathcal{K}(\mathbf{r}_i, \mathbf{r}_j) + b = \mathbf{w}\phi(\mathbf{r}_j) + b \qquad (3)$$

where the bias b is chosen so that $q_i f(\mathbf{r}_i) = 1$ for any i with $0 < \alpha_i < \gamma$ and the weight vector $\mathbf{w} = \sum_{i=1}^{n} q_i\alpha_i\phi(\mathbf{r}_i)$.

The protein-protein interface a_j corresponding to the residue r_j is determined by

$$a_j = \begin{cases} I & \text{if } f(\mathbf{r}_j) \geq 0; \\ N & \text{otherwise.} \end{cases} \qquad (4)$$

4   Experiments and Results

4.1   Dataset

Firstly, the individual proteins from a set of 70 protein-protein heterocomplexes in the study of Chakrabarti and Janin [23] were used. A dataset of 77 individual proteins with sequence identity < 30% was subsequently obtained after removing redundant proteins and molecules, consisting of fewer than 10 residues [5]. According to the scheme of Chakrabarti and Janin [23], these proteins include six different categories of protein-protein interfaces. The categories and the number of representatives in each category are antibody antigen (13), protease-inhibitor (11), enzyme complexes (13), large protease complexes (7), G-proteins, cell cycle, signal transduction (16) and miscellaneous (17). Since the level of sequence identity is low, the resulting dataset is more challenging than the datasets used in previous studies [9] as well as by other authors [8]. The list of the 77 proteins is available at *http://www.public.iastate.edu/~chhyan/ISMB2004/list.html.*

4.2   Definition of Surface Residue and Interface Residues

The reduction of ASA upon complex formation is used in the definition of interface residues. The DSSP program [24] is used to calculate ASA for each residue in the unbound molecule (MASA) and in the complex (CASA). Rost and Sander define a residue to be a surface residue if its MASA is at least 25% of its nominal maximum area [25]. A surface residue is defined to be an interface residue if its calculated ASA in the complex is less than that in the monomer by at least $1A^{\circ 2}$ [26]. By using structural information from Protein Data Bank (PDB) files, surface residues were extracted and divided into interface residues and non-interface residues. Finally, a total of 2340 positive examples corresponding to interface residues and 5091 negative examples corresponding to non-interface residues were obtained [5].

4.3   Features

The selected features of each amino acid consist of (A) one feature representing amino acid composition, (B) one representing for hydrophobic character, (C) three features representing secondary structure propensity, (D) one representing accessible surface area, and (E) twenty features representing evolutionary information derived from the position-specific scoring matrices generated from PSI-BLAST profiles. There were together 26 features in all. All the features were applied and the backward elimination procedure was used for feature selection. Please contact the author for information on the features and the algorithm used for producing these features.

4.4   Prediction Accuracy

To ensure a fair comparison with earlier prediction methods [5,9] on this dataset of 77 individual proteins, a five-fold cross-validation was employed. We have used several measures to evaluate the prediction accuracy: the prediction accuracy is measured by the percentage of correctly predicted classes of protein-protein interface residues [5]; the sensitivity score is the fraction of correctly predicted positive protein-protein interface residues; and the specificity is the fraction of true positives in the protein-protein interface residues predicted as positive [5].

4.5 Results

The SVM method was implemented using LIBSVM library [27] which usually leads to fast convergence in large optimization problems. The Gaussian kernel $\mathcal{K}(\mathbf{x}, \mathbf{y}) = e^{-\sigma \|\mathbf{x}-\mathbf{y}\|^2}$ showed superior performance over the linear and polynomial kernels for predicting protein secondary structure [13,14], relative solvent accessibility [15], and ASA of amino acids [16]. The parameters of the Gaussian kernel and SVM, as $\sigma = 2$, $\gamma = 1$, and the neighborhood window $w = 15(h = 7)$ were determined empirically for optimal performances within the ranges [0.1, 3.0], [0.5, 2.0], and [7, 19] respectively.

Table 1. The performance of SVM approach based on different features with backward elimination, using a five-fold cross-validation on the dataset of 77 proteins

Method	Accuracy (%)	Standard deviation
A, B, C, D, E	72.7	0.65
A, B, C, D	70.3	0.82
A, B, C	68.8	0.88
A, C, D	70.7	0.76
A, D, E	**74.2**	0.58
A, D	71.7	0.76
C, D	70.4	0.85
D, E	73.8	0.62
A	68.9	0.65
B	68.8	0.63
C	68.8	0.63
D	71.4	0.66
E	72.2	0.50

The performance of all the features developed in this study with SVM on the dataset of 77 proteins is shown in Table 1. The SVM using amino acid composition (A) predicted protein-protein interface residues with higher accuracy (68.9%) in comparison with the hydrophobic characters (B) and secondary structure propensity (C) of amino acids. In the case of the ASA of amino acid residues (D), the performance of SVM was 2.5% better than using the amino acid composition (referred in Table 1). Therefore, accessible surface area, which provided information about the degree of the residues in contact with the solvent molecules, is a better feature for predicting protein-protein interaction sites. The SVM using evolutionary information generated from PSI-BLAST profiles (E) predicted protein-protein interface residues with best accuracy (72.2%) in comparison with using amino acid composition, hydrophobic characters, secondary structure propensity, and ASA of amino acids. As shown in Table 1, the

Table 2. Comparison of performances of SVM approach in protein-protein interface residue prediction based on amino acid composition, ASA of amino acid residues, and PSSMs generated by PSI-BLAST, with other methods on the dataset of 77 proteins

Method	Accuracy	Specificity	Sensitivity
Yan *et al.* 2003 (SVM)	66.0	0.44	0.43
Yan 2004 (SVM & Bayesian)	72.0	0.58	0.39
K-nearest neighbor (KNN)	65.7	-	-
Neural network (NN)	67.9	-	-
The present method	**74.2**	**0.78**	**0.72**

best prediction accuracy of 74.2% was achieved by using the hybrid module with the features of amino acid composition, ASA of amino acids, and evolutionary information generated by PSI-BLAST profiles. The results from Table 1 indicate that the features of ASA and evolutionary information from PSI-BLAST profiles are more important for the prediction of protein-protein interaction sites than those of hydrophobic characters and secondary structure propensity.

Table 2 shows the performances of different predictors for protein-protein interface residues on the dataset. Compared to the SVM method of Yan *et al.* using single sequence inputs [9], the present method significantly improved accuracy with 8.2%. The accuracy was improved by 2.2% compared to results of a two-stage classifier consisting of an SVM and a Bayesian network classifier of Yan *et al.* [5]. As shown in Table 2, the sensitivity and specificity of the present SVM approach outperformed the previous SVM methods of Yan *et al.* [5,9].

We also implemented the K-nearest neighbors (KNN) and neural network (NN) methods based on amino acid composition by using Weka software [28]. We used the same five-fold cross-validation to provide an objective comparison of these methods. For KNN classifier, the appropriate value of K is selected in [1, 5] based on cross-validation. The accuracy of the present approach is significantly higher than the results of KNN and NN methods using the amino acid composition (see Table 2). Zhou and Shan [6] reported an accuracy value of 69%-70% for a set of 35 test proteins when the input was calulated from the bound and unbound structures. Recently, a consensus neural network method proposed by Chen and Zhou [4] obtained the prediction accuracy of 70% with 47% coverage of native interface residues.

5   Discussion and Conclusion

In this paper, we investigated the combination of input features that could be effectively used with SVM for protein-protein interface residue prediction. We explore the amino acid composition, hydrophobic characters, secondary structure propensity, accessible surface area (ASA) of amino acids, and evolutionary information generated from PSI-BLAST profiles. Experimental results confirmed the optimal features: amino acid composition, ASA of amino acid residues, and evolutionary information with PSI-BLAST profiles.

Performance comparison results confirmed that the SVM approach achieved better accuracy than the earlier reported accuracies in predicting protein-protein interface residues on the tested dataset. The method should be tested on more datasets before making strong conclusions on the optimal features or its efficacy. The prediction of protein-protein interface residues by the present approach could facilitate protein-protein interactions and the functions of amino acid sequences, which applications are worthwhile for further investigating.

Acknowledgement

The work is partly supported by a grant to J. C. Rajapakse, by the Biomedical Research Council (grant no. 04/1/22/19/376), of Agency of Science and Technology Research, administered through the National Grid Office, Singapore.

References

1. Lichtarge, O., Sowa, M. E., and Philippi,A.: Evolutionary traces of functional surfaces along the G protein signaling pathway. Methods Enzymol **344** (2001) 536–556
2. Sowa, M. E., He, W., Slep, K. C., Kercher, M. A., Lichtarge, O., and Wensel, T. G.: Prediction and confirmation of a site critical for effector regulation of RGS domain activity, Nat Struct Biol **8** (2001) 234–237
3. Zhou, H. X.: Improving the understanding of human genetic disease through predictions of protein structures and protein-protein interaction sites. Curr Med Chem **11** (2004) 539–549
4. Chen, H. and Zhou, H. X.: Prediction of interface residues in protein-protein complexes by a consensus neural network method: Test against NMR data. Proteins **61** (2005) 21–35
5. Yan, C., Dobbs, D., and Honavar, V.: A two-stage classifier for identification of protein-protein interface residues. Bioinformatics **20** (2004) i371–i378
6. Zhou, H. X. and Shan, Y.: Prediction of protein interaction sites from sequence profile and residue neighbor list. Proteins **44** (2001) 336–343
7. Fariselli, P., Pazos, F., Valencia, A., and Casadio, R.: Prediction of protein-protein interaction sites in heterocomplexes with neural networks. Eur. J. Biochem. **269** (2002) 1356–1361
8. Ofran, Y. and Rost, B.: Predicted protein-protein interaction sites from local sequence information. FEBS Lett. **544** (2003) 236–239
9. Yan, C., Dobbs, D., and Honavar, V.: Identification of residues involved in protein-protein interaction from amino acid sequencea support vector machine approach. In A. Abraham, K. Franke, and M. Kppen, (eds), Intelligent Systems Design and Applications, Springer-Verlag, Berlin, Germany (2003) 53–62
10. Vapnik, V.: Statistical Learning Theory. Wiley and Sons, Inc., New York (1998)
11. Cristianini, N. and Shawe-Taylor, J.: An Introduction to Support Vector Machines and other kernel-based learning methods. Cambridge University Press (2000)
12. Nguyen, M. N. and Rajapakse, J. C.: Two-stage support vector machines for protein secondary structure prediction. Neural, Parallel and Scientific Computations **11** (2003) 1–18

13. Nguyen, M. N. and Rajapakse, J. C.: Two-stage multi-class SVMs for protein secondary structure prediction. Pacific Symposium on Biocomputing (PSB), Hawaii (2005)
14. Nguyen, M. N. and Rajapakse, J. C.: Prediction of protein secondary structure with two-stage multi-class SVM approach. International Journal of Data Mining and Bioinformatics **1**(3) (2007) 248–269
15. Nguyen, M. N. and Rajapakse, J. C.: Prediction of protein relative solvent accessibility with a two-stage SVM approach. Proteins: Structure, Function, and Bioinformatics **59** (2005) 30–37
16. Nguyen, M. N. and Rajapakse, J. C.: Two-stage support vector regression approach for predicting accessible surface areas of amino acids. Proteins: Structure, Function, and Bioinformatics **63** (2006) 542–550
17. Rajapakse, J. C., Duan, K.-B., and Yeo, W. K.: Proteomic cancer classification with mass spectra data. American Journal of Pharmacology **5**(5) (2005) 228–234
18. Duan, K.-B., Rajapakse, J. C., Wang, H., and Azuaje, F.: Multiple SVM-RFE for gene selection in cancer classification witn expression data. IEEE Transactions on Nanobioscience **4**(3) (2005) 228–234
19. Thornton, J. and Taylor, W. R.: Structure prediction. Protein Sequencing, J.B.C. Findlay and M.J. Geisow eds., IRL Press, Oxford (1989) 147–190
20. Wang, L. H., Liu, J., Li, Y. F., and Zhou, H. B.: Predicting protein secondary structure by a support vector machine based on a new coding scheme. Genome Informatics **15** (2004) 181–190
21. Jones, D. T.: Protein secondary structure prediction based on position-specific scoring matrices. Journal of Molecular Biology **292** (1999) 195–202
22. Altschul, S. F., Madden, T. L., Schaffer, A. A., Zhang, J. H., Zhang, Z., Miller, W., and Lipman, D. J.: Gapped BLAST and PSI-BLAST: a new generation of protein database search programs. Nucleic Acids Res **25** (1997) 3389-3402
23. Chakrabarti, P. and Janin, J.: Dissecting protein-protein recognition sites. J. Mol. Biol. **272** (2002) 132–143
24. Kabsch, W. and Sander, C.: Dictionary of protein secondary structure: pattern recognition of hydrogen bonded and geometrical features. Biopolymers **22** (1983) 2577–2637
25. Rost, B. and Sander, C.: Conservation and prediction of solvent accessibility in protein families. Proteins **20** (1994) 216–226
26. Jones, S. and Thornton, J. M.: Principles of protein-protein interactions. Proc. Natl Acad. Sci., USA **93** (1996) 13–20
27. Hsu, C. W. and Lin, C. J.: A comparison on methods for multi-class support vector machines. IEEE Transactions on Neural Networks **13** (2002) 415–425
28. Witten, I. H. and Frank, E.: Data Mining: Practical machine learning tools and techniques. 2nd Edition, Morgan Kaufmann, San Francisco (2005)

Predicting HIV Protease-Cleavable Peptides by Discrete Support Vector Machines

Carlotta Orsenigo[1] and Carlo Vercellis[2]

[1] Dip. di Scienze Economiche, Aziendali e Statistiche, Università di Milano, Italy
carlotta.orsenigo@unimi.it
[2] Dip. di Ingegneria Gestionale, Politecnico di Milano, Italy
carlo.vercellis@polimi.it

Abstract. The Human Immunodeficiency Virus (HIV) encodes an enzyme, called HIV protease, which is responsible for the generation of infectious viral particles by cleaving the virus polypeptides. Many efforts have been devoted to perform accurate predictions on the HIV-protease cleavability of peptides, in order to design efficient inhibitor drugs. Over the last decade, linear and nonlinear supervised learning methods have been extensively used to discriminate between protease-cleavable and non cleavable peptides. In this paper we consider four different proteins encoding schemes and we apply a discrete variant of linear support vector machines to predict their HIV protease-cleavable status. Empirical results indicate the effectiveness of the proposed method, that is able to classify with the highest accuracy the cleavable and non cleavable peptides contained in two publicly available benchmark datasets. Moreover, the optimal classification rules generated are characterized by a strong generalization capability, as shown by their accuracy in predicting the HIV protease cleavable status of peptides in out-of-sample datasets.

Keywords: HIV protease, cleavable peptides prediction, discrete support vector machines.

1 Introduction

The Human Immunodeficiency Virus (HIV) encodes an enzyme, called HIV protease, which is responsible for the generation of infectious viral particles. HIV protease function is to cleave virus polypeptides at defined susceptible sites. This cut gives rise to new viral proteins which are able to spread from the native cell and infect other cells. Thus, HIV protease plays a fundamental role in enabling the replication of the virus.

In molecular biology, many efforts have been devoted to investigating the HIV protease problem specificity. The interaction between the enzyme and the virus polyprotein is based upon the following paradigm, also known as "lock and key" model (Chou, 1996; Rögnvaldsson and You, 2004). The active peptide in the protein, which is generally composed by a sequence of eight amino acids around the cleavage site, must fit as a key for binding to the HIV protease

E. Marchiori, J.H. Moore, and J.C. Rajapakse (Eds.): EvoBIO 2007, LNCS 4447, pp. 197–206, 2007.

active region. This means that new infectious particles are generated if the HIV protease-cleavable site properly satisfies the substrate specificity of the active enzyme region (Chou, 1996). If the chemical combination is not verified, the bounding to the active protease site can still be accomplished but no cleavage is performed or the production of immature noninfectious viral proteins takes place.

According to this paradigm, effective HIV protease drugs can be designed with the aim of smoothing the cleavage ability of the enzyme. This task may require to identify inhibitors that, binding to the protease site, are able to prevent any further binding (competitive inhibition) or change the structure of the enzyme (non-competitive inhibition) (Narayanan et al., 2002). In both cases, the prediction of which peptides can be cleaved and which can instead act as protease inhibitors is of great importance.

The discrimination between cleavable and non cleavable sequences can be naturally formulated as a binary classification problem. For this reason, several supervised learning techniques have been applied to provide fast and accurate predictions. In order to explain the complex relationship between peptides and cleavability, many authors have resorted to nonlinear classification models; see (Rögnvaldsson and You, 2004) and the references therein. However, there are situations in which the problem of devising protease-cleavable sequences is linear in nature, and can be efficiently solved by means of linear supervised learning methods (Rögnvaldsson and You, 2004).

Recently, support vector machines (SVM) have been successfully applied to predict the HIV protease-cleavable status of two sets of peptides. The first dataset has been extensively used for the comparison of different classification approaches (Cai et al., 2002; Yang and Chou, 2004; Rögnvaldsson and You, 2004; Nanni, 2006). It contains amino acid sequences that can be cleaved or not by the protease from the Human Immunodeficiency Virus of type 1 (HIV-1 protease). The second dataset, used in (Poorman et al., 1991; Chou, 1996), contains sequences known cleavable by the enzyme from the Human Immunodeficiency Virus of type 2 (HIV-2 protease).

In this paper, we perform the classification of these two datasets by means of an alternative method based on *discrete support vector machines* (DSVM). By this term we denote SVM in which the empirical classification error is represented by a discrete function, called *misclassification rate*, counting the number of misclassified examples, in place of a proxy of the misclassification distance considered by traditional SVM approaches (Orsenigo and Vercellis, 2003, 2004). The inclusion of the discrete term leads to the formulation of a mixed-integer programming problem, whose objective function is composed by the weighted sum of three terms, expressing a trade-off between accuracy and potential of generalization.

The empirical results obtained on the benchmark datasets show the effectiveness of the proposed method that, compared to traditional support vector machines, is able to discriminate between cleavable and non cleavable peptides with the highest accuracy. Moreover, the optimal classification rules generated are characterized by a strong generalization capability, as shown by their accuracy in predicting the HIV protease cleavable status of peptides in out-of-sample datasets. Since the new method performs a linear separation of the examples, the computational tests seem to support the intuition that, in this case, the protease cleavage site specificity problem can be efficiently investigated by means of linear classification models.

2 Classification and Discrete Support Vector Machines

The problem of discerning between peptides according to their HIV protease-cleavability likeliness can be cast in the form of a binary classification problem. In a binary classification problem we are provided with a set of examples (i.e. amino acids sequences) whose class (protease-cleavable or non cleavable status) is known, and we are required to devise a function that performs an optimal discrimination of the examples with respect to their class values. From a mathematical point of view, the problem can be stated as follows. Suppose we are given a set $\mathcal{S}_m = \{(\mathbf{x}_i, y_i),\ i \in \mathcal{M} = \{1, 2, \ldots, m\}\}$ of training examples, where $\mathbf{x}_i \in \Re^n$ is an input vector of explanatory variables and $y_i \in \mathcal{D} = \{-1, +1\}$ is the categorical output value associated to $\mathbf{x}_i$. Let $\mathcal{H}$ denote a set of functions $f(\mathbf{x}) : \Re^n \mapsto \mathcal{D}$ that represent hypothetical relationships between $\mathbf{x}_i$ and y_i. A classification problem consists of defining an appropriate hypotheses space $\mathcal{H}$ and a function $f^* \in \mathcal{H}$ which optimally describes the relationship between inputs and outputs. Actually, a vast majority of binary classifiers proceed by defining a *score function* $g(\mathbf{x}) : \Re^n \mapsto \Re$ and then inducing a classification rule in the form $f(\mathbf{x}) = \mathrm{sgn}(g(\mathbf{x}))$. In particular, many classifiers build a surface $g(\mathbf{x}) = 0$ in $\Re^n$ that attempts to spatially separate the examples of opposite class value. For instance, when the space $\mathcal{H}$ is based on the set of separating hyperplanes in $\Re^n$, we have $g(\mathbf{x}) = \mathbf{w}'\mathbf{x} - b$, so that a generic hypothesis is given by $f(\mathbf{x}) = \mathrm{sgn}(\mathbf{w}'\mathbf{x} - b)$.

In order to choose the optimal parameters $\mathbf{w}$ and b, support vector machines resort to the minimization of the following risk functional (Vapnik, 1995; Cristianini and Shawe-Taylor, 2000)

$$R(f) = \frac{1}{m} L(y, f(\mathbf{x})) + \lambda \|f\|_K^2, \qquad (1)$$

where $K(\cdot, \cdot)$ is a given symmetric positive definite function named *kernel*; $\|f\|_K^2$ denotes the norm of f in the reproducing kernel Hilbert space induced by K and plays a regularization role; $L(y, f(\mathbf{x}))$ is a loss function that measures the accuracy by which the predicted output $f(\mathbf{x})$ approximates the actual output y; λ is a parameter that controls the trade-off between the empirical error and the regularization term.

In the theory of SVM, the loss function measures the distance of the misclassified examples from the separating hyperplane, and is given by

$$L(y, f(\mathbf{x})) = \sum_{i \in \mathcal{M}} |1 - y_i g(\mathbf{x}_i)|_+, \qquad (2)$$

where g is a score function such that $f(\mathbf{x}) = \operatorname{sgn}(g(\mathbf{x}))$ and $|t|_+ = t$ if t is positive and zero otherwise. A different family of classification models, introduced in (Orsenigo and Vercellis, 2003, 2004) and termed *discrete support vector machines* (DSVM), is motivated by an alternative loss function that counts the number of misclassified examples, given by

$$L(y, f(\mathbf{x})) = \sum_{i \in \mathcal{M}} c_i \theta(-y_i g(\mathbf{x}_i)), \tag{3}$$

where $\theta(t) = 1$ if t is positive and zero otherwise, while c_i is a penalty for the misclassification of the example $\mathbf{x}_i$. This leads to the formulation of a mixed-integer programming problem that corresponds to the minimization of (1) using the loss function in (3), with the inclusion of an additional regularization term counting the number of attributes which define the separating hyperplane. The minimization of this term is aimed at reducing the dimension of the space $\mathcal{H}$ in order to derive optimal hypotheses of lower complexity and higher generalization capability.

The problem of determining an optimal separating hyperplane is formulated as follows in the DSVM framework. The number of misclassified points is computed by means of the binary variables

$$p_i = \begin{cases} 0 & \text{if } \mathbf{x}_i \text{ is correctly classified} \\ 1 & \text{if } \mathbf{x}_i \text{ is misclassified} \end{cases}. \tag{4}$$

The count of the number of attributes defining the separating hyperplane is instead based on the binary variables

$$q_j = \begin{cases} 0 & \text{if } w_j = 0 \\ 1 & \text{if } w_j \neq 0 \end{cases}. \tag{5}$$

Let $h_j, j \in \mathcal{N}$, be the penalty cost of using attribute j. Let S and R be sufficiently large constant values, and α, β, γ the parameters to control the trade-off among the objective function terms. The following *discrete support vector machines* model can be formulated

$$\min_{\mathbf{w}, b, \mathbf{p}, \mathbf{u}, \mathbf{q}} \frac{\alpha}{m} \sum_{i=1}^{m} c_i p_i + \beta \sum_{j=1}^{n} u_j + \gamma \sum_{j=1}^{n} h_j q_j \tag{DSVM}$$

$$\text{s.t.} \quad y_i (\mathbf{w}' \mathbf{x}_i - b) \geq 1 - S p_i \qquad i \in \mathcal{M} \tag{6}$$

$$u_j \leq R q_j \qquad j \in \mathcal{N} \tag{7}$$

$$-u_j \leq w_j \leq u_j \qquad j \in \mathcal{N} \tag{8}$$

$$\mathbf{u} \geq \mathbf{0}, \quad \mathbf{p}, \mathbf{q} \text{ binaries},$$

where the family of bounding variables $u_j, j \in \mathcal{N}$, and the constraints (8) are introduced in order to linearize the norm of f in the risk functional (1).

The first term in the objective function represents the regularization term used in the support vector machines framework in order to restore the well-posedness of the optimization problem and to increase the predictive capability of the classifier. The second term evaluates the prediction accuracy on the training sample by means of the misclassification rate. The third term supports the generalization capability of the solution by minimizing the number of attributes used in the definition of the separating hyperplane, hence reducing the complexity of the classification rule generated.

A feasible suboptimal solution to model DSVM can be derived by a heuristic procedure based on a sequence of linear programming (LP) problems. The heuristic starts by considering the LP relaxation of problem DSVM. Each LP problem $DSVM_{t+1}$ in the sequence is obtained by fixing to zero the relaxed binary variable with the smallest fractional value in the optimal solution of the predecessor $DSVM_t$. Notice that, if problem $DSVM_t$ is feasible and its optimal solution is integer feasible, the procedure is stopped, and the solution generated at iteration t is retained as an approximation to the optimal solution of problem DSVM. Otherwise, if problem $DSVM_t$ is unfeasible, the procedure modifies the previous LP problem $DSVM_{t-1}$ by fixing to one all its fractional variables. The problem $DSVM_{t-1}$ defined in this way is feasible and any of its optimal solutions is integer. Thus, the procedure is stopped and the solution found for $DSVM_{t-1}$ is retained as an approximation to the optimal solution of DSVM.

3 Encoding Schemes for Amino Acid Sequences

The active region in the protein is usually composed by an octapeptide, that is a sequence of eight amino acids around the scissible bond. According to the representation introduced in (Schechter and Berger, 1967), each sequence can be expressed as $P_4, P_3, P_2, P_1, P_{1'}, P_{2'}, P_{3'}, P_{4'}$, where the protease active site, if it exists, is located between positions P_1 and $P_{1'}$. Each amino acid P_i can take one of the values contained in the following amino acid alphabet $\mathcal{B} = \{s_1, s_2, \ldots, s_{20}\} = \{A,C,D,E,F,G,H,I,K,L,M,N,P,Q,R,S,T,V,W,Y\}$.

The majority of the classification methods, such as those considered in this paper, require to perform a numerical encoding of the categorical predictors. Hence, an appropriate encoding scheme must be adopted to convert amino acid sequences into numeric values. Several encoding schemes have been proposed and tested for classification tasks involving amino acids in general, and for signal peptide cleavage site prediction in particular; for a summary of some popular encoding methods refer to (Maetschke et al., 2005).

In this paper four encoding schemes are considered. The first, termed *orthonormal* encoding, represents the most common method used in the cleavage site prediction for the treatment of the categorical values. According to this method, each amino acid symbol s_i in the sequence is replaced by an orthonormal vector $\mathbf{d}_i = (\delta_{i1}, \delta_{i2}, \ldots, \delta_{i20})$, where δ_{ij} is the Kronecher delta symbol. Then, a sequence of K amino acids can be encoded by means of K orthonormal vectors, leading to a binary encoding of the original sequence of a total length of $20K$ attributes. Hence, the main drawback of the orthonormal encoding is the growth of the number of attributes by a 20 factor.

The second encoding scheme is based on combining the orthonormal representation with *substitution matrices*, such as the popular BLOSUM50 matrix that exhibited good performances in the classification of cleavable peptides (Nanni, 2006). Substitution matrices estimate the probability at which each amino acid in a protein is replaced by a different amino acid over time. These matrices are usually derived by the amino acids exchanges observed in *homologous* sequences, which are proteins alike because of shared ancestry. Here the idea is to build a 20×20 matrix whose entry in row i and column k expresses a frequency score of observing amino acids s_i and s_k at the same position in homologous sequences. Then, the orthonormal encoding is modified by multiplying the vector $\mathbf{d}_i$ by the entry at row i and column i of the substitution matrix. Again, this encoding results in a growth of the number of attributes by a 20 factor.

The third method, termed *frequency based* encoding, is obtained by replacing the amino acid s_i with the conditional probability of observing the symbol s_i given the positive cleavable status of the peptide; of course, the same might be done for the non cleavable class value, depending on which one is less likely to occur. The rational behind this encoding is to replace each amino acid with a value that takes into account its individual impact on the target class. If no prior guess on the conditional probabilities can be derived from the knowledge of the domain experts, one can wisely approximate the probabilities with relative contingency tables frequencies evaluated on the HIV-protease training dataset. The frequency encoding has the advantage of preserving the original number of attributes.

Finally, a further encoding is derived by combining the orthonormal representation with the frequency based scheme, by multiplying the vector $\mathbf{d}_i$ by the frequency of the symbol s_i, conditioned on the class value. In this case too, the encoding leads to a 20 factor growth in the number of attributes.

4   Computational Results

The predictive ability of the proposed method has been evaluated on two datasets composed by sequences of eight amino acids that can be cleaved or not by the HIV protease.

The first dataset, originally proposed in (Thompson et al., 1995) and extended in (Cai et al., 1998), has been widely used in order to compare alternative linear and nonlinear classification approaches. It contains 362 examples where 114 are cleavable by the HIV-1 protease and 248 represent non cleavable peptides. The best known accuracy achieved on this dataset appears to be 97.5% (Nanni, 2006), yet obtained through rather involved data manipulations.

The second dataset, that has been less investigated in the past, contains 149 examples where 22 peptides are cleavable by the HIV-2 protease and 127 consist of non cleavable sequences. In this dataset each positive example is given by a substrate of the HIV-2 protease. Of the 127 non cleavable peptides, 122 have been derived from the sequence of hen egg lysozyme, that did not exhibit any

Table 1. Accuracy and sensitivity results for the HIV-1 protease dataset

| | Classifier | | | |
| | Accuracy - (Sensitivity) | | | |
Encoding	DSVM	SVM$_L$	SVM$_G$	SVM$_{RBF}$
freq	**97.6%**	92.8%	91.7%	92.8%
	(95.7%)	(88.4%)	(92.1%)	(86.8%)
orth	97.0%	90.6%	90.9%	90.6%
	(95.9%)	(87.5%)	(81.4%)	(88.4%)
orth+freq	96.3%	90.4%	89.9%	91.4%
	(92.7%)	(87.5%)	(88.4%)	(85.7%)
orth+BLOSUM50	97.0%	91.7%	91.2%	91.7%
	(**97.3%**)	(86.6%)	(86.8%)	(85.7%)

cleavage site even after complete denaturation, whereas the remaining five negative examples have been obtained by replacing the first amino acid of a specific octapeptide with five different amino acids, as described in (Chou, 1996). The best accuracy achieved on this dataset is 97.3%.

The effectiveness of the new classifier has been compared with the results provided by linear SVM, which have proven to perform quite efficiently in the classification of the first dataset, and by SVM with gaussian and radial basis functions as kernels. In order to evaluate the accuracy of these classifiers k-fold cross-validation was used for both datasets; more specifically, ten-fold cross-validation was applied to the HIV-1 protease dataset, whereas five-fold cross-validation was implemented for the HIV-2 dataset, due to its limited size.

Table 1 and 2 show the accuracy values exhibited by the competing methods according to the different schemes adopted for encoding the amino acid sequences: the frequency based encoding (*freq*), the orthonormal representation (*orth*), and the two forms derived by combining the orthonormal representation with the frequency based scheme (*orth+freq*) and the substitution matrix BLOSUM50 (*orth+BLOSUM50*). The accuracy of method DSVM has been derived using the sequential LP-based heuristic described in section 2. The results for SVM methods with linear, gaussian and radial basis kernels have been computed by means of the LIBSVM library (Chang and Lin, 2001). For all the methods, a preliminary grid search was applied in order to select the most promising combination of the parameters. Besides the accuracy values, tables 1 and 2 provide the mean rate of the true positive responses (TPR), indicated also as *sensitivity*, returned by the competing methods on the two datasets. The sensitivity can be used to evaluate the effectiveness of a classifier in providing a correct prediction for the cleavable peptides. In fact, the rate $(1 - TPR)$ measures the error of the first kind that, in this case, is represented by the error of assigning to the cleavable peptides the non cleavable label.

The results presented in table 1 and 2 point out a number of interesting issues. The classifier based on discrete support vector machines outperforms, in

Table 2. Accuracy and sensitivity results for the HIV-2 protease dataset

Encoding	Classifier Accuracy - (Sensitivity)			
	DSVM	SVM$_L$	SVM$_G$	SVM$_{RBF}$
freq	**98.7%**	96.7%	96.7%	96.1%
	(100%)	(80.0%)	(83.0%)	(82.0%)
orth	96.0%	91.4%	90.0%	90.0%
	(88.0%)	(56.0%)	(46.0%)	(60.0%)
orth+freq	96.7%	96.7%	92.0%	91.3%
	(59.0%)	(70.0%)	(70.0%)	(65.0%)
orth+BLOSUM50	95.4%	90.0%	87.9%	90.7%
	(79.0%)	(47.0%)	(42.0%)	(37.0%)

terms of accuracy, the competing classification techniques considered for these tests, with a mild dependence from the specific scheme adopted for encoding the protein sequences. This empirical conclusion can be drawn for both HIV-1 and HIV-2 protease datasets, and the best performance is obtained using the frequency based encoding scheme. In particular, the accuracy provided by the DSVM classifier is higher than the best prediction accuracy value ever obtained for the two datasets.

Moreover, the comparison of the linear classifiers DSVM and SVM$_L$ with SVM based on gaussian and radial basis kernels confirms the hypothesis that, for these datasets, the problem of discerning between cleavable and non cleavable peptides is linear in nature, and can be efficiently solved by means of linear classification models.

Finally, we notice that DSVM consistently dominates the other methods referring also to the sensitivity, although the best result is in this case achieved by means of the *orth+BLOSUM50* representation. However, considering the overall performance on the two datasets, it appears that the frequency based scheme is more robust than the other encoding methods.

In order to further support the usefulness of the proposed approach, the optimal classification rules generated were applied to predict the HIV-1 protease cleavable status of new peptides, gathered after the publication of the first dataset. In particular, three new samples were considered: the first, described in (Beck et al., 2000), contains 45 new protein sequences known cleavable by the enzyme. The second, provided in (Beck et al., 2001), lists 38 sequences, where 25 are cleavable and 13 are non cleavable by the HIV-1 protease. The third dataset, presented in (Tözsr et al., 2000), consists of 49 peptides where 38 are cleavable, 3 are non cleavable and 8 are left undetermined. As noticed in (Rögnvaldsson and You, 2004), due to the way non cleavable sequences were collected, these datasets are hard to classify, since cleavable and non cleavable peptides are rather similar.

Table 3. Accuracy prediction for the HIV-1 protease out-of-sample datasets

| | Classifier | | | |
| | N. of correct predictions - (Accuracy) | | | |
Dataset	DSVM	SVM_L	SVM_G	SVM_{RBF}
Beck-00	43/45	32/45	34/45	38/45
	(93.3%)	(71.1%)	(75.5%)	(84.5%)
Beck-01	35/38	34/38	32/38	34/38
	(92.1%)	(89.5%)	(84.2%)	(89.5%)
Tözsr-00	39/41	35/41	36/41	35/41
	(95.1%)	(85.3%)	(87.8%)	(85.3%)

The prediction results on this datasets are reported in table 3. For each dataset, the table indicates the number of correct predictions returned by applying the optimal classification rule derived for each method, using the frequency based representation. The highest performance achieved by DSVM seems to indicate the robustness and effectiveness of this class of methods, that are able to generate optimal classification rules characterized by a stronger generalization capability.

References

Beck, Z.Q., Hervio, L., Dawson, P.E., Elder, J.E. and Madison, E.L.: Identification of efficiently cleaved substrates for HIV-1 protease using a phage display library and use in inhibitor development. Virology **274** (2000) 391–401

Beck, Z.Q., Lin, Y.-C. and Elder, J.E.: Molecular basis for the relative substrate specificity of human immunodeficiency virus type 1 and feline immunodeficiency virus proteases. Journal of Virology **75** (2001) 9458–9469

Cai, Y., Chou, K.: Artificial neural network model for predicting HIV protease cleavage sites in protein. Advances in Engineering Software **29** (1998) 119–128

Cai, Y., Liu, X., Xu, X., Chou, K.: Support vector machines for predicting HIV protease cleavage sites in protein. Journal of Computational Chemistry **23** (2002) 267–74

Chang, C.C., Lin, C.J.: LIBSVM: a library for support vector machines. (2001)

Chou, K.C.: Prediction of human immunodeficiency virus protease cleavage sites in proteins. Analytical Biochemistry **233** (1996) 1–14

Cristianini, N., Shawe-Taylor, J.: An introduction to support vector machines and other kernel-based learning methods. Cambridge University Press, Cambridge, U.K. (2000)

Yang, Z.R., Chou, K.C.: Bio-support vector machines for computational proteomics. Bioinformatics **20** (2004) 735–741

Nanni, N.: Comparison among feature extraction methods for HIV-1 protease cleavage site prediction. Pattern Recognition **39** (2006) 711–713

Maetschke, S., Towsey, M., Boden, M.: Blomap: An encoding of amino acids which improves signal peptide cleavage prediction. In Chen, Y., L.W., ed.: Proceedings of the 3rd Asia-Pacific Bioinformatics Conference. (2005) 141–150

Narayanan, A., Wu, X., Yang, Z.R.: Mining viral protease data to extract cleavage knowledge. Bioinformatics **18** (2002) 15–13

Orsenigo, C., Vercellis, C.: Multivariate classification trees based on minimum features discrete support vector machines. IMA Journal of Management Mathematics **14** (2003) 221–234

Orsenigo, C., Vercellis, C.: Discrete support vector decision trees via tabu-search. Journal of Computational Statistics and Data Analysis **47** (2004) 311–322

Poorman, R., Tomasselli, A., Heinrikson, R., Kezdy, F.: A cumulative specificity model for proteases from human immunodeficiency virus types 1 and 2, inferred from statistical analysis of an extended substrate data base. The Journal of Biological Chemistry **266** (1991) 14554–14561

Rögnvaldsson, T., You, L.: Why neural networks should not be used for HIV-1 protease cleavage site prediction. Bioinformatics **20** (2004) 1702–1709

Schechter, I., Berger, A.: On the size of the active site in proteases. I. Papain. Biochemical and Biophysical Research Communications **27** (1967) 157–162

Thompson, T., Chou, K., Zheng, C.: Neural network prediction of the hiv-1 protease cleavage sites. Journal of Theoretical Biology **177** (1995) 369–379

Tözsr, J., Zahuczky, G., Bagossi, P., Louis, J.M., Copeland,T.D., Oroszlan,S., Harrison,R.W. and Weber,I.T.: Comparison of the substrate specificity of the human T-cell leukemia virus and human immunodeficiency virus proteinases. European Journal of Biochemistry **267** (2000) 6287–6295

Vapnik, V.: The nature of statistical learning theory. Springer Verlag, New York (1995)

Inverse Protein Folding on 2D Off-Lattice Model: Initial Results and Perspectives

David Pelta and Alberto Carrascal

Models of Decision and Optimization Research Group
Depto. de Ciencias de la Computación e I.A.
Universidad de Granada, 18071 Granada, Spain
dpelta@decsai.ugr.es

Abstract. Inverse protein folding or protein design stands for searching a particular amino acids sequence whose native structure or folding matches a pre specified target.

The problem of finding the corresponding folded structure of a particular sequence is, *per se*, a hard computational problem.

We use a genetic algorithm for searching the space of potential sequences, and the fitness of each individual is measured with the output of a second GA performing a minimization process in the space of structures.

Using an off-lattice protein-like 2D model, we show how the implemented techniques are able to obtain a variety of sequences attaining the target structures proposed.

1 Introduction

In very simple terms, a protein consists of a chain of amino acids' residues. The chemical properties and forces between the amino acids residues are such that, whenever the protein is left in its natural environment, it folds to a specific 3-dimensional structure, called its native state, which minimizes the total free energy. This 3D structure is specially relevant because it determines completely how the protein functions and interacts with other molecules. Most biological mechanisms at the protein level are based on shape-complementarity, so that proteins present particular concavities and convexities that allow them to bind to each other and form complex structures, such as skin, hair and tendon [5].

The field of structural bioinformatics is plenty of very interesting problems, and two of them are computationally hard: the protein structure prediction problem and the inverse protein folding problem (or protein design) Both attract the attention of researchers since a long time ago.

Protein structure prediction is one of the most significant technologies pursued by computational structural biology and theoretical chemistry. It has the aim of determining the three-dimensional structure of proteins from their amino acid sequences. In more formal terms, this is expressed as the prediction of protein tertiary structure from primary structure. Given the usefulness of known protein

E. Marchiori, J.H. Moore, and J.C. Rajapakse (Eds.): EvoBIO 2007, LNCS 4447, pp. 207–216, 2007.

structures in such valuable tasks as rational drug design, this is a highly active field of research.

It is assumed that: a) all the information needed for a protein to fold, is coded in the sequence [1] and, b) the corresponding structure is the one that minimizes the free energy of the system. Under this situation, the problem can be re-stated as minimizing an energy function that may have several definitions as in the all-atom model [16], HP lattice models [12,18], those only considering the backbone [9] or atomic models where torsion (usually restricted) angles are considered [3].

Ab initio protein folding methods seek to build three-dimensional protein models "from scratch", i.e., based on physical principles rather than (directly) on previously solved structures. Under simple models, this problem was shown to be NP-hard [6,17].

The progress of the field can be tracked through the results of the *Critical Assessment of Techniques for Protein Structure Prediction* (CASP) wide community contest. See for example, [13] for a review of the last decade.

Inverse Protein Folding (or Protein Design) is the design of new protein molecules from scratch. The number of possible amino acid sequences is infinite, but only a subset of these sequences will fold reliably and quickly to a single native state. Protein design involves identifying such sequences, in particular those with a physiologically active native state.

Inverse Protein Folding requires an understanding of the process by which proteins fold. In a sense it is the reverse of structure prediction: a tertiary structure is specified, and a primary sequence is identified which will fold to it.

Some simplifications are usually made to approach this problem. The set of aminoacids is limited, the fitness of a particular sequence is measured using a threading like approach, target structures are restricted to lattices, etc. [7,8,2]. Pierce and Winfree [14] showed that optimizing the set of rotamers for a specified backbone conformation is NP-Hard. The reader should note that under this model, there is no "folding process". A good overview of the field can be seen in [15,11] and the references therein.

In this work, we focus on this second problem using a genetic algorithm to explore the sequences space (*GA-seq*). In order to measure the fitness of each individual, we need to use a second genetic algorithm that explores the space of structures (*GA-struct*) looking for the one that minimizes an energy function. A remarkably aspect is that, although a 2D space is used, structures are not restricted to a particular lattice geometry.

The hypothesis of this work is that using *GA-seq* and given a particular target structure, *it is possible to obtain a sequence that, when folded with* GA-struct, *the corresponding structure resembles the target one.*

The paper is organized as follows: in Section 2, the models for sequences and structures are presented, and the main characteristics of both GAs implemented. Next, Section 3 is devoted to experiments and results. We divide the experiments in two parts: firstly, we analyze *GA-struct* in terms of convergence and suitability; secondly, *GA-seq* is tested for several target structures. Final comments and future lines of research are outlined in Section 4.

2   Model Definition

The inverse protein folding model is composed by a set of parameters determining the nature and interaction degree of amino acids, and the type of possible protein structures. More formally, the model is defined with a 4-tuple $M(S, T, A, F)$ where:

- S: is the topological space used to embed the conformations. Here, we use $\Re^2$ with Euclidean distance
- T: is the set of available amino acids, $|T| = N$
- A: a $N \times N$ symmetric interaction matrix. Each cell $a_{i,j}$ indicates the degree of interaction between an amino acid of type i and another of type j.
- F: is the interaction function. It may be defined as $F_{ij} = a_{i,j}/(d_{i,j})^2$, where $d_{i,j}$ is the Euclidean distance between amino acids i, j. Another possible definition for F is as a potential function $P_{ij} = -a_{i,j}/d_{i,j}$. Through this function, different folding forces may be represented, like electrostatic forces or hydrophobic-hydrophilic characteristics [10].

A structure E is defined as a pair (L, G), where L is the sequence of amino acids ($l_i \in T$), and G is a set of angles that are used to determine the spatial coordinates of every amino acids in S. If $S = \Re^3$ then $|G| = 2 * (m - 1)$; when $S = \Re^2$ then $|G| = m - 1$.

2.1   GAs for Structure Prediction and Inverse Folding

Now, we will briefly describe the GAs employed for searching the space of sequences (GA-seq) and the space of associated structures (GA-$struct$).

Each individual in the GA-seq used to explore the space of sequences L, represents a fixed size list of amino acids. Uniform crossover and standard mutation are used. Other parameters are specified later. The fitness of each individual consists of two steps: to "fold" the sequence and to measure how good is the fit against the target structure. The first part is also done with a genetic algorithm named GA-$struct$.

Every individual in GA-$struct$, which explores the space of conformations, represents a set of angles coded using floating point numbers. In this way, we avoid the constraints of using a particular lattice geometry. Standard mutation, and MMX [4] crossover operator are employed. The fitness of each individual (the energy of the folded sequence) is calculated as:

$$Energy = \sum_{i<j} P_{i,j} \tag{1}$$

for every pair of amino acids. Once the GA-$struct$ is finished, the ever best individual, representing a particular structure A is taken and the goodness of fit against the target structure B is measure as follows:

$$RMSD_d(A, B) = \frac{1}{m}\sqrt{\sum_{i=1}^{m}\sum_{j=1}^{m}(d_{i,j}^A - d_{i,j}^B)^2} \tag{2}$$

where $d_{i,j}^X$ stands for the Euclidean distance between amino acid i and j in structure X. Essentially, we measure how similar are the internal distances between both structures. This value represents the fitness of the individual in the *GA-seq*.

3 Computational Experiments and Results

Given the model M, we design two experiments with different aims

1. Given a particular sequence, *is it possible to obtain the lowest energy structure?*. Here we want to verify if the *GA-struct* is working fine solving a protein structure prediction problem.
2. Given a particular target structure, *is it possible to obtain a sequence that, when folded with* GA-struct, *the corresponding structure resembles the target one?*. This is the inverse protein folding or protein design problem.

3.1 Protein Structure Prediction

In a first test, we try to minimize the energy of a simple sequence that may lead to a membrane-like structure. It is composed by 14 amino acids ($m = 14$) and just two types are available ($N = 2, T = \{0, 1\}$). The interaction matrix A is defined as:

$$A = \begin{pmatrix} 1 & -1 \\ -1 & 1 \end{pmatrix}$$

In this way, amino acids of different types are attracted, while those of the same type are repelled with the same energy.

The sequence to be folded is $L = (00000001111111)$. After preliminary tests, the following parameters were determined for *GA-struct*: population size 100, mutation probability (per angle): 0.03, for the crossover: 5 parents lead to 2 child.

The convergence of the *GA-struct* was analyzed in the following way: we ran 10^3 repetitions for each value of maximum fitness evaluations available that were fixed in $10^2, 10^3, 10^4, 10^6$. So, for every stopping condition, we have 10^3 values corresponding to the best solution found. Histograms are calculated from such values and they are displayed in Fig. 1.

When just 10^2 evaluations are allowed, a wide range of energy values are obtained (values are calculated with Eq. 1). Moreover, such values are centered around energy = -0.5, thus contradicting the principles of structural molecular stability.

As more evaluations are available, there is a clear tendency to a minimum value of energy around -1.4, while the dispersion is diminished, thus meaning

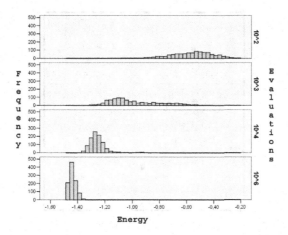

Fig. 1. Energy levels of 10^3 best individuals for different stopping limits

Table 1. Values obtained after 10^3 repetitions for every value of "Evals"

Evals	Avg	SD	Min	Max
10^2	-0,562	0,148	-1,058	-0,203
10^3	-0,978	0,186	-1,296	-0,422
10^4	-1,254	0,065	-1,381	-0,824
10^6	-1,436	0,025	-1,481	-1,339

Fig. 2. Example of a good energy structure for sequence $L = (00000001111111)$

that the probability of "misfolding" is also decremented. This is clearer if we look at Table 1. Figure 2 shows a typical structure associated with the test sequence.

Then, we design a new sequence that may fold as a globular structure (like a spiral). We slightly change the interaction matrix to promote packing of amino acids. The new matrix is:

$$A = \begin{pmatrix} 0.001 & -1 \\ -1 & 0.001 \end{pmatrix}$$

Now, amino acids of different types are attracted as before, but the repulsion among those of the same type is minimum.

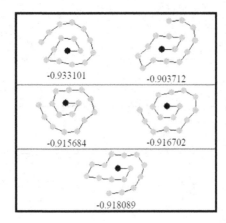

Fig. 3. Five best energy structures for sequence $L = (0111111111111111)$

The test sequence, with $m = 16$ is $L = (0111111111111111)$. One would expect that the first amino acid attracts the other ones, lending to a spiral-like structure. Using the same parameters as before for the *GA-struct*, and setting a limit of 10^4 energy evaluations, we performed 5 repetitions. The five best structures obtained are shown in Fig. 3. As expected the spiral-like structure arose. Intuitively, this is a natural way to put closer those amino acids that make a greater contribution to the energy function.

3.2   Inverse Protein Folding

The experiments done in the previous part, enable us to use *GA-struct* as a fitness calculator in *GA-seq* used for solving the inverse protein folding problem.

The first experiment to perform uses as target structure the globular one obtained before and we try to find other sequences of length $m = 16$ that may fold as a spiral. As stated before, the fitness of each sequence will be measured through the $RMSD_d$. *GA-struct* is run just once for each individual and uses 10^4 fitness evaluations. *GA-seq* uses 50 individuals, mutation probability of 0.03 (per amino acid) and uniform crossover. In this experiment, we allow for 10^3 iterations. The model M is kept as before.

The results obtained are shown in Fig. 4, where different folded sequences are displayed with their energy value as calculated with Eq.2. This example matches what happens in Nature: several sequences may share the same fold.

The next experiment is done with a different target structure, shown in Fig. 5 while the rest of the model is kept.

The results obtained are shown in Fig. 6 (a). Looking at the corresponding structures, it seems that the sequences that can be designed with the model defined, have no potential to fold as the target structure. Just one of them (the lower one) is reasonably good.

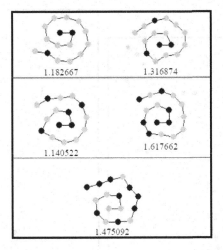

Fig. 4. Five sequences with best agreement to the target structure

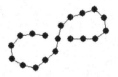

Fig. 5. Target structure

To verify this fact, we modify the previous model, allowing for an additional type of amino acid, and modifying the interaction matrix. Now, $N = 3, T = \{0, 1, 2\}, m = 19$ and the new interaction matrix A is:

$$A = \begin{pmatrix} 0 & 0 & -1 \\ 0 & 0.01 & 0 \\ -1 & 0 & 0 \end{pmatrix}$$

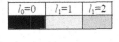

Under this new model, the space of sequences grows to $|L| = 3^{19}$, so we extend the population size to 100 individuals and each run is given 7500 iterations. Results are shown in Fig. 6 (b). It can be seen that the values of RMSDd are lower with this extended model, while the sequences obtained give folded structures clearly better than before.

The sequences have a central region where attractive amino acids are located, while the outer regions are filled with repulsive ones. It can also be noted that particular regions of the sequences are invariant. These "conserved" regions may be associated with evolutionary stability of the solutions.

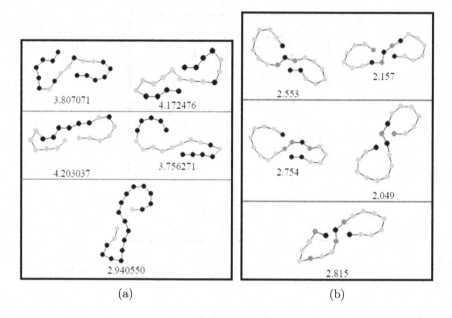

3.807071

4.172476

4.203037

3.756271

2.940550

(a)

2.553

2.157

2.754

2.049

2.815

(b)

Fig. 6. Sequences with best agreement to the target structure using a model with (a) two and (b) three types of aminoacids

In order to make a deeper analysis of this fact, we constructed a "profile" using a multiple alignment of the best 15 sequences found. The profile Q is, in this case, a 3×19 matrix where each cell $q_{i,j}$ stands for the number of times the amino acid i appeared at position j in the best solutions found.

Using this information, we measure the entropy of each position in the sequence as follows:

$$S(i) = -\sum_{i=1}^{N} p(a,i) \times log(p(a,i)) \qquad (3)$$

where $p(a,i)$ is the probability of finding an amino acid of type a at position i (as calculated using the frequencies stored in the profile Q).

The values of $S(i)$, $\forall\ i = 1, 2, \ldots, 19$ are shown in Fig. 7. The values of $S(i)$ are lower for those positions that, when the sequence is folded, are located in the outer regions of the structure. These regions are usually composed of amino acids of type l_1. This is the only "self"-repulsive amino acid, so it is the unique way to construct such soft curves.

Regions with higher values of entropy are composed with amino acids of type l_0 or l_2: both are neutral against l_1 while have certain attraction between them.

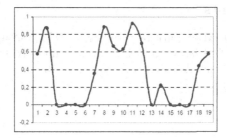

Fig. 7. Entropy values per position of the best sequences found

4 Final Comments

In this contribution we approach the inverse protein folding problem over 2D off-lattice model with genetic algorithms. The GA used for searching the space of sequences has a straight design and implementation. However, the fitness calculation of each individual is the hardest part because it implies the resolution of the protein structure prediction problem. This is an NP-Hard problem on simpler models than the one used here.

The first part was devoted to show the good behavior of the GA developed for structure prediction and then, we focus on the problem of protein design.

The experiments are encouraging and proved and justified the usefulness of genetic algorithms for these problems in two senses:

1. the search mechanism allowed for the target structures tested, to obtain sequences that, when folded, lead to native structures that showed an extremely good agreement with the target one;
2. the use of a population-based technique allowed to obtain a set of potential sequences, folding to the same structure, that can be further analyzed on a post-processing steps that may consider other elements not included in the fitness function like physical feasibility, stability against mutations, etc.

Admittedly, further studies on bigger structures, extended amino acids alphabets and GAs parametrization should be conducted and we think the simplicity of the model used here is more a benefit than a problem to perform them. Moreover, we consider that such studies should be made in 2D models before going into more complex 3D ones.

The fitness calculation will be surely a potential bottleneck for more complex studies. However, the parallel evaluation of solutions is a clear and practical way to speed up computations at a quite low cost. This step is being done.

Another research line, fits the scope of the automatic design of self-assembly systems, where proteins are a paradigmatic case. Under this topic, we have made the initial steps with this work, but we can extend the search process to look also for the interaction matrix and potentially, for the set of amino acids available. We hope to get some results in the near future.

Acknowledgments

This work is supported in part by Projects TIN2005-08404-C04-01 from the Spanish Ministry of Science and Education and TIC-00129-PE from the Andalusian Government. Authors wish to thank the anonymous reviewers for their useful comments.

References

1. C. B. Anfinsen. Principles that govern the folding of protein chains. *Science*, 181(96):223–230, Jul 1973.
2. P. Berman, B. DasGupta, D. Mubayi, R. Sloan, G. Turan, and Y. Zhang. The inverse protein folding problem on 2d and 3d lattices. *Discrete Applied Mathematics*, 2007. to appear.
3. N. Budin, S. Ahmed, N. Majeux, and A. Caflisch. An evolutionary approach for structure-based design of natural and non-natural peptidic ligands. *Comb Chem High Throughput Screen*, 4(8):661–673, Dec 2001.
4. D. M. D. Barrios, A. Carrascal and J. Rios. Optimisation with real coded genetic algorithms based on mathematical morphology. *International Journal of Computer Mathematics*, 80(3):275–293, 2003.
5. M. M. C. David L. Nelson, editor. *Lehninger Principles of Biochemistry*. W. H. Freeman, 2004.
6. A. S. Fraenkel. Complexity of protein folding. *Bulletin of Mathematical Biology*, 6, 1993.
7. A. Gupta, J. Manuch, and L. Stacho. Structure-approximating inverse protein folding problem in the 2d hp model. *J. of Comp. Biology*, 12(10):1328–1345, 2005.
8. D. JONES. De novo protein design using pairwise potentials and a genetic algorithm. *Protein Sci*, 3(4):567–574, 1994.
9. F. Koskowski and B. Hartke. Towards protein folding with evolutionary techniques. *J Comput Chem*, 26(11):1169–1179, Aug 2005.
10. N. Krasnogor, W. Hart, J. Smith, and D. Pelta. Protein structure prediction with evolutionary algorithms. In W. Banzhaf et al.(Eds), *Proceedings of GECCO*, pages 1596–1601. Morgan Kaufman, 1999.
11. E.-W. Lameijer, T. Bäck, J. N. Kok, and A. P. Ijzerman. Evolutionary algorithms in drug design. *Natural Computing*, 4:177–243, 2005.
12. K. Lau and K. Dill. Theory for protein mutability and biogenesis. *PNAS*, 87(2):638–642, 1990.
13. J. Moult. A decade of casp: progress, bottlenecks and prognosis in protein structure prediction. *Curr Opin Struct Biol.*, 15(3):285–9, 2005.
14. N. A. Pierce and E. Winfree. Protein design is np-hard. *Protein Engineering*, 15(10):779–782, 2002.
15. N. Pokala and T. M. Handel. Review: protein design–where we were, where we are, where we're going. *J Struct Biol*, 134(2-3):269–281, 2001.
16. R. Unger. The genetic algorithm approach to protein structure prediction. *Structure and Bonding*, 110:153–175, 2004.
17. R. Unger and J. Moult. Finding the Lowest Free Energy Conformation of a Protein is an NP–hard Problem: Proof and Implications. *Bulletin of Mathematical Biology*, 55(6):1183–98, 1993.
18. R. Unger and J. Moult. Genetic algorithms for protein folding simulations. *Journal of Molecular Biology*, 231(1):75–81, 1993.

Virtual Error: A New Measure for Evolutionary Biclustering

Beatriz Pontes[1], Federico Divina[2], Raúl Giráldez[2], and Jesús S. Aguilar–Ruiz[2]

[1] Department of Computer Science, University of Seville
Avenida Reina Mercedes s/n, 41012 Sevilla, Spain
`bepontes@lsi.us.es`
[2] School of Engineering, Pablo de Olavide University
Ctra. de Utrera, km. 1, 41013, Sevilla, Spain
`{fdivina,giraldez,aguilar}@upo.es`

Abstract. Many heuristics used for finding biclusters in microarray data use the mean squared residue as a way of evaluating the quality of biclusters. This has led to the discovery of interesting biclusters. Recently it has been proven that the mean squared residue may fail to identify some interesting biclusters. This motivates us to introduce a new measure, called *Virtual Error*, for assessing the quality of biclusters in microarray data. In order to test the validity of the proposed measure, we include it within an evolutionary algorithm. Experimental results show that the use of this novel measure is effective for finding interesting biclusters, which could not have been discovered with the use of the mean squared residue.

1 Introduction

Nowadays, technological advances offer the possibility of completely sequentialize the genome of some living species. This constitutes a great source of information which needs to be analyzed. Microarray techniques allow us to study genomes on their own or also to combine some of them in order to extract relational knowledge [12].

Microarray data are usually transformed into a numerical matrix which could then be analyzed. There exist various techniques to extract relevant information from a microarray, depending on the specific application in study. These techniques include clustering methods [4], where the goal is to cluster together genes that have a similar behaviour under all the experimental conditions. This grouping is carried out by means of any specific algorithm or mathematical formula based on genes similarity over all conditions [13]. It may be interesting, however, to analyze whether several genes in a microarray show the same behaviour under a subset of the experimental conditions. This has motivated a recent line of research named biclustering. Biclustering techniques aim at individuating subsets of genes that present the same behaviour under a subset of experimental conditions. This problem has been proven to be even much more complex than clustering [8].

E. Marchiori, J.H. Moore, and J.C. Rajapakse (Eds.): EvoBIO 2007, LNCS 4447, pp. 217–226, 2007.

Biclustering was first applied to genomic data in [6], where the authors present a greedy search method for finding biclusters. In the same work, a measure for assessing the quality of biclusters, named *Mean Squared Residue* (MSR), is proposed. This measure has been used by many researches who have proposed different heuristics for biclustering biological data, e.g., [2,14]. Some other authors have established a search model to detect significant biclusters, without using a specific formula to optimize [11]. For a review of different biclustering techniques, we refer the reader to [9,10]. Among the used techniques, it is interesting to emphasize the application of evolutionary computation to the problem of finding biclusters in microarray data [5,8]. In these work the search was biased towards biclusters with low MSR.

The use of MSR to guide the search for biclusters in microarray data has led to the discovery of interesting biclusters. However, it has been proven that MSR may fail to recognize some interesting biclusters as quality biclusters [1]. This motivates us to introduce a new measure, called *Virtual Error* (VE), for assessing the quality of biclusters in microarray data. In order to evaluate the validity of the proposed measure, we include it within an evolutionary algorithm (EA). In a previous version of this EA, the search was guided mainly by the MSR. Experimental results show that the so modified EA is capable of finding interesting biclusters, which could not have been discovered with the use of MSR.

This paper is organized as follows: in Section 2 we present the motivations for this paper, we therefore describe the quality measure we propose in Section 3. Section 4 describes the used EA and how VE has been included into the algorithm, while some experimental results are shown in Section 5. Finally, Section 6 summarizes the main conclusions.

2   Motivation

One of the most used quality measures for biclusters is the *Mean Squared Residue*, MSR [6]. MSR tries to evaluate the coherence of the genes and conditions of a bicluster $\mathcal{B}$ consisting of I rows and J columns. MSR is defined as:

$$\text{MSR}(\mathcal{B}) = \frac{1}{I \cdot J} \sum_{i=1}^{i=I} \sum_{j=1}^{j=J} (b_{ij} - b_{iJ} - b_{Ij} + b_{IJ})^2 \tag{1}$$

where b_{ij}, b_{iJ}, b_{Ij} and b_{IJ} represent the element in the *ith* row and *jth* column, the row and column means, and the mean of the submatrix, respectively. If the gene expression levels fluctuate in unison under the conditions contained in a bicluster $\mathcal{B}$, then $\text{MSR}(\mathcal{B}) = 0$. In general, the lower the MSR, the stronger the coherence exhibited by the bicluster, hence the better the quality. It follows that a trivial or constant bicluster where there is no fluctuation is characterized by a very low value of MSR. In order to reject these kind of biclusters, most heuristics combine the MSR with some other measures, e.g., the row variance [8,6].

As demonstrated in [1], MSR may not be the optimal measure for assessing the quality of some kinds of biclusters. In this work, the author makes a further study on the main characteristics inherent to biclusters, extracting from them two main principles, shifting patterns and scaling patterns. Genes in a bicluster might present either one of these patterns or both of them simultaneously. It is demonstrated that the MSR value is useful to recognize shifting behaviours in the biclusters, while it may fail to recognize a bicluster presenting scaling patterns.

Figure 1 shows two biclusters whose genes fluctuate in unison under the conditions contained in the bicluster. Each line in the graphs represents the expression levels of a gene under different conditions. This figure also presents the numerical values for each bicluster in a matrix, where columns correspond to genes and rows to experimental conditions. Despite the fact the the genes present the same behaviour under the experimental conditions, the MSR value for the two biclusters does not seem to indicate that they are equally good biclusters. The MSR for the two biclusters is 236.25 and 385, respectively. As it can be seen in Figure 1, the only difference between these two biclusters is represented by the value assumed by the genes under the third condition. Comparing these two biclusters graphically, we cannot conclude that the left one is better than the right one, as it would be unfair to claim that genes presenting lower values for a certain condition are preferable to higher values.

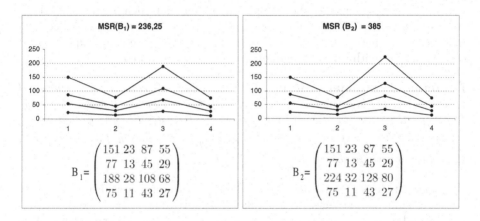

Fig. 1. Examples of similar biclusters with different MSR values

This motivates us to propose a novel measure for assessing the quality of bicluster in microarray data. This measure should avoid taking into account the numerical similarities in the submatrix. Instead, it should quantify the behaviour of the genes under all the conditions contained in the bicluster. We therefore propose a novel criterion, called *Virtual Error* (VE), based on the concept of behavioural patterns.

3 Virtual Error

The main idea behind VE is to create a pattern from each bicluster in order to represent the general trends within it. This pattern will try to capture the overall behaviour of the genes over the conditions in the bicluster, checking if the expression levels of the genes vary in unison, with independence on the specific values and slopes. VE is based on the use of a tendency pattern for each bicluster. Therefore, this quality value will depend on the way in which the pattern is built.

Next, we formally define how the pattern is created, starting from a bicluster $\mathcal{B}$, consisting of I conditions (rows) and J genes (columns), and where each element in the bicluster is represented as b_{ij}, where $1 \leq i \leq I$ and $1 \leq j \leq J$.

Definition 1 (Tendency Pattern). *Given a bicluster $\mathcal{B}$ containing I conditions and J genes, we define the tendency pattern as a collection of I elements P_i, each of them given by: $P_i = \frac{\sum_{j \in J} b_{ij}}{J}, b_{ij} \in \mathcal{B}, 1 \leq i \leq I, 1 \leq j \leq J$.*

Thus, each of the points of the pattern will represent a significative value for all genes under a specific condition.

Once the pattern has been built, the aim is to examine to what extent the genes are similar to it. In this sense, we need a mechanism in order to do an appropriate comparison between each gene and the pattern. This mechanism would be responsible for smoothing every gene behaviour, since the most important issue is to characterize their conduct but not their numerical values. An example of this is represented by scaling patterns (see section 2), where two different genes may present the same behaviour under the same experimental conditions, but with different magnitude of expression values.

Definition 2 (Standardization). *Let $\mathcal{B}$ be a bicluster containing J genes and I conditions. Let b_{ij} denotes the elements of $\mathcal{B}$, for $1 \leq i \leq I$ and $1 \leq j \leq J$. We then define the standardization of $\mathcal{B}$ as the bicluster $\mathcal{B}'$, whose element b'_{ij} are $b'_{ij} = \frac{b_{ij} - b_{Ij}}{\sigma_{g_j}}, 1 \leq i \leq I, 1 \leq j \leq J$, where σ_{g_j} is the standard deviation of all the expression values of gene j.*

By means of the standardization, two distinct tasks are carried out. The first one is to shift all the genes to a similar range of values (near 0 in this case). The second one is to homogenize the expression values for each gene, modifying in this way their values under all the conditions, and smoothing their graphical representation.

It is important to notice that in order to fairly compare genes values to pattern values, all of them must be enclosed in the same range of values. Thus, the pattern must be also standardized, generating a so called virtual pattern. This is shown in equation 2, where P_i refers to the pattern value for condition i, and $\overline{P}$, σ_P refer to the mean and the deviation of all the values in the pattern, respectively.

$$P'_i = \frac{P_i - \overline{P}}{\sigma_P} \tag{2}$$

Definition 3 (Virtual Error). *Given a bicluster $\mathcal{B}$ containing I conditions and J genes, and a pattern P containing I values, we define VE as the mean of* *the numerical differences between each standardized gene and pattern values for each condition:*

$$VE(\mathcal{B}) = \frac{1}{I \cdot J} \sum_{i=1}^{i=I} \sum_{j=1}^{j=J} (b'_{ij} - P'_i)$$

$VE(\mathcal{B})$ corresponds to our measure proposal, which states that biclusters with lower levels of VE are considered to be better than those with higher values. This is due to the fact that VE computes the differences between the standardized genes and the pattern, therefore, the more similar the genes are, the lower the value for VE. It then is important to note that shifting patterns do not increase the value of VE, since standardizing genes allows the VE to compare their behaviour within the same range of values. In the case of scaling patterns, it has a minimal effect on our measure, as the standardization decreases the numerical differences among genes. As an instance, biclusters shown in Figure 1 have VE values practically equal to zero ($VE(B_1) = 2,77 \times 10^{-17}$ and $VE(B_2) = -1,39 \times 10^{-17}$). These values indicate that VE considers both biclusters as equally good. VE owes its name to the fact that the error is not computed using the original genes and pattern, but with virtual ones, once the original data has been standardized.

In the whole, this new measure provides a value for each bicluster, quantifying the similarities among genes by means of comparing their behaviour to a pattern. This comparison is carried out in such a way that shifting and scaling trends are minimally penalized, while behavioural differences among genes notably increase the quality value.

4   Description of the Algorithm

In order to assess the effectiveness of the VE as a measure for establishing the quality of biclusters, we have incorporated it in the EA SEBI [8]. In order to use the VE within SEBI, we have modified the fitness function of SEBI, as explained next.

SEBI adopts a sequential covering strategy: an EA, called EBI (for Evolutionary BIclustering), is called n times, where n is an user-defined parameter. EBI takes as input the expression matrix and returns a bicluster, which is stored in a list called *Results*, and EBI is called again.

In order to avoid too much overlapping among the found biclusters, we associate a weight to each element of the expression matrix. After a bicluster is returned, these weights are adjusted. The weight of an element depends on the number of biclusters in *Results* containing the element. The more biclusters cover an element, the higher the weight of the element will be (see [8] for more details).

In [8], the fitness of an individual X was:

$$f(X) = \frac{MSR(X)}{\delta} + \frac{1}{row\_variance(X)} + +w\_d + penalty \qquad (3)$$

where $MSR(X)$ represents the mean squared residue of X, δ is a user supplied threshold, $row\_variance(X)$ is the row variance of X, w_d is used for penalizing smaller biclusters and *penalty* is the sum of the weights assigned to the element of the expression matrix belonging to the bicluster X. Notice that the fitness has to be minimized. It follows that the aim of EBI was to find biclusters with mean squared residue lower than δ, with high volume, with a relatively high row variance, and minimizing the effect of overlapping among biclusters.

In the version of EBI we use in this paper, we have modified the fitness function defined in equation 3 in the following way:

$$f(X) = \text{VE}(X) + w\_d + penalty \tag{4}$$

where VE(X) is the virtual error of X, w_d and *penalty* are defined as in [8], but are scaled to adapt to VE. Also this fitness has to be minimized, hence we prefer biclusters characterized by a low VE, high volume and minimum overlapping with biclusters contained in *Results*. In this fitness function, we do not use the row variance, as it happened in equation 3. This is because with the use of VE we do not need this factor to reject trivial biclusters, as it happened when the MSR was used.

As in [8], the initial population consists of biclusters containing only one element of the expression matrix. Tournament selection is used for selecting parents. Selected pairs of parents are recombined with a crossover operator with a given probability p_c (default value 0.9), and the resulting offspring is mutated with a probability p_m (default value 0.1). Elitism is applied with a probability p_e (default value 0.75). At the end of the evolutionary process, EBI returns the best individual, according to the fitness.

Each individual of the population encodes one bicluster. Biclusters are encoded by means of binary strings of length $N + M$, where N and M are the number of rows (genes) and of columns (conditions) of the expression matrix, respectively. Each of the first N bits of the binary string is related to the rows, in the order in which the bits appear in the string. In the same way, the remaining M bits are related to the columns. If a bit is set to 1, it means that the relative row or column belongs to the encoded bicluster; otherwise it does not.

5   Experiments

In order to show the quality of our approach, we run SEBI on two well known datasets. The first one, the yeast *Saccharomyces cerevisiae* cell cycle expression dataset [7], is a microarray which contains 2884 genes and 17 conditions. The second dataset is the human B-cells expression data [3], that consists of 4026 genes and 96 conditions.

With regard to the EA parameters, we used the same parameter setting as in [8]. Thus, we can compare the results with those obtained in the previous EA version, where the MSR was used as main term of the fitness function. Specifically, we used a population of 200 individuals and a number of generations of 100. The crossover probability was of 0.85 and the mutation probability was 0.2. The number of biclusters was set to 100, that is, SEBI generated one hundred biclusters for each dataset.

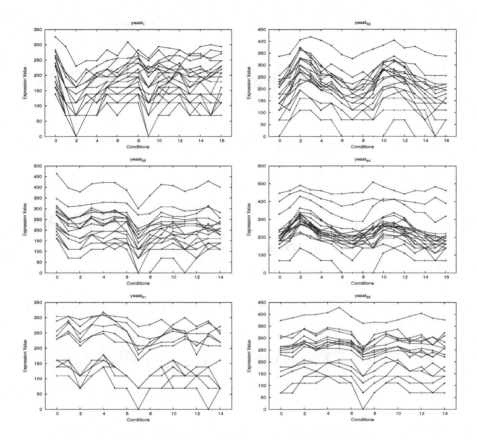

Fig. 2. Biclusters found on the Yeast dataset

Figure 2 shows six biclusters out of the one hundred found on the yeast dataset (see Table 1 for numerical results). The bicluster labelled $yeast_1$ is found with the fist call of SEBI. As we can see, this bicluster is visually interesting. Furthermore, it has low VE of 0.38, but high residue of 535.8. This is a remarkable result, since first biclusters found with VE were interesting, while this is not the case with MSR, as it can be seen in [8] and [6].

In general, we can notice from a visual inspection of all the biclusters that the genes present a similar behaviour under the set of selected conditions. All the VE of the biclusters are lower than 0.38. We find especially interesting the fact that, some genes in the bicluster are distant from the rest of it but they show a similar trend. For example, bicluster $yeast_{44}$ is interesting because it differentiates three genes at the top of the graph from the others, although all the genes seem to have the same behaviour. This points out that VE is not sensitive to the scale or magnitude difference in the expression values of the genes. Furthermore, this is a kind of bicluster difficult to find by using the MSR [1].

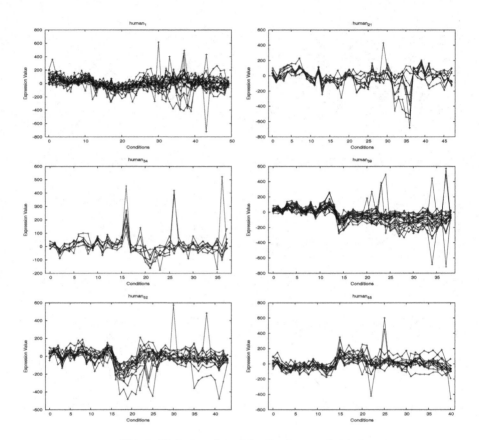

Fig. 3. Biclusters found for the human dataset

Concerning the size of the biclusters, many biclusters contain all the seventeen conditions. A similar result was also obtained in the previous version of SEBI [8], where MSR was used as main term in fitness function. However, the number of genes was higher in this case than those obtained in the aforementioned work. Thus, the use of VE allows SEBI to include more genes without damaging the bicluster quality.

The human dataset is larger and more complex than the yeast dataset. Therefore, it is also more complex to find good biclusters with low VE. Six out of one hundred biclusters found on such dataset are shown in Figure 3 (see Table 1 for numerical results).

The bicluster $(human_1)$, that SEBI with VE finds in the first execution of EBI, is interesting. However, it does not happen the same with MSR. Although it has also low VE (0.57), the most remarkable aspect is that the residue of this bicluster is very high (7173.5). This strengthens our conclusion that SEBI can find interesting biclusters in the first iterations when VE is used as quality measure instead of MSR.

Table 1. Information about biclusters of Figures 2 and 3

Yeast Dataset				Human Dataset				
Bicluster	VE	MSR	#Genes	#Cond.	Bicluster	VE	MSR	#Genes	#Cond.
$yeast_1$	0.38	535.8	23	17	$human_1$	0.57	7173.5	21	50
$yeast_{32}$	0.29	408.9	19	17	$human_{21}$	0.39	6405.4	9	48
$yeast_{35}$	0.28	380.6	18	15	$human_{34}$	0.43	3278.8	7	38
$yeast_{44}$	0.30	583.5	21	17	$human_{39}$	0.44	5786.1	21	39
$yeast_{51}$	0.34	346.7	12	15	$human_{52}$	0.42	5660.7	15	44
$yeast_{85}$	0.36	232.1	16	15	$human_{55}$	0.46	4069.5	14	41

Regarding the shape of the biclusters, all of them present a similar trend, although the genes are closer in this case than in yeast dataset case. Also in the human dataset, VE is not sensitive to the magnitude difference in the values of the genes. This aspect can be observed in the bicluster $human_{52}$, where there is a decrease of the expression level between 15 and 25 for all the genes but with different scale. Finally, the number of genes and conditions are similar to those produced in the previous version of SEBI with MSR.

Table 1 summarizes the numerical results for Figures 2 and 3. The left table corresponds to the yeast dataset, while the right one to the human dataset. For each table, the first column indicates the name of bicluster. The second column gives the VE value and the third ones the MSR measurement for each bicluster. The last two columns report the number of genes and conditions of the bicluster, respectively.

The most interesting result that can be extracted from these tables is that the biclusters present low VE but high MSR for both datasets. Taking into account that these bicluster are interesting, they could be rejected by the other approaches which use the MSR as main quality measure. For instance, the version of SEBI proposed in [8] used a threshold for rejecting biclusters with MSR higher than this threshold. For the yeast dataset this threshold was set to 300, and for the human dataset to 1200, as in [6]. Therefore, all the biclusters shown in Table 1, with the exception of $yeast_{85}$, would have been rejected.

6   Conclusions

In this work, we have proposed a novel measure for assessing the quality of biclusters in microarray data, called *Virtual Error* (VE). VE is based on the concept of tendency pattern. The majority of the existing biclustering methods are based on the well-known *Mean Squared Residue* (MSR) as the quality measure. However, MSR may fail to recognize some interesting biclusters.

In order to test the goodness of our proposal, we have used VE as main term in the fitness function of an EA. We have then applied the resulting EA to two well-known datasets. From experimental results we can draw the following three main conclusions.

By using VE, the EA found very interesting biclusters with very low VE values. The same biclusters would not have been considered as high-quality biclusters if evaluated with MSR. The row variance is not needed in order to reject trivial or constant biclusters as it happens when MSR is the main term in the fitness function. Another result is that VE is not sensitive to the scale or magnitude difference in the expression values of the genes, as long as they present the same behaviour. It has been proven that this kind of biclusters is difficult to find by using the MSR.

References

1. J. S. Aguilar-Ruiz. Shifting and scaling patterns from gene expression data. *Bioinformatics*, 21:3840–3845, 2005.
2. J. S. Aguilar-Ruiz, D. S. Rodriguez, and D. A. Simovici. Biclustering of gene expression data based on local nearness. In *1Proceedings of EGC 2006,*, pages 681–692, Lille, France, 2006.
3. A. A. Alizadeh, M. B. Eisen, R. E. Davis, and et al. Distinct types of diffuse large b-cell lymphoma identified by gene expression profiling. *Nature*, 403:503–511, 2000.
4. A. Ben-Dor, R. Shamir, and Z. Yakhini. Clustering gene expression patterns. *Journal of Computational Biology*, 6(3-4):281–297, 1999.
5. S. Bleuler, A. Prelić, and E. Zitzler. An EA framework for biclustering of gene expression data. In *Congress on Evolutionary Computation (CEC-2004)*, pages 166–173, Piscataway, NJ, 2004. IEEE.
6. Y. Cheng and G. M. Church. Biclustering of expression data. In *In Proceedings of the 8th International Conference on Intellingent Systemns for Molecular Biology*, pages 93–103, La Jolla, CA, 2000.
7. R. Cho, M. Campbell, E. Winzeler, and et al. A genome-wide transcriptional analysis of the mitotic cell cycle. *Molecular Cell*, 2:65–73, 1998.
8. F. Divina and J. S. Aguilar-Ruiz. Biclustering of expression data with evolutionary computation. *IEEE Transactions on Knowledge & Data Engineering*, 18(5):590–602, 2006.
9. S. C. Madeira and A. L. Oliveira. Biclustering algorithms for biological data analysis: A survey. *IEEE Transactions on Computational Biology and Bioinformatics*, 1:24–25, 2004.
10. A. Prelić, S. Bleuler, P. Zimmermann, and et al. A systematic comparison and evaluation of biclustering methods for gene expression data. *Bioinformatics*, 22:1122–1129, 2006.
11. A. Tanay, R. Sharan, and R. Shamir. Discovering statistically significant biclusters in gene expression data. *Bioinformatics*, 18:136–144, 2002.
12. C. Tilstone. Dna microarrays: Vital statistics. *Nature*, 424:610–612, 2003.
13. H. Wang, W. Wang, J. Yang., and P. S. Yu. Clustering by pattern similarity in large data sets. In *ACM SIGMOD International Conference on Management of Data*, page 394–405, Madison, WI, 2002.
14. J. Yang, H. Wang, W. Wang, and P. S. Yu. An improved biclustering method for analyzing gene expression profiles. *International Journal on Artificial Intelligence Tools*, 14:771–790, 2005.

Characterising DNA/RNA Signals with Crisp Hypermotifs: A Case Study on Core Promoters

Carey Pridgeon[1] and David Corne[2]

[1] Department of Computer Science, University of Exeter, Exeter EX4 4QF, UK
[2] School of MACS, Heriot-Watt University, Edinburgh EH14 8AS, UK
carey.pridgeon@gmail.com, d.w.corne@macs.hw.ac.uk

Abstract. A common way to characterise important and conserved signals in nucleotide sequences, such as transcription factor binding sites, is via the use of so-called *consensus* sequences or consensus patterns. A well-known example is the so-called "TATA-box" commonly found in eukaryotic core promoters. Such patterns are valuable in that they offer an insight into basic molecular biology processes, and can support reasoning regarding the understanding, design and control of these processes. However it is rare for such patterns to be accurate; instead they represent a very approximate characterisation of the signal under study. At the opposite extreme, we may instead characterise such a signal via a neural network, or a high-order Markov model, and so on. These have better sensitivity and specificity, but are unreadable, and consequently unhelpful for conveying an understanding of the underlying molecular biology processes that could support insight or reasoning. We describe a simple pattern language, called crisp hypermotifs (CHMs), that leads to highly readable patterns that can support understanding and reasoning, yet achieve greater sensitivity and specificity than the commonly used approaches to crisply characterise a signal. We use evolutionary computation to discover high-performance CHMs from data, and we argue that CHMs be used in place of classical consensus motifs, and justify that by presenting examples derived from a large dataset of mammalian core promoters. We provide CHM alternatives to the well-known core promoter TATA-box and Initiator patterns that have better sensitivity and specificity than their classical counterparts.

1 Introduction

It is common in molecular biology to seek models that discriminate between different groups of nucleotide sequences. For example, we need to be able to discriminate between intron/exon splice sites and control sequences, between core promoters and control sequences, between different classes of microRNA, and so forth. The model used for discrimination can take many forms, but at a high level we can categorise them as occupying a continuum from 'readable/approximate' to 'opaque/accurate'. At the 'readable/approximate' end, we have straightforward sequences or simple patterns that characterise the positive (i.e. not the control) sequences. For example, such a model might be a simple consensus sequence, from the alphabet {A,C,G,T}, which is commonly observed in the positive sequences and rarely observed in the controls. Or,

E. Marchiori, J.H. Moore, and J.C. Rajapakse (Eds.): EvoBIO 2007, LNCS 4447, pp. 227–235, 2007.
© Springer-Verlag Berlin Heidelberg 2007

we may have a pattern from the IUPAC[1] alphabet, such as YYCARR, which matches several of the positive cases but fewer or none of the controls. Meanwhile, at the 'opaque/accurate' end of this continuum, we may have models such as hidden Markov models (e.g. Henderson et al, 97), higher-order Markov models,(e.g. Salzberg et al, 98) neural networks (e.g. Reese, 01), support vector machines (e.g. Zien et al, 00), and so on. These tend to provide more accurate classification than the simple patterns, and are consequently used for genome annotation and similar tasks, however they are opaque to analysis – i.e. it is very difficult or impossible to glean knowledge and insight concerning the precise sequences and patterns of nucleotides that form the signal of interest.

Conceivably, there are DNA and RNA tasks for which consensus or simple patterns are sufficient, since the signal of interest is highly conserved. But for the majority of such tasks, simple patterns made from the IUPAC alphabet (see Table 1) poorly capture the variability of nucleotide composition in the signal, and score quite low in sensitivity and specificity.

Table 1. The IUPAC alphabet for nucleotide subsets, used in motifs

Nucleotide subset	IUPAC Symbol
A	A
C	C
G	G
T	T
A, C	M
A, G	R
A, T	W
C, G	S
C, T	Y
G, T	K
A, C, G	V
A, C, T	H
A, G, T	D
C, G, T	B
A, C, G, T	N

An interesting challenge is to find ways that can provide accuracy competitive with 'opaque' techniques, yet which have enough readability to support fruitful insight and conjecture concerning underlying molecular biology mechanisms. The idea of using a *hypermotif* (Pridgeon & Corne, 05) was motivated by this challenge. In the latter paper, we focused on the use of *weighted* hypermotifs for use in conjunction with neural networks, to provide discrimination at the 'opaque' end of the model spectrum. In this paper we concentrate only on beginning to examine 'crisp' (i.e. unweighted)

---

[1] IUPAC: International Union of Pure and Applied Chemistry – a source of standards, include common abbreviations for sets of nucleotides.

hypermotifs, focusing on whether we can find HM patterns that can readably capture highly variable biological signals.

The remainder of this paper is set out as follows. Section 2 describes and discusses crisp hypermotifs, section 3 describe the straightforward evolutionary algorithms we use to discover crisp hypermotifs for characterizing core promoter regions, section 4 shows results, and section 5 finishes with some brief discussion and consideration of future work.

2   Crisp Hypermotifs

It is common to use a pattern of symbols from the IUPAC alphabet (Table 1) to form patterns (which we will call IUPAC-motifs), to represent a family of nucleotide sequences. Such patterns have the benefit of 'human readability'. Given an IUPAC motif M, say, which discriminates well between positive and control examples for some significant DNA binding site, as well as the use of M as a (although possibly weak) predictive model for annotation of new and unknown sites, M also readably and directly expresses a pattern of nucleotides that constitute the signal. This in turn can lead to new biological insight concerning the molecular biology mechanism involved, such as potential binding molecules, or potential docking sites for suspected molecules.

However, IUPAC-based motifs are quite impoverished in their ability to characterize signals. To see this, it will be useful to first set out some simple notation. Suppose we have a signal that is present in a positive dataset P, and known not to be present in a negative or control dataset N. We desire to find a pattern that matches only the sequences in P, and does not match any of the sequences in N. Whether or not this is possible in the first place depends on the kind of pattern structure in use. To understand this point, that we call 'discriminatory power', consider how many ways our dataset (the set D, consisting of the union of N and P) can be partitioned into 2 sets. For a total of k sequences, the number of distinct such partitions is $2^{k-1} - 1$ (excluding the partition involving D and the empty set). Since this number represents all possible ways that D can be divided into two sets (e.g. P and N), it represents, the 'ideal' level of discriminatory power. Now consider motifs formed from a sequence of l symbols in the IUPAC alphabet. There are 15^l possible such motifs, and therefore at most 15^l ways in which such a motif can lead to a partition of a set of such sequences. If our sequence set contained all l-base sequences, of which there are 4^l, this represents a fraction $(15^l / (2^{(4^l - 1)} - 1)) \approx 2^{-4^l}$ of all possible partitions, suggesting that straightforward IUPAC alphabet motifs are highly unlikely to be able to discriminate between the two sets in an arbitrarily chosen partition of sequences. To see a simple example, consider a sequence signal which is present in these eight 3-nucleotide sequences:

ACG, CAG, ACT, CAT, ACA, CAA, ACC, CAC

but which is present in no other 3-nucleotide sequence. The pattern which captures this must recognise that the first and second positions may be AC or CA, but not AA and not CC; meanwhile, the third position can be any nucleotide. This simply cannot be expressed by a single sequence motif. The attempt `MMN' is fine in terms of the third position, and matches all eight of the positive examples, however it also matches four negatives which start with AA, and four more which start with CC.

A crisp hypermotif (CHM) extends the notion of a motif by using an extended alphabet that covers all possible subsets of a 2-nucleotide segment (rather than, as with standard motifs, all possible subsets of a 1-nucleotide segment). A CHM consists of a number of 16-bit vectors in sequence. Each 16-bit vector (or chunk), represents a subset of 2-nucleotide sequences for a 2-nucleotide window. In a single chunk, each bit represents the presence or absence of a given dinucleotide; bit positions are mapped to dinucleotides in alphabetical order, so that the first bit represents AA and the last represents TT. For example, the chunk:

$$1000100000000001$$

represents the subset {AA, CA, TT}. The following 3-chunk hypermotif:

$$0011000000010 \quad 000000001111 \quad 100000000000$$

represents the set of 6-base sequences in which the first pair can be AG or AT or TG, the second pair can be TA, TC, TG or TT, and the third pair can only be AA. Note that we do not allow a chunk to be all zeroes. Meanwhile, we can express a CHM readably by simply listing ithe set of allowed nucleotide pairs. For example, the three-chunk CHM above can be readably expressed as:

$$\{AG\text{-}AT\text{-}TG\} \mid \{TA\text{-}TC\text{-}TG\text{-}TT\} \mid \{TA\}$$

In practice, we note that it can be simplified further (and we make use of this in our results section) by employing the IUPAC alphabet where possible. Also, we omit the curly braces when only one pair is present for a particular chunk – hence, the above CHM is equivalent to this:

$$\{AK\text{-}TG\} \mid TN \mid TA$$

Now let's compare the discriminatory power of CHMs and IUPAC-motifs. When representing l-base sequences (where we will assume l is even), we have seen that there are 15^l motifs, and hence up to 15^l partitions of a set of sequences can be encoded. For CHMs this number is $(2^{16} - 1)^{l/2}$ and they are therefore a ratio $(2^{16} - 1)^{l/2} / 15^l \approx 10^{2.41l} / 10^{1.18l} \approx 10^{1.23l}$ times more discriminatory than motifs. For reasonable sequence lengths, this suggests enormously better discriminatory power, yet retaining a potential level of readability.

3 Evolving Crisp Hypermotifs

In this initial study of CHMs, we use a straightforward evolutionary algorithm to evolve CHMs, assigning fitness using the Matthews Correlation Coefficient (MCC) (Matthews, 75) indicated below.

$$MCC = \frac{TP \cdot TN - FP \cdot FN}{\sqrt{(TP + FP)(TP + FN)(TN + FP)(TN + FN)}}$$

where TP means true positives, TN means true negatives, and so forth. The MCC provides a convenient measure of discriminative ability, and is commonly used in bioinformatics, but is not an ideal measure (indeed there is no ideal single-objective fitness measure in this case). The evolutionary algorithm (EA) (Fogel et al, 66; Schwefel, 81; Goldberg, 89) uses binary tournament selection (Goldberg, 89), steady-sate replacement (Syswerda, 90) with crowding (de Jong, 75), a population size of 200, crowding factor of 20, and the only operator is mutation, with a rate of 1, and a strength of precisely 1 gene-flip per operation. In the runs that produced the results discussed in the next section, populations were initialized with 10% mutated (flipped) versions of the CHM corresponding to either TATAWAW, YYCARR or YYANWYY. Each individual in the population is a CHM of a given fixed number of chunks (fixed per experiment). So, for example, in a case of 4-chunk CHMs, the individuals were 64-bit binary strings. Fitness was assigned by sliding the CHM pattern along each sequence in the data to record whether or not it matched that sequence. If it matched once or more within a core promoter sequence, this was recorded as a single true positive, and a match to a control sequence was recorded as a false positive.

4 Experiments and Results

We used the well-known fairly comprehensive dataset of mammalian promoters and control sequences collected by Reese (2001), available at the following URL: http://www.fruitfly.org/seq_tools/datasets/Human/promoter/. These data consist of 565 vertebrate promoters and 5,235 control sequences. The controls are either sequences from coding regions or sequences that are known to be introns. Following Reese (2003) and other authors, we extracted the *core promoter* region from each sequence (and the corresponding region in the control sequences), by extracting the 40 bases upstream of the transcription start site and 11 bases downstream of the TSS.

It is important to note here that in the current paper we treat the data as authoritative, and are interested in the specific CHM patterns that we can find. That is, we are not at the moment interested in using CHMs as predictive tools, and hence we are not investigating and evaluating a method for promoter classification. Rather, we are interested in understanding what patterns are in known promoter data. Hence there is no requirement for separating the data into training, test and validation sets, since this is an investigation of patterns in the full dataset itself. Consequently, we do not make any claims about the 'performance' of the discovered patterns on unseen data, but we simply imply tentative generalization from the fact that (as we will see) CHM patterns seem to capture core promoter signals in these data more effectively than the standard consensus patterns, and hence may be candidates for replacement of such patterns in molecular biologists' intuitions.

4.1 Performance of IUPAC Based Motifs

First, it is instructive to view the discriminatory ability of the well-known motifs on the Reese dataset. This is shown in Table 2.

Table 2. Discriminatory ability of well-known core promoter motifs

Motif	Sensitivity	Specificity	MCC
TATAWAW	25.1327	1.98	0.340557
YYCARR	0.0	0.0	-0.014
YYANWYY	0.0	0.0	-0.058

The first motif is the standard TATA-box consensus, which is found in the core region in just over 25% of Reese mammalian promoter data, but is also found in marginally below 2% of the control (non-promoter) sequences. This TATA-box motif is clearly highly representative and indicative of a core promoter region (roughly, if we find this in a sequence, it is over 12 times more likely to be a core promoter sequence than not), however it is also well-known not to occur in many genes. It is interesting to ask what is the 'next best' motif, that covers core-promoters that do not contain the TATA box. However there is no clear candidate for this in the molecular biology literature, and it seems clear that any other such pattern (at least, when using simple consensus and IUPAC based motifs) has much less sensitivity than TATAWAW. Our feeling, however, is that a CHM-based pattern could better capture the variability inherent in the TATA-box signal, achieving greater sensitivity, and reducing the proportion of promoters that are believed not to contain a TATA-box. Meanwhile, in later CHM-based studies (not done here) we aim to discover patterns for non-TATA genes, which may perhaps not be representable with appreciable sensitivity or MCC by IUPAC-based patterns.

The second and third motifs are textbook patterns for the initiator (Inr) region that is believed to often occupy a region across and downstream of the transcription start site in eukaryotic genomes. Notably, neither of these patterns matches any core promoters in the Reese dataset (and they also match no controls). The Inr region is well-known to be highly variable, yet it is perhaps surprising that the consenus patterns match none of the Reese data. However this becomes less surprising when we remember the potential representative inadequacy of these signals. For example, if we have the following positive signals in a dataset: CAT, GGT, CTT, CGA, then their simple consensus is CGT, however this does not match any of them.

4.2 Discovering Crisp Hypermotifs

Some preliminary study was first done to test how an EA (of the design in section 3) compared with standard hillclimbing, using the same operator and initialization strategy, but for different numbers of CHM chunks. Without exception, we find that the EA is able to achieve a better MCC than HC, with p-values never above 0.05 using randomization testing (Edgington, 95), and showing increasing significance for larger CHMs. We intend to also test simulated annealing on this task, but for the present study we chose to directly move to using EAs. Next, we examined the difference between the performance of standard IUPAC based motifs and CHMs, using EAs to evolve 4-chunk CHMs and 8-base IUPAC-motifs in paired tests, each initialized with TATAWAW. In these tests, fitness was a simple linear combination of promoter matches and non-promoter matches, using different weightings per pair, to obtain a spread of coverage over sensitivity values. Again, we found evidence of statistical significance for the superiority of CHMs over IUPAC-Motifs in all cases.

Meanwhile, in this article our purpose is to demonstrate discovered CHMs that improve upon IUPAC-motifs for core promoters in terms of sensitivity and specificity. We show in Table 3 a selection of the results from several runs of the EA producing CHMs, with fitness as MCC, and for varying numbers of chunks. In every case (except one) we note a better MCC than achieved on these data by TATAWAW. In the case of the one with lower MCC than that, this remains interesting because it is a short signal from a 3-chunk run initialized with YYCARR; hence it is an initiator region signal, but which scores immensely better in discriminatory power than YYCARR, and is highly readable (note that S means C or G). A clear correlation is noted of MCC against length of signal, suggesting both that core-promoter signal material is to be found across a wide region of the core promoter itself, and that CHMs are able to capture at least some of the material variability therein. TATA itself is preserved in one of the signals, and this is an interesting one:

$$\{AS-CN-KV\}| TA | TA | AS$$

This is an example of a short, readable CHM which is clearly a TATA-box signal, but is appreciably more discriminative than TATAWAW, dominating it in sensitivity and specificity on these data.

Table 3. Discriminatory ability of crisp hypermotifs discovered by evolutionary computation

Motif	Sensitivity	Specificity	MCC
HT | {AW-TA} | {AR-TA}|{AV-CS-GN-TR}|{AG-CN-GV-TC}|{AS-SN-TS}|{AS-SN-TS}|{AN-CS-GB-TN}|{AV-SN-TB}|{AS-CV-GN-TV}	42.65	1.49	0.535407
{CD-GH-TW} | TA | WA | {AV-TA}|{AV-SN-TG} | {AS-SN-TS} | {AB-SN-TS}	42.65	2.02	0.507981
{CD-GR-TA}| TA | WA | {AV-CS- GS-TD} | {AV-SN-TC} | {AS-SN-TS}	40.17	1.89	0.492901

Table 3. (*Continued*)

YA I TA I {AR-TA} I {AV-CG-GS-TH} I {AV-SN-TS}	41.24	2.31	0.482169
{AT-TA}I {AW-TA} I AR I {AS-CV-GN-TV} I {AS-SN-TC} I {MS-GB}	34.69	1.2	0.480743
{AT-TA} I WA I {AV-TA} I {AS-SN-TA} I {AG-CS-CV-TC} I {AS-SN-TB}	39.115	1.99	0.478691
{CW-KA} I TA I WA I {AV-CC-GS-TH}I {AR-SN-TS}	43.01	2.87	0.473613
{CH-GA-TA} I TA I WA I {AV-SS-TD} I {AV-SN-TS}	44.78	3.24	0.473608
TATAI{AR-TA}I{AV-CT-GN-TD}	36.99	2.48	0.436179
{AT-CK-GY}I AT I AW I {AD-CG-GA}	35.93	2.29	0.434937
{AS-CN-KV}I TA I TA I AS	31.50	1.47	0.434379
{AA-CA-CT-GN} I SS I SS	39.3	8.2	0.292

5   Conclusions and Future Work

It is vital for biologists and related scientists to learn a full understanding of low-level molecular biology operations, since this is critical to reasoning about the design and control of bioactive molecules for a very wide range of activities, ranging through medicine, nanotechnology, and crop engineering. Crisp hypermotifs represent a clear potential advance on IUPAC-based motifs for characterizing signals in a readable yet discriminatory form, and we have shown here that EAs can be efficiently used to find discriminatory CHMs. Further we show evidence that EAs are more suitable than HC for this task, although there is much more experimental study to do before we can advise on the most appropriate search method, and we expect this to vary anyway from task to task.

Interestingly, CHM-based replacements for the well-known TATA-box and Inr region have been discovered, which indicate clearly, since they cannot be 'simplified' to IUPAC-alphabet signals, that the CHM representation is more appropriate for capturing the type of variability present in core promoter signals. Most notably, we find that the two standard Inr region patterns in the literature do not match any promoter sequences at all. However, by moving to CHM space we are able to express patterns that validly represent sequences in this region, and with reasonable discriminatory power.

There are three areas of further work we envisage for the study of CHM-based patterns. First, and obviously, we intend to extend the current work to additional datasets and additional bioinformatics tasks, as well as further examine the algorithm and operator variants. Second, we suspect that evolving 'decision-lists', or simple sets of, CHMs may enable us to find classification models for tasks such as promoter prediction that are competitive with methods at the accurate/opaque end of the model spectrum. Finally, we note that an inherent aspect of the CHM model we have described here is that nucleotides are associated in adjacent pairs. However, this need not be fixed, and it may not be ideal for many bioinformatics tasks. For example, many DNA binding related signals may be better captured by CHMs that capture associations between nucleotides and their associated one helix turn apart. Specifically, we note that we can encode a more variable CHM pattern by associating with the binary string a permutation of the n nucleotides, such as "261435", which could indicate, for example, that the first chunk expresses the subset of dinucleotides in positions 2 and 6, the second chunk expresses the subset in positions 1 and 4, and so on. In further work we will explore the use of EAs to discover CHMs in this larger space.

References

Edgington, E.S. (1995) *Randomisation Testing*, Marcel Dekker, New York

Eskin, E., U. Keich, M. S. Gelfand, and P. Pevzner (2003) Genome-wide analysis of bacterial promoter regions. In Proc. 8th Pac. Symp. Biocomp., pages 29–40, Kauai, Hawaii, January 3-7 2003. ISCB.

Fogel, L.J., A.J. Owens, and M.J. Walsh (1966) *Artificial Intelligence Through Simulated Evolution*, John Wiley, New York.

Goldberg. D.E. (1989) Genetic algorithms in search, optimization and machine learning, Addison-Wesley.

Henderson, J. Salzberg, S. Fasman, K. H. (1997) Finding Genes in DNA with a Hidden Markov Model, Journal of Computational Biology, 4(2):127—142.

Holland, J. H. (1975) *Adaptation in Natural and Artificial Systems*, University of Michigan Press, Ann Arbor, MI.

Kenneth Alan De Jong. (1975) An analysis of the bevavior of a class of genetic adaptive systems. PhD thesis, University of Michigan.

Kanhere, A. and Bansal, M. (2005) A novel method for prokaryotic promoter prediction based on DNA stability, *BMC Bioinformatics*, 6:1

Matthews, B.W. (1975) *Biochim. Biophys. Acta,* 405, 442-451.

Ohler U, Niemann H, Liao G, Rubin GM (2001) Joint modeling of DNA sequence and physical properties to improve eukaryotic promoter recognition, *Bioinformatics*, 17(Suppl 1):S199-206

Pridgeon, C., Corne, D. (2005) Hypermotifs: Novel Discriminatory Patterns for Nucleotide Sequences and their Application to Core Promoter Prediction in Eukaryotes, in *Proc. CIBCB* 05, IEEE CS Press, pp. 1—7.

Reese MG: (2001) Application of a time-delay neural network to promoter annotation in the Drosophila melanogaster genome. *Comput Chem* 2001, 26:51-56

Salzberg, S. L. Delcher, A. L. Kasif, S. White, O.(1998) Microbial gene identification using interpolated Markov models, *Nucleic Acids Research*, 26(2):544—548.

Schwefel, H.-P. (1981) *Numerical Optimization of Computer Models*, John Wiley, Chichester, U.K.

Gilbert Syswerda: (1990) A Study of Reproduction in Generational and Steady State Genetic Algorithms. FOGA 1990: 94-101

Zien, A. Ratsch, G. Mika, S. Scholkopf, B. Lengauer, T. Muller, K.-R. (2000) Engineering support vector machine kernels that recognize translation initiation sites, *Bioinformatics*, 16(9):799—807.

Evaluating Evolutionary Algorithms and Differential Evolution for the Online Optimization of Fermentation Processes

Miguel Rocha[1], José P. Pinto[1], Isabel Rocha[2], and Eugénio C. Ferreira[2]

[1] Departament of Informatics / CCTC - University of Minho
Campus de Gualtar, 4710-057 Braga - Portugal
mrocha@di.uminho.pt, josepedr@gmail.com
[2] IBB - Institute for Biotechnology and Bioengineering
Center of Biological Engineering - University of Minho
Campus de Gualtar, 4710-057 Braga - Portugal
irocha@deb.uminho.pt, ecferreira@deb.uminho.pt

Abstract. Although important contributions have been made in recent years within the field of bioprocess model development and validation, in many cases the utility of even relatively good models for process optimization with current state-of-the-art algorithms (mostly offline approaches) is quite low. The main cause for this is that open-loop fermentations do not compensate for the differences observed between model predictions and real variables, whose consequences can lead to quite undesirable consequences. In this work, the performance of two different algorithms belonging to the main groups of *Evolutionary Algorithms* (EA) and *Differential Evolution* (DE) is compared in the task of online optimisation of fed-batch fermentation processes. The proposed approach enables to obtain results close to the ones predicted initially by the mathematical models of the process, deals well with the noise in state variables and exhibits properties of graceful degradation. When comparing the optimization algorithms, the *DE* seems the best alternative, but its superiority seems to decrease when noisier settings are considered.

Keywords: Fermentation processes, Online optimization, Differential Evolution, Real-valued Evolutionary Algorithms.

1 Introduction

In recent years, many efforts have been devoted to the optimization of processes in bioengineering as a number of valuable products such as recombinant proteins, antibiotics and amino-acids are produced using fermentation techniques. A problem that has received special attention is the dynamic optimization of fed-batch bioreactors. This process has traditionally been conducted on the substrate feed rate as key manipulated variable in operation. The optimization problem is therefore solved before the beginning of the fermentation process (open-loop optimal control) and consists on finding an expression or a sequence of values for

E. Marchiori, J.H. Moore, and J.C. Rajapakse (Eds.): EvoBIO 2007, LNCS 4447, pp. 236–246, 2007.
© Springer-Verlag Berlin Heidelberg 2007

the feeding rate that maximizes an objective function that represents the process productivity, subject to the constraints represented by a dynamical model.

Several optimization methods have been applied to solve this kind of problem. It has been shown that for relatively simple bioreactor systems, which are expressed in differential equations models, the optimization problem can be solved analytically from the Hamiltonian function by applying the Minimum Principle of Pontryagin [14]. However, in the majority of the cases reported, determination of the optimal feed rate profile has a problem of singular control.

Numerical methods make a distinct approach to dynamic optimization. The gradient algorithms are used to adjust the control trajectories in order to iteratively improve the objective function [3]. In contrast, dynamic programming methods discretize both time and control variables to a predefined number of values. A systematic backward search method in combination with the simulation of the system model equations is used to find the optimal path through the defined grid. However, in order to achieve a global minimum, the computational burden is very high [3].

An alternative comes from the use of algorithms from the *Evolutionary Computation (EC)* field, which have been used in the past to optimize nonlinear problems with a large number of variables. These techniques have been applied with success to the optimization of feeding or temperature trajectories [8][1], and, when compared with traditional methods, usually perform better [12][5].

However, even when the mathematical models used for open-loop optimization are reliable and validated by experimentation, in a real environment several sources of noise can contribute to changes in the observed values of the state variables. These issues are of particular importance when dealing with recombinant high-cell density fermentations, as the process, besides the nonlinearities exhibited, tends to change dramatically upon some events, like induction. Also, it is likely that there exists a time-variance of both yield and kinetic parameters not contemplated in most process models. These scenarios have an important impact on the experimental results that end up being worse than the ones predicted after running the offline optimization.

An alternative to cope with model inaccuracies is the use of online optimization algorithms that periodically generate new solutions as the process is running, making use of the measurement of relevant state variables for update of the internal model. Indeed, unlike the previously stated alternatives where the optimization is conducted prior to the experimental process, in this case, the optimization is performed simultaneously, taking into account values of the state variables measured by sensors within the fermentation process.

In this work, the performance of two different algorithms belonging to the main groups of *Evolutionary Algorithms* (EA) and *Differential Evolution* (DE) is compared in this task of online optimization. These methods were the ones that performed better in offline optimization, in a previous study [6]. Three case studies were taken from literature in order to test the performance of both algorithms. These are used to perform an offline optimization and then a simulation of a real-world fermentation is conducted. The relevant state variables are, in each case, disturbed by adding a small noise value, at regular periods of time. The behavior of both algorithms is compared, as well as the performance of the initial optimization results given the perturbations considered.

The paper is organized as follows: firstly, the fed-batch fermentation case studies are presented; next, the algorithms of DE and a real-valued EA are described; next, the results of the application of the different algorithms to the case studies are presented; finally, the paper presents a discussion of the results, conclusions and further work.

2   Case Studies: Fed-Batch Fermentation Processes

The case studies used in this work are related to the simulation of fed-batch fermentation processes. In these processes there is an addition of certain nutrients along the process, in order to prevent the accumulation of toxic products, allowing the achievement of higher product concentrations. During this process the system states change considerably, from a low initial to a very high biomass and final product concentrations. This dynamic behavior motivates the development of optimization methods able to find the optimal input feeding trajectories in order to improve the process performance. For the optimization of the process white box mathematical models are developed, based on differential equations that represent the mass balances around the relevant state variables.

2.1   Case Study I

This case study is related to a fed-batch recombinant *Escherichia coli* fermentation process which was previously optimized in [10][11].

The dynamical model can be described by the following equations:

$$\frac{dX}{dt} = (\mu_1 + \mu_2 + \mu_3)X - DX \tag{1}$$

$$\frac{dS}{dt} = (-k_1\mu_1 - k_2\mu_2)X + \frac{F_{in,S}S_{in}}{W} - DS \tag{2}$$

$$\frac{dA}{dt} = (k_3\mu_2 - k_4\mu_3)X - DA \tag{3}$$

$$\frac{dO}{dt} = (-k_5\mu_1 - k_6\mu_2 - k_7\mu_3)X + OTR - DO \tag{4}$$

$$\frac{dC}{dt} = (k_8\mu_1 + k_9\mu_2 + k_{10}\mu_3)X - CTR - DC \tag{5}$$

$$\frac{dW}{dt} \simeq F_{in,S} \tag{6}$$

where X, S, A, O, C represent biomass, glucose, acetate, dissolved oxygen and carbon dioxide concentrations (in g/kg); μ_1 to μ_3 are time variant specific growth rates and k_i are constant yield coefficients. D is the dilution rate, $F_{in,S}$ the substrate feeding rate (in kg/h), W the fermentation weight (in kg), OTR and

CTR are the oxygen and carbon dioxide transfer rates. The kinetic behaviour is given by the specific growth rates μ_1 to μ_3, which are non-linear functions of the state variables. This dependence can be found in [9].

The purpose of the optimization is to determine the feeding rate profile $(F_{in,S}(t))$ that maximizes the productivity of the process, defined as the units of product (recombinant protein) formed per unit of time. In this case, this is usually related with the final biomass obtained. Thus, a *performance index (PI)* is defined by the following expression:

$$PI = \frac{X(T_f)W(T_f) - X(0)W(0)}{T_f} \tag{7}$$

The state variables are initialized with the values: $X(0) = 5$, $S(0) = 0$, $A(0) = 0$, $W(0) = 3$. Due to physical limitations the value of $F_{in,S}(t)$ must be in the range $[0.0; 0.4]$ and $W(t) \leq 5$. The final time (T_f) is set to 25 hours.

2.2 Case Study II

This case study handles a hybridoma reactor described by the equations [12]:

$$\frac{dX_v}{dt} = (\mu - k_d)X_v - \frac{F_1 + F_2}{V}X_v \tag{8}$$

$$\frac{dGlc}{dt} = \frac{F_1}{V}Glc_{in} - \frac{F_1 + F_2}{V}Glc - q_{Glc}X_v \tag{9}$$

$$\frac{dGln}{dt} = \frac{F_2}{V}Gln_{in} - \frac{F_1 + F_2}{V}Gln - q_{Gln}X_v \tag{10}$$

$$\frac{dLac}{dt} = q_{Lac}X_v - \frac{F_1 + F_2}{V}Lac \tag{11}$$

$$\frac{dAmm}{dt} = q_{Amm}X_v - \frac{F_1 + F_2}{V}Amm \tag{12}$$

$$\frac{dMab}{dt} = q_{Mab}X_v - \frac{F_1 + F_2}{V}Mab \tag{13}$$

$$\frac{dV}{dt} = (F_1 + F_2) \tag{14}$$

where the state variables X_v, Glc, Gln, Lac, Amm, Mab, V are the concentrations of viable cells, glucose, glutamine, lactate, ammonia, monoclonal antibodies and culture volume, respectively. The control variables F_1 and F_2 are the volumetric feed rates. The complete kinetic expressions are given in [12].

A single feed problem can be obtained by considering $F = F_1 + F_2$. The target of the optimization process, in this case, is to increase the total amount of monoclonal antibodies produced. So, the PI is given by:

$$PI = \int_0^{T_f} -q_{Mab}X_v(t)V(t) \tag{15}$$

Initialization values for the state variables are the following: $X_v = 2.0 \times 10^8 cells/L$, $Glc = 25g/L$, $Gln = 4g/L$, $Lac = 0g/L$, $Amm = 0g/L$, $Mab = 0g/L$, $V = 0.8L$. T_f is $10days$ and the value of $V(t)$ is constrained by $V(t) \leq V_{max}$.

2.3   Case Study III

This system is a fed-batch bioreactor for the production of ethanol by *Saccharomyces cerevisiae*, firstly studied by Chen and Huang [4]. The aim is to find the substrate feed rate profile that maximizes the final amount of ethanol.

The model equations are the following:

$$\frac{dx_1}{dt} = g_1 x_1 - u\frac{x_1}{x_4} \tag{16}$$

$$\frac{dx_2}{dt} = -10g_1 x_1 + u\frac{150 - x_2}{x_4} \tag{17}$$

$$\frac{dx_3}{dt} = g_2 x_1 - u\frac{x_3}{x_4} \tag{18}$$

$$\frac{dx_4}{dt} = u \tag{19}$$

where x_1, x_2 and x_3 are the cell mass, substrate and ethanol concentrations (g/L), x_4 the volume of the reactor (L) and u) the feeding rate (L/h).

On the other hand, the kinetic variables g_1 and g_2 are given by:

$$g_1 = \frac{0.408}{1 + \frac{x_3}{16}}\frac{x_2}{0.22 + x_2} \tag{20}$$

$$g_2 = \frac{1}{1 + \frac{x_3}{71.5}}\frac{x_2}{0.44 + x_2} \tag{21}$$

The *performance index (PI)* is given by: $PI = x_3(T_f)x_4(T_f)$.

The final time is set to $T_f = 54$ hours, and the initial values for the state variables are the following: $x_1(0) = 1$, $x_2(0) = 150$, $x_3(0) = 0$ and $x_4(0) = 10$. Additionally, there are physical constraints over the variables, namely: $0 \leq x_4(t) \leq 200$ for all time points and the feeding rate $0 \leq u(t) \leq 12$.

3   The Optimization Algorithms

The aim of the offline optimization is to find the feeding trajectory, represented as an array of real-valued variables, that yields the best performance index. Each variable will encode the amount of substrate to be introduced into the bioreactor, in a given time interval, and the solution will be given by the temporal sequence of such values. In this case, the size of the solution will be determined based on the final time of the process (T_f) and the discretization step (d) considered in the numerical simulation of the model, given by the expression: $\frac{T_f}{d}$. To reduce the solution size, feeding values were defined only at certain equally spaced points, and the remaining values are linearly interpolated. The size of the genome (G) becomes $G = \frac{T_f}{dI} + 1$, where I stands for the number of points within each interpolation interval. The value of d used in the experiments was $d = 0.005$, for all case studies and the value of I is 200, 100 and 500 in case studies I, II and III, respectively.

The evaluation process, for each solution, is achieved by running a numerical simulation of the defined model, giving as input the feeding values in the solution. The numerical simulation is performed using a linearly *implicit-explicit (IMEX) Runge-Kutta* scheme, used for stiff problems, included in package *OdeToJava* [2]. The fitness value is then calculated from the values of the state variables according to the *PI* defined for each case.

3.1 Differential Evolution

Differential Evolution (DE) is a population-based approach to function optimization that generates trial individuals by calculating vector differences between other randomly selected members of the population. In this work, a variant of the *DE* algorithm called *DE/rand/1* was considered that uses a binomial crossover [13]. In this case, the following scheme is followed, in every generation, for each individual i in the population:

1. Randomly select 3 individuals r_1, r_2, r_3 distinct from i;
2. Generate a trial vector based on: $t = r_1 + F \cdot (r_2 - r_3)$
3. Incorporate coordinates of this vector with probability CR;
4. Evaluate the candidate and use it in the new generation if it is at least as good as the current individual.

3.2 Real-Valued EA

In this work, a real-valued *EA* was adopted that provided good results in previous work [11][6]. A brief presentation of the algorithm is given in this section. A more detailed description can be found in those references. The *EA* uses real-valued representations and a number of mutation and crossover operators to create new solutions, namely:

- *Random Mutation*, which replaces one gene by a new randomly generated value, within the allowed range [7];
- *Gaussian Mutation*, which adds to a given gene a value taken from a Gaussian distribution, with a zero mean.
- the standard *Two-Point crossover*;
- the *Arithmetical* crossover [7];
- the *Sum* crossover, where the offspring genes denote the sum or the subtraction of the genes in the parents.

The mutation operators are applied to a number of genes randomly generated between 1 and N (value of N was 10 in the experiments). In each generation half of the population is kept from the previous generation and the remaining is replaced by the new individuals created using the above reproduction operators. Selection is performed by using a ranking of the individuals in the population and applying a roulette wheel scheme.

4   Online Optimization

During the fermentation process, some of the state variables can be measured, but its values are scarcely used for closed-loop optimization purposes, and are rather employed to evaluate qualitatively the performance of the process. However, it is possible to develop dynamic optimization algorithms capable of timely reacting to this new knowledge generated by updating the corresponding internal model and generating new solutions.

EAs and *DE* are promising approaches to this real-time optimization task, since they keep a population of solutions that can be easily adapted to perform re-optimization. Indeed, a population of solutions previously obtained can be evaluated under the new scenario and better adapted solutions can be created through the use of reproduction operators. The fact that a set of solutions is kept, and not only the best solution, makes a faster adaptation to new conditions possible, while taking advantage of previous optimisation efforts.

In this work, an online optimization strategy based on *EAs* and *DE* is proposed, working in two stages: before the fermentation process starts, an offline optimization is conducted with the algorithms described in the previous sections. After this preliminary optimization, online optimization algorithms use information gathered by measuring the value of relevant state variables to improve the PI during the real fermentation process. These algorithms react by updating its internal model and reaching a new solution, that is available to be sent back to the fermentation monitoring software.

The version of the *EA/DE* used to perform online optimization is similar to the ones described before. When new information regarding the state variables is received, the following steps are followed by both *EA* and *DE*:

1. a starting point (in time) is determined for the re-optimized solution, by adding the time label of the received data with the predicted time necessary to compute a new solution (since it is impossible to reach and therefore apply a solution before that time).
2. the last available population is adapted by removing from the genome of each individual the genes that encode feeding values for elapsed time periods.
3. half of the individuals in the population are replaced by new randomly generated solutions (these individuals are chosen randomly, although the best individual is always kept). This helps in maintaining genetic diversity, a specially needed feature for the optimization in changing landscapes.
4. the internal model of the fermentation used by the *EA/DE* is updated with the new information available from the real process and each of the individuals is re-evaluated taking this new knowledge into consideration.

5. the normal process of the *DE* or *EA* proceeds for a given number of iterations.
6. the best solution obtained is sent to the fermentation process and can be used in the real process.

5   Experiments

5.1   Setup

In this study, and given time and physical constraints, real fermentations were not conducted and instead these were replaced by simulating the fermentation process and adding noise to state variables. This process is implemented by considering two interacting components: an *optimizer*, that implements the optimization algorithm (*DE* or *EA*), and a *noise simulator (NS)*, that simulates the real fermentation process adding noise to the state variables.

This is performed by considering that there is a deviation between the model prediction and the behaviour of the process due to inaccuracies of the model. Therefore, for each sampling time, the state variables that represent the real process are obtained from the simulated variables by adding white noise. These new values of the state variables would originate a major deviation of the process from its optimal behavior, which had been defined during offline optimization. To partially compensate for this deviation, the new values. of the state variables will be used by the optimization algorithm to calculate a new feeding profile.

The following sequence of events takes place:

1. an offline optimization is performed by the *optimizer* and its results are passed on to the *NS*, used to compute the predicted values of the state variables. The *optimizer* stops and waits for new information.
2. the variable t, which stores the simulated time in the *NS* is set to $t = 0$.
3. while $t < T_f$ (where T_f denotes the final time of the fermentation process) the following steps are executed:
   (a) the values of all state variables at time t are disturbed by the *NS* by adding/ subtracting noise, given by the original value multiplied by a value taken from an uniform distribution with range $[0, U]$. The new values of the state variables are sent to the *optimizer*.
   (b) the *optimizer* receives this information and runs the steps for online optimization listed in the previous section. The best solution reached is sent to the *NS* that updates its model accordingly.
   (c) the *NS* updates $t = t + \Delta t$

Each run for the initial optimization is stopped after $100,000$ function evaluations. For the *DE*, the population size is set to 20, F was set to 0.5, CR to 0.6. In terms of the real-valued EA, the population size is set to 200. The value of Δt was set to 1 (h.) in case study I, 0.5 (d.) in II and 2 (h.) in III.

5.2 Results

The results will be presented in terms of the mean of the *PI* values obtained in 30 runs, as well as 95% confidence intervals. The Tables 1, 2 and 3 show the results of the algorithms obtained on case studies I, II and III, respectively. In every case, the first column represents the parameter U of the uniform representation used to generate noise (an increase in this parameter implies noisier setups). The next two columns show the results for the *DE* and the *EA* during offline optimization; columns 4 and 5 show the results obtained for the same algorithms, but applying the noise disturbances without changing the solutions of offline optimization (simulating the case where there are discrepancies between model predictions and real processes but without intervention of online optimizers) and, finally, the last two columns show the results obtained by the online optimization.

Table 1. Results obtained by the *DE* and *EAs* in case study I

U	Initial optim.		Initial+noise		Online opt.	
	DE	EA	DE	EA	DE	EA
0.01	9.47 ± 0.00	8.85 ± 0.04	4.67 ± 0.70	4.79 ± 0.73	9.11 ± 0.14	8.72 ± 0.14
0.02	9.47 ± 0.00	8.83 ± 0.05	4.41 ± 0.75	4.69 ± 0.78	8.80 ± 0.24	8.53 ± 0.25
0.03	9.47 ± 0.00	8.81 ± 0.05	4.20 ± 0.76	4.35 ± 0.81	8.47 ± 0.34	8.17 ± 0.35

Table 2. Results obtained by the *DE* and *EAs* in case study II

U	Initial optim.		Initial+noise		Online opt.	
	DE	EA	DE	EA	DE	EA
0.01	394.7 ± 0.2	386.3 ± 0.8	371.7 ± 8.5	367.9 ± 7.1	386.2 ± 4.8	379.8 ± 3.8
0.02	394.7 ± 0.2	385.2 ± 0.7	353.9 ± 14.9	351.2 ± 12.3	374.1 ± 9.2	371.8 ± 8.3
0.03	394.7 ± 0.2	386.1 ± 0.9	330.0 ± 23.5	343.0 ± 15.4	364.5 ± 13.0	367.6 ± 11.0

Table 3. Results obtained by the *DE* and *EAs* in case study III

U	Initial optim.		Initial+noise		Online opt.	
	DE	EA	DE	EA	DE	EA
0.01	20405 ± 4	20374 ± 9	20097 ± 133	20236 ± 108	20421 ± 115	20408 ± 119
0.02	20407 ± 3	20379 ± 7	19832 ± 305	19986 ± 244	20404 ± 243	20392 ± 242
0.03	20405 ± 5	20376 ± 9	19711 ± 357	19938 ± 393	20282 ± 317	20236 ± 335

The first conclusion to draw from the results in that, in every case study, even a low level of noise is enough to clearly disturb the results, although that effect is clearly more visible in case study I. In fact, the levels of noise studied are certainly within the range of the differences observed between model predictions and experimental results in biotechnological processes. However, the consequences in terms of process performance when an open-loop fermentation (without online optimization) is performed are quite extreme, implying that in many cases the utility of even relatively good models for process optimization with current state-of-the-art optimization techniques (mostly offline approaches) is quite low. Therefore, the results obtained with online optimization strategies indicate that the reward obtained in terms of process productivity is probably enough to justify its implementation and the corresponding costs.

In fact, the results obtained for all 3 case studies are quite close to the ones predicted by offline optimization without added noise, thus implying that the optimization scheme is robust to the levels of noise studied in this work. Furthermore, the degradation of the results that is caused by the increase of U is quite graceful, as an increase in U does not cause dramatic effects in the PI.

A comparison of the results obtained by both optimization algorithms show that DE seems to be more effective than the EAs. The difference is very clear when offline optimization is performed, but decreases when the level of noise increases. In fact, the differences for $U = 0.02$ and 0.03 are not significant from a statistical perspective and in case study II, the EA even outperforms the DE for $U = 0.03$. Nevertheless, if an alternative has to be chosen the DE still has an advantage, since it shows the best results (mean) in almost all scenarios.

6   Conclusions and Further Work

In this work, the task of optimizing feed profiles for fed-batch fermentation problems was approached by proposing optimization algorithms, such as EAs and DE, that are able to implement online optimization strategies, i.e., to perform the optimization simultaneously with the real process. The proposed approach was validated by conducting a number of experiments that used a noise simulator to emulate the differences between the values predicted by the mathematical model and the real values in the fermentation project. The results of the experiments show that even small differences lead to important disruptions in the behavior that was predicted by offline optimization.

The proposed approach to online optimization deals well with the noise and exhibits properties of graceful degradation. When comparing the optimization algorithms, the DE seems the best alternative, but its superiority that seems to decrease when noisier settings are considered.

In future work, the priority is to validate these results by implementing this approach with a real fed-batch fermentation process. Furthermore, other case studies will be tested and distinct optimization algorithms will be taken into account.

Acknowledgment

This work was supported by the Portuguese Foundation for Science and Technology under project POSC/EIA/59899/2004, partially funded by FEDER.

References

1. P. Angelov and R. Guthke. A Genetic-Algorithm-based Approach to Optimization of Bioprocesses Described by Fuzzy Rules. *Bioprocess Engin.*, 16:299–303, 1997.
2. Ascher, Ruuth, and Spiteri. Implicit-explicit runge-kutta methods for time-dependent partial differential equations. *Applied Numerical Mathematics*, 25:151–167, 1997.
3. A.E. Bryson and Y.C. Ho. *Applied Optimal Control - Optimization, Estimation and Control*. Hemisphere Publication Company, New York, 1975.
4. C.T. Chen and C. Hwang. Optimal Control Computation for Differential-algebraic Process Systems with General Constraints. *Chemical Engineering Communications*, 97:9–26, 1990.
5. J.P. Chiou and F.S. Wang. Hybrid Method of Evolutionary Algorithms for Static and Dynamic Optimization Problems with Application to a Fed-batch Fermentation Process. *Computers & Chemical Engineering*, 23:1277–1291, 1999.
6. R. Mendes, I. Rocha, E. Ferreira, and M. Rocha. A comparison of algorithms for the optimization of fermentation processes. In *2006 IEEE Congress on Evolutionary Computation*, pages 7371–7378, Vancouver, BC, Canada, jul 2006.
7. Z. Michalewicz. *Genetic Algorithms + Data Structures = Evolution Programs*. Springer-Verlag, USA, third edition, 1996.
8. H. Moriyama and K. Shimizu. On-line Optimization of Culture Temperature for Ethanol Fermentation Using a Genetic Algorithm. *Journal Chemical Technology Biotechnology*, 66:217–222, 1996.
9. I. Rocha. *Model-based strategies for computer-aided operation of recombinant E. coli fermentation*. PhD thesis, Universidade do Minho, 2003.
10. I. Rocha and E.C. Ferreira. On-line Simultaneous Monitoring of Glucose and Acetate with FIA During High Cell Density Fermentation of Recombinant E. coli. *Analytica Chimica Acta*, 462(2):293–304, 2002.
11. M. Rocha, J. Neves, I. Rocha, and E. Ferreira. Evolutionary algorithms for optimal control in fed-batch fermentation processes. In G.Raidl et al., editor, *Proceedings of the Workshop on Evolutionary Bioinformatics - EvoWorkshops 2004, LNCS 3005*, pages pp.84–93. Springer, 2004.
12. J.A. Roubos, G. van Straten, and A.J. van Boxtel. An Evolutionary Strategy for Fed-batch Bioreactor Optimization: Concepts and Performance. *Journal of Biotechnology*, 67:173–187, 1999.
13. R. Storn and K. Price. Differential Evolution - a Simple and Efficient Heuristic for Global Optimization over Continuous Spaces. *Journal of Global Optimization*, 11:341–359, 1997.
14. V. van Breusegem and G. Bastin. Optimal Control of Biomass Growth in a Mixed Culture. *Biotechnology and Bioengineering*, 35:349–355, 1990.

The Role of a Priori Information in the Minimization of Contact Potentials by Means of Estimation of Distribution Algorithms

Roberto Santana, Pedro Larrañaga, and Jose A. Lozano

Department of Computer Science and Artificial Intelligence
University of the Basque Country, Donostia-San Sebastian, Spain
rsantana@si.ehu.es, pedro.larranaga@ehu.es, ja.lozano@ehu.es

Abstract. Directed search methods and probabilistic approaches have been used as two alternative ways for computational protein design. This paper presents a hybrid methodology that combines features from both approaches. Three estimation of distribution algorithms are applied to the solution of a protein design problem by minimization of contact potentials. The combination of probabilistic models able to represent probabilistic dependencies with the use of information about residues interactions in the protein contact graph is shown to improve the efficiency of search for the problems evaluated.

Keywords: estimation of distribution algorithm, protein design, energy minimization algorithms.

1 Introduction

The goal of protein design is to find sequences of aminoacids with desired structural and functional properties. The problem has been approached by the application of directed search methods which cast the search as an optimization method. The approach requires the definition of a simplified model of the proteins, a fitness function that associates a value to each solution according to its 'quality', and a search procedure to efficiently sample the search space. In the field of protein design, these methods have been called "directed approaches to protein design".

Another class of methods has been covered under the umbrella of "probabilistic approaches to protein design" [14]. They use site-specific aminoacid probabilities rather than specific sequences and are usually employed in domains where the information available about the problem is incomplete. Probabilistic approaches include the use of consensus sequences [8] to determine low energy sequences and other methods where the probabilities learned can be used to guide search algorithms.

In this paper, we present a different approach which is based on the use of estimation of distribution algorithms (EDAs) [7,9,13]. EDAs are evolutionary algorithms that construct an explicit probability model of a set of selected solutions.

E. Marchiori, J.H. Moore, and J.C. Rajapakse (Eds.): EvoBIO 2007, LNCS 4447, pp. 247–257, 2007.

EDAs have been used for protein structure prediction in simplified models [17], protein side chain placement [18] and *de novo* peptide design [2]. Their suitability to deal with protein problems is given by the incorporation of machine learning techniques in the construction of the models. These learning algorithms automatically extract relevant regularities and complex structural patterns shared by promising solutions. The information learned can be compactly stored in the probabilistic model, which is later used to guide the exploration of the search space. EDAs are also different from probabilistic approaches that use probabilities to bias the search (e.g. Monte Carlo based techniques [21]) and where probabilities are unchanged during the search.

The paper is organized as follows. In the following section we present the energy function and introduce the problem of finding the aminoacid sequence with the lowest energy. Section 3 describes the main characteristics of EDAs and introduces the EDAs based on tree models used in our application. Section 4 gives a description of the experimental framework. The numerical results are shown in Section 5. Section 6 presents the main conclusions of our work and discuss future work.

2   Approach to Protein Design: Finding the Sequence with the Lowest Energy

In this section, we introduce the problem of finding the aminoacid sequence with the lowest energy for a given energy function. We use X_i to represent a discrete random variable. A possible value of X_i is denoted x_i. Similarly, we use $\mathbf{X} = (X_1, \ldots, X_n)$ to represent an n-dimensional random variable and $\mathbf{x} = (x_1, \ldots, x_n)$ to represent one of its possible values.

We will approach the protein design problem following a strategy that is based on the optimization of contact functions. Contact potentials or scoring functions [19,20] measure how likely it is for a sequence to fold to a given structure. Although the potential functions have been mainly used to distinguish native from decoy structures [19,20], they can also be employed to study the distribution of native-like features in sequence space [11].

In [10,11], the sequence evolutionary selection mechanisms are analyzed focusing on the stability energy of sequences. Although the 'survival probability' of a protein sequence depends on a number of other factors such as protein function and protein flexibility, the sequence-structure relationship can be analyzed in terms of energy. The analysis assumes that native sequences were selected because they were highly probable as a function of energy.

We will denote the native sequence corresponding to the structure σ as $\mathbf{x}^\sigma$. $E(\mathbf{x}, \sigma)$ is the energy of sequence $\mathbf{x}$ in structure σ and $E_\sigma = E(\mathbf{x}^\sigma)$ is the native energy of sequence $\mathbf{x}^\sigma$ in structure σ. The quantity $N(E_\sigma)$ is the number of sequences whose energy in σ would be no greater than that of the actual native sequence. $N(E_\sigma)$ is called the *evolutionary capacity* of structure σ because it reflects how far the current state of molecular evolution σ is from the possible optimum in terms of energy [11].

Given a protein structure σ and an energy function defined on the space of aminoacid sequences with cardinality 20^n, where n is the number of aminoacids, we address the problem of finding the lowest energy sequence among the 20^n possible solutions.

2.1   Energy Function

The $TE13$ potential function was introduced in [19] to correctly rate the native structure in relation to a set of decoy structures. This potential function was calculated using a linear programming approach.

Before presenting the function, let us to introduce some notation. $u_{\alpha\beta}(r)$ will denote a step potential between a pair of aminoacids α and β, and $p_{\alpha\beta}$ its asociated parameter calculated from solving linear inequalities. The distance between the geometric centers of two aminoacid side chains, r, is divided into 13 steps between 2 and 9 $\mathring{A}$. The first step along r is between 2 and 3 $\mathring{A}$, and the rest of the 12 steps are 0.5 $\mathring{A}$ each. Each of the $u_{\alpha\beta}(r)$ (as a function of the index β) is 1 only at one of the windows (steps) and zero elsewhere.

The total potential energy is:

$$E(\mathbf{x}, \sigma) = \sum_{\alpha,\beta} p_{\alpha\beta} n_\alpha u_{\alpha\beta}(r) \equiv \sum_\lambda p_\lambda n_\lambda \tag{1}$$

where index α parameterizes the type of the two interacting aminoacids. n_α is the number of contacts of a specific type found when threading the complete sequence $\mathbf{x}$ into the known shape σ. n_α and $u_{\alpha\beta}(r)$ are combined together to form n_λ, the number of contacts of a specific type and at a specific distance, λ, of structure σ. For each of the λ-s, there is a corresponding independent parameter p_λ calculated from solving linear inequalities. The total number of parameters is $((21 \times 20)/2) \times 13 = 2730$.

$TE13$ is defined for the distances between the side chain centers that have to be given. Therefore, the side chain center has to be given and from this information we calculate the distance r for each pair of residues. We construct the contact graph of each protein considering the existence of edges between two vertices if the corresponding residues have contact distances below $9\mathring{A}$.

3   Estimation of Distribution Algorithms

We will work with positive probability distributions denoted by $p(\mathbf{x})$. Similarly, $p(\mathbf{x}_S)$ will denote the marginal probability distribution for $\mathbf{X}_S$, where $S \subset \{1, \ldots, n\}$.

In EDAs, each individual represents one possible solution and it is encoded using the vector representation introduced above. A key characteristic and crucial step of EDAs is the construction of the probabilistic model.

The simplest EDA is the univariate marginal distribution algorithm (UMDA) which uses a probabilistic model where all variables are considered independent. The probabilistic model used by UMDA is described by Equation (2).

$$p_{UMDA}(\mathbf{x}) = \prod_{i=1}^{n} p(x_i) \tag{2}$$

The probability of each solution is equal to the product of the variables' univariate probabilities.

In this paper, we will apply UMDA to the minimization of contact potentials. We also propose the application to the protein design problem of a model that captures bivariate dependencies between the variables. This probabilistic model is based on a tree where each variable may depend on at most another variable that is called the parent.

A probability distribution $p_{Tree}(\mathbf{x})$ that is conformal with a tree is defined as:

$$p_{Tree}(\mathbf{x}) = \prod_{i=1}^{n} p(x_i|pa(x_i)) \tag{3}$$

where $pa(x_i)$ denotes a configuration of $Pa(X_i)$, the parent of X_i in the tree, and $p(x_i|pa(x_i)) = p(x_i)$ when $Pa(X_i) = \emptyset$, i.e. X_i is the root of the tree. The distribution $p_{Tree}(\mathbf{x})$ itself will be called a tree model when no confusion is possible. Probabilistic trees are represented by acyclic connected graphs.

The construction of the tree structure from data implies the detection of the most important bivariate interactions between the variables. This can be done applying statistical independence tests [15] or methods based on the analysis of the mutual information between variables as in [1]. We follow the second approach. The pseudocode of the tree-based EDA (Tree-EDA) is shown in Algorithm 1.

Algorithm 1: **Tree-EDA**

---

1 $D_0 \leftarrow$ Generate M individuals randomly

2 $l = 1$

3 **do** {

4 $D_{l-1}^s \leftarrow$ Select $N \leq M$ individuals from D_{l-1} according to a selection method

5   Compute the univariate and bivariate marginal frequencies $p_i^s(x_i|D_{l-1}^s)$ and $p_{i,j}^s(x_i, x_j|D_{l-1}^s)$ of D_{l-1}^s

6   Calculate the matrix of mutual information using univariate and bivariate marginals

7   Calculate the maximum weight spanning tree from the matrix of mutual information

8   Compute the parameters of the model

9 $D_l \leftarrow$ Sample M individuals (the new population) from the tree

10 } **until** A stop criterion is met

As presented in Algorithm 1, the bivariate probabilities are initially calculated for every pair of variables. From these bivariate probabilities, the mutual information between variables is found. To construct the tree structure, an algorithm introduced in [3], that calculates the maximum weight spanning tree from the matrix of mutual information between pairs of variables, is used. To sample the solutions from the tree, we have used probabilistic logic sampling (PLS) [5].

One problem faced by the algorithms that learn probabilistic models within EDAs is the arousal of spurious correlations between variables. This is due, among other factors, to the small datasets used to estimate the probabilities. Spurious correlations deteriorate the accuracy of the models and negatively influence the efficiency of the search. Our initial experiments showed that the quality of the learned trees could be improved when the interactions represented in the tree structure were constrained to those between residues that are making contacts in the contact graph of the protein as it was defined in Section 2.1. Therefore, one variant of the tree learning algorithm constrains the calculation of bivariate probabilities and mutual information to those pairs of variables corresponding to residues that are in contact in the contact graph. The variant of Tree-EDA that restricts the interactions represented by the tree structure to interacting pairs of variables is called Tree-EDA^r.

Tree-EDA^r can be considered as an example of the class of EDAs that use a priori problem information in order to improve the search. However, the information about the structure does not completely determine the final structure of the model as it is common in other EDAs that employ a priori problem information [12,16]. Instead, the mutual information between variables in the current selected population is taken into account to define the final structure of the probabilistic model.

4   Experiments

The objectives of our experiments are:

- To determine the ability of the probability model used by EDAs to capture relevant features of the protein design problem considered.
- To evaluate the capacity of UMDA, Tree-EDA and Tree-EDA^r to solve the energy minimization problem.
- To establish the influence of using information about the contact graph of the protein in the quality of the solutions obtained.

To search sequences with the lowest energy, we selected a set of 61 protein instances[1] from an initial set of 3901 protein instances[2]. This is a reduced and non-redundant set of protein shapes used for fold recognition. It is a good representative of the known folds of the protein databank [10]. For each protein, information about the side chain geometric centers of the protein is available[3].

---

[1] The list of the selected sequences is available from http://www.sc.ehu.es/ccwbayes/ EDA/EDAProteinProblems.html

[2] These instances have been obtained from Prof. Leonid Meyerguz's page: http://www.cs.cornell.edu/~leonidm/counting/protein_list.txt

[3] http://www.cs.cornell.edu/~leonidm/counting/pdb_sample.tar

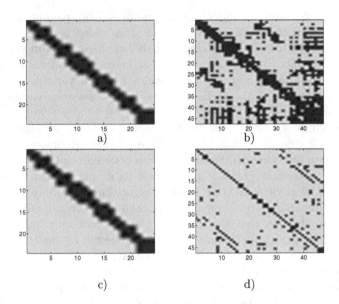

Fig. 1. Contact matrices of proteins pdb4clg-A (a) and pdb1aoo (b). Most frequent interactions found by Tree-EDA in the first generation for proteins pdb4clg-A (c) and pdb1aoo (d).

4.1 Parameters of the Algorithms

The parameters of the EDAs have been set as follows. The population size was set at 5000. The maximum number of generations is 500. Truncation selection with parameter $T = 0.15$ has been used. In this selection scheme, the best $T \cdot N$ individuals of the population are selected to construct the probabilistic model. We apply a replacement strategy called best elitism in which the selected population at generation t is incorporated into the population of generation $t + 1$, keeping the best individuals found so far and avoiding to revaluate their fitness function. The algorithm stops when the maximum number of generations is reached or the selected population has become too homogeneous (no more than 10 different individuals).

4.2 Design of the Experiments

To compare the results of the algorithms we conducted 50 experiments for each instance and algorithm. The performance of the algorithms was evaluated considering the fitness of the best solution found in each experiment, the best fitness among all the best solutions found, and the number of experiments in which the best fitness overall the 50 experiments was found.

To determine whether differences between the fitness of the solutions found by the algorithms are statistically significant the Kruskal-Wallis test [6] was employed. The test significance level was 0.05.

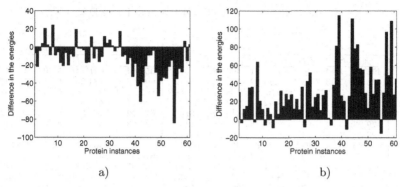

Fig. 2. Improvement in the energies of the solutions found by Tree-EDA and Tree-EDA^r with respect to those found by UMDA for the 61 protein instances. a) Difference between the average energies of Tree-EDA and UMDA. b) Difference between the average energies of Tree-EDA^r and UMDA.

5   Numerical Results

As an initial experiment, we investigated if there was any mapping between the statistical dependencies learned by the tree learning algorithm and the structure of the problem. We stored the tree structures learned in the first generation of Tree-EDA for protein instances pdb4clg-A and pdb1aoo (see footnote 1). The algorithm was run 1000 times for each instance. We counted the number of times each edge appeared in the trees. Figure 1a) and 1b) show the contact matrices constructed for edges that were in at least 50 of the 1000 trees learned. The similarity between the structures learned and the original contact matrices of the graphs (Figures 1c) and 1d)) is remarkable.

The decision of analyzing only the structures learned in the first generation was motivated by the fact that, as the EDA evolves, diversity in the population is lost and spurious correlations due to sample arise. The possibility of capturing many of the interactions that are in the original contact graph shows that EDAs are able to recover problem information from the protein structure from a statistical analysis of the data generated along the evolution.

We evaluated the performance of EDAs to sample the space of low energy sequences for function $TE13$. We do not have information about the actual lowest energy sequences. Therefore, the comparison between algorithms can be done only in relative terms. UMDA, Tree-EDA and Tree-EDA^r were run on the 61 instances of the protein benchmark. In Table 1, the results of Tree-EDA and Tree-EDA^r for each of the instances are presented. The table shows the best value of the fitness[4] found in all the experiments $(-f)$, the number of times the best solution has been found (S) and the average fitness $(\bar{f})$ of the solutions. It can seen from the table that, in terms of best solution found and the average

---

[4] Original fitness values are negative, however notice that the table actually shows their opposite values.

Table 1. Results of the Tree-EDA and Tree-EDA^r

pdb	Tree-EDA $-f$	S	$-f$	Tree-EDA^r $-f$	S	$-f$	pdb	Tree-EDA $-f$	S	$-f$	Tree-EDA^r $-f$	S	$-f$
1afp	1493.46	1	1458.46	1494.83	1	1468.78	1zwd	658.84	1	621.14	663.07	2	642.17
1apq	1426.39	1	1367.53	1442.43	2	1405.37	2mrb	806.87	1	784.90	808.44	4	797.63
1bba	800.31	7	775.36	800.31	20	788.67	2pta	882.57	1	837.70	903.44	1	867.30
1bh4	758.44	3	734.41	759.42	4	743.75	3ins-A	446.54	2	418.86	446.54	12	428.22
1bkv-A	425.23	25	411.48	425.23	56	419.71	3znf	679.72	1	668.07	682.03	7	670.64
1bqf-A	479.03	5	458.60	479.03	18	466.56	8tfv-A	368.23	80	367.04	368.23	85	367.42
1btd-A	703.33	1	678.32	726.62	1	694.10	1aoj-A	1252.12	1	1163.01	1258.01	1	1226.78
1ciq-B	452.90	45	449.99	452.90	62	451.01	1b7d-A	1750.34	1	1666.33	1822.28	1	1743.25
1clv-I	926.78	1	892.70	933.72	4	913.21	1bbr-H	3615.33	1	3458.34	3947.21	1	3798.36
1dec	947.68	1	907.02	955.11	3	924.31	1bf0	1665.05	1	1596.53	1711.79	1	1643.32
1dfn-A	695.52	1	668.45	703.26	2	676.09	1cdq	2122.12	1	2043.20	2192.00	1	2100.08
1eiu	907.32	6	898.07	908.19	1	903.53	1doy	3324.07	1	3266.49	3373.96	1	3325.90
1gnf	1167.75	1	1116.09	1172.37	1	1147.62	1ehd-A	3243.84	1	3159.73	3282.18	1	3236.33
1gps	1323.88	1	1284.45	1329.07	1	1301.12	1eo0-A	2242.45	1	2070.38	2389.50	1	2227.81
1iva	1300.78	1	1282.35	1301.56	3	1294.18	1fxr-A	1924.04	1	1831.44	1988.22	1	1910.87
1ktx	1002.84	1	941.52	1016.46	1	960.75	1gam-A	2356.16	1	2266.02	2465.97	1	2386.30
1mct-I	766.62	1	745.84	771.84	2	757.00	1gat-A	1495.89	1	1401.05	1513.75	1	1457.24
1mea	691.47	2	664.44	691.47	7	679.30	1hd0-A	2282.86	1	2179.34	2331.39	1	2252.34
1mhu	907.43	1	880.05	907.43	4	889.83	1if1-A	2984.11	1	2840.86	3051.07	1	2993.47
1myn	1289.52	1	1230.85	1299.16	1	1263.31	1imp	2478.64	1	2374.79	2513.36	1	2458.20
1pnh	687.83	1	669.68	784.83	1	691.89	1ivl-A	3338.91	1	3195.89	3470.71	1	3407.85
1pyc	1101.69	1	1050.11	1123.14	1	1083.41	1kst	2166.32	1	2080.83	2203.51	1	2138.72
1qdp	1104.75	1	1060.80	1114.81	2	1083.88	1nra	1888.56	1	1806.72	1914.91	1	1863.79
1qfn-B	338.55	28	330.84	338.55	65	335.45	1pba	2154.58	1	2033.58	2199.66	1	2112.74
1qk6-A	816.63	1	780.12	817.42	2	797.95	1qd9-A	3929.16	1	3762.24	4026.28	1	3952.83
1res	1038.58	1	981.14	1069.92	1	1026.72	1vfy-A	1857.82	1	1797.28	1912.55	1	1844.27
1roo	975.74	32	969.57	977.36	1	971.67	1whf	2443.19	1	2293.68	2487.02	1	2424.10
1sh1	1405.55	1	1348.21	1436.10	1	1386.09	2hgf	2923.62	1	2848.07	3046.69	1	2960.59
1sp2	673.46	1	608.36	673.46	18	652.37	2r63	1829.48	1	1678.30	1912.22	1	1802.58
1ter	528.31	4	503.82	528.31	26	513.96	4mt2	1864.31	1	1809.06	1884.13	2	1849.47
5cro	1527.11	1	1449.72	1605.88	1	1516.00							

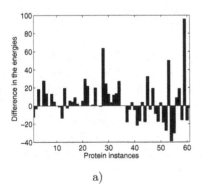

a)

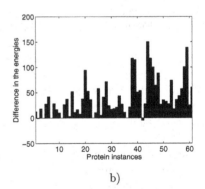

b)

Fig. 3. Improvement in the energies of the solutions found by Tree-EDA and Tree-EDA^r with respect to those found by UMDA for the 61 protein instances. a) Difference between the energies of the best solutions overall run found by Tree-EDA and UMDA. b) Difference between the energies of the best solutions found by Tree-EDA^r and UMDA.

fitness, Tree-EDA^r consistently finds better results than Tree-EDA. However, for most of instances, the best solution is found only once.

Figures 2 and 3 show the improvements achieved by Tree-EDA and Tree-EDA^r with respect to UMDA. Negative values mean that UMDA outperformed the other algorithm. Figures 2 a), b) show the improvement taking into account the average energies calculated from the best solution found in each of the 50 experiments conducted. In Figure 3 a), b), the improvement was calculated considering the absolute best solution found among the 50 experiments.

The application of the Kruskal-Wallis test found significant statistical differences between UMDA and Tree-EDA for 24 of the 61 instances. For 19 of these instances, UMDA was better than Tree-EDA. Significant statistical differences between UMDA and Tree-EDA^r were found for 52 of the 61 instances. For 49 of these instances, Tree-EDA^r outperformed UMDA. The statistical tests confirmed what can be seen from the figures: Considering the average energy, Tree-EDA was not able to improve results found by UMDA. For the best solutions, Tree-EDA found better solutions than UMDA in only 32 of the 61 instances. Nevertheless, the results of the statistical tests showed that the performance of Tree-EDA^r was consistently and clearly superior to UMDA both in average and best solutions. This example shows that the use of problem information can be critical for successful application of EDAs that consider interactions. The failure of Tree-EDA to improve (on average) results achieved by UMDA may be due to a small population size, insufficient to accurately compute the statistics needed to detect the correct interactions.

6   Conclusions and Future Work

EDAs belong to a new class of stochastic optimization methods that are able to capture, by the application of machine learning techniques, relevant features about the problem domain. The results presented in this paper show how solutions achieved in protein design problems can be improved by the use of probabilistic models able to represent interactions between the variables. This general procedure can be applied to other problems in protein design. Furthermore, as it has been shown in this paper, the information stored in the probabilistic models learned during the search could be useful to reveal important characteristics about the problem domain.

As a future research trend we envision the application of EDAs to more complex models used for protein design. In particular, we consider EDAs could be applied to the solution of the protein design problem using backbone dependent rotamer libraries [4] similarly to the way they have been used for protein design in [18].

Acknowledgements

The authors thank Leonid Meyerguz for providing the set of instances used in our experiments. This work was supported by the SAIOTEK-Autoinmune (II)

2006 and Etortek research projects from the Basque Government. It has been also supported by the Spanish Ministerio de Ciencia y Tecnología under grant TIN 2005-03824.

References

1. S. Baluja and S. Davies. Using optimal dependency-trees for combinatorial optimization: Learning the structure of the search space. In *Proceedings of the 14th International Conference on Machine Learning*, pages 30–38. Morgan Kaufmann, 1997.
2. I. Belda, S. Madurga, X. Llorá, M. Martinell, T. Tarragó, M. Piqueras, E. Nicolás, and E. Giralt. ENPDA: An evolutionary structure-based de novo peptide design algorithm. *Journal of Computer-Aided Molecular Design*, 19(8):585–601, 2005.
3. C. K. Chow and C. N. Liu. Approximating discrete probability distributions with dependence trees. *IEEE Transactions on Information Theory*, 14(3):462–467, 1968.
4. R. L. Dunbrack. Rotamer libraries in the 21st century. *Current Opinion in Structural Biology*, 12:431–440, 2002.
5. M. Henrion. Propagating uncertainty in Bayesian networks by probabilistic logic sampling. In J. F. Lemmer and L. N. Kanal, editors, *Proceedings of the Second Annual Conference on Uncertainty in Artificial Intelligence*, pages 149–164. Elsevier, 1988.
6. J. C. Hsu. *Multiple Comparisons: Theory and Methods*. Chapman and Hall, 1996.
7. P. Larrañaga and J. A. Lozano, editors. *Estimation of Distribution Algorithms. A New Tool for Evolutionary Computation*. Kluwer Academic Publishers, Boston/Dordrecht/London, 2002.
8. M. Lehmann, D. Kostrewa, M. Wyss, R. Brugger, A. D'Arcy, L. Pasamontes, and A. van Loon. From DNA sequence to improved functionality: Using protein sequence comparisons to rapidly design a thermostable consensus phytase. *Protein Engineering*, 13:49–57, 2000.
9. J. A. Lozano, P. Larrañaga, I. Inza, and E. Bengoetxea, editors. *Towards a New Evolutionary Computation: Advances on Estimation of Distribution Algorithms*. Springer-Verlag, 2006.
10. L. Meyerguz, C. Grasso, J. Kleinberg, and R. Elber. Computational analysis of sequence selection mechanisms. *Structure*, 12(4):547–557, 2004.
11. L. Meyerguz, D. Kempe, J. Kleinberg, and R. Elber. The evolutionary capacity of protein structures. In *Proceedings of the Eighth Annual International Conference on Research in Computational Molecular Biology*, pages 290–297, San Diego, California, 2004. Morgan Kaufmann Publishers, San Francisco, CA.
12. H. Mühlenbein, T. Mahnig, and A. Ochoa. Schemata, distributions and graphical models in evolutionary optimization. *Journal of Heuristics*, 5(2):213–247, 1999.
13. H. Mühlenbein and G. Paaß. From recombination of genes to the estimation of distributions I. Binary parameters. In *Parallel Problem Solving from Nature - PPSN IV*, pages 178–187, Berlin, 1996. Springer Verlag. LNCS 1141.
14. S. Park, H. Kono, W. Wang, E. T. Boder, and J. G. Saven. Progress in the development and application of computational methods for probabilistic protein design. *Computers and Chemical Engineering*, 29:407–421, 2005.
15. M. Pelikan and H. Mühlenbein. The bivariate marginal distribution algorithm. In R. Roy, T. Furuhashi, and P. Chawdhry, editors, *Advances in Soft Computing - Engineering Design and Manufacturing*, pages 521–535, London, 1999. Springer-Verlag.

16. R. Santana, E. P. de León, and A. Ochoa. The edge incident model. In *Proceedings of the Second Symposium on Artificial Intelligence (CIMAF-99)*, pages 352–359, Habana, Cuba, March 1999.

17. R. Santana, P. Larrañaga, and J. A. Lozano. Protein folding in 2-dimensional lattices with estimation of distribution algorithms. In *Proceedings of the First International Symposium on Biological and Medical Data Analysis*, volume 3337 of *Lecture Notes in Computer Science*, pages 388–398, Barcelona, 2004. Springer Verlag.

18. R. Santana, P. Larrañaga, and J. A. Lozano. Side chain placement using estimation of distribution algorithms. *Artificial Intelligence in Medicine*, 39(1):49–63, 2006.

19. D. Tobi and R. Elber. Distance-dependent, pair potential for protein folding: Results from linear optimization. *Proteins*, 41(1):40–46, 2000.

20. J. Zhu, Q. Zhu, Y. Shi, and H. Liu. How well can we predict native contacts in proteins based on decoy structures and their energies? *Proteins: Structure, Function, and Genetics*, 52(4):598–608, 2003.

21. J. Zou and J. G. Saven. Using self-consistent fields to bias Monte Carlo methods with applications to designing and sampling protein sequences. *The Journal of Chemical Physics*, 118(8):3843–3854, 2003.

Classification of Cell Fates with Support Vector Machine Learning

Ofer M. Shir[1,*], Vered Raz[2,*], Roeland W. Dirks[2], and Thomas Bäck[1,**]

[1] Natural Computing Group, Leiden University
Niels Bohrweg 1, 2333 CA Leiden, The Netherlands
{oshir, baeck}@liacs.nl
http://natcomp.liacs.nl
[2] Department of Molecular Cell Biology,
Leiden University Medical Center
2300 RC Leiden, The Netherlands
{v.raz, r.w.dirks}@lumc.nl

Abstract. In human mesenchymal stem cells the envelope surrounding the nucleus, as visualized by the nuclear lamina, has a round and flat shape. The lamina structure is considerably deformed after activation of cell death (apoptosis). The spatial organization of the lamina is the initial structural change found after activation of the apoptotic pathway, therefore can be used as a marker to identify cells activated for apoptosis. Here we investigated whether the spatial changes in lamina spatial organization can be recognized by machine learning algorithms to classify normal and apoptotic cells. Classical machine learning algorithms were applied to classification of $3D$ image sections of nuclear lamina proteins, taken from normal and apoptotic cells. We found that the Evolutionary-optimized Support Vector Machine (SVM) algorithm succeeded in the classification of normal and apoptotic cells in a highly satisfying result.

This is the first time that cells are classified based on lamina spatial organization using the machine learning approach. We suggest that this approach can be used for diagnostic applications to classify normal and apoptotic cells.

1 Introduction

The nuclear envelope separates nuclear and cytoplasmic compartments in metazoan cells. The nuclear envelope is composed of outer and inner nuclear membranes and the nuclear lamina, which is connected to the inner nuclear membrane. The nuclear lamina is a filamentous protein network that gives the nucleus structure. Two major types of lamin proteins can be distinguished. B-type lamins are ubiquitously expressed and essential for cell viability. The expression of A-type lamins is developmentally regulated, and mutations in the human genes result in

---

* These authors contributed equally.
** NuTech Solutions, Martin-Schmeisser-Weg 15, 44227 Dortmund, Germany.

E. Marchiori, J.H. Moore, and J.C. Rajapakse (Eds.): EvoBIO 2007, LNCS 4447, pp. 258–269, 2007.

a variety of hereditary diseases with premature aging syndromes [1]. From the dynamic interactions between the lamina and chromatin domains, an active role of the nuclear lamina in gene regulation and chromatin organization was suggested [1][2][3]. Thus, the organization of the nuclear lamina can contribute to cellular function.

In *human mesenchemyal stem cells* (hMSCs) the nuclear lamina shows a round and flat shape after $3D$ image reconstruction. This distinct nuclear lamina shape is dramatically changed after caspase-8 activation and cell death. Activation of the caspase-8 pathway leads to an increase in intranuclear organization of the nuclear lamina, followed by massive degradation of the lamina proteins [4]. Intranuclear structures of the nuclear lamina were observed in interphase cells [5][6], and senescent cells [7], but the function of lamina reorganization is yet unclear. An increase in lamina intranuclear structures is one of the first events in nuclear remodeling after caspase-8 activation, and it precede the wholemarks of apoptosis, such as chromatin fragmentation and nuclear breakdown [4]. As the spatial organization of the nuclear lamina is initially changed when cells undergo deactivation, such as senescence or apoptosis, it can serve as a marker to distinguish between an active or non-active cells.

Here we considered classical machine-learning algorithms as a tool to classify nuclear lamina organization of normal and caspase-8-activated cells. The emphasis is on exploring the feasibility of the task, rather than comparing state-of-the-art classification algorithms. After the appropriate training, combined with an evolutionary optimization, we show that the Support Vector Machines (SVM) algorithm resulted in highly-satisfying classification.

Related Work. Machine learning algorithms have been widely used in biological applications for the tasks of pattern-recognition, classification, prediction and others. However, the application of machine learning techniques to the task of classifying $3D$ images has been very limited, as reported so far. A recent study [8], which originates in the field of brain imaging (fMRI), claimed to be the first research to apply machine learning techniques to a clinical diagnosis. This study reported successful classification of drug addicted human subjects and controls, based on $3D$ brain images, using state-of-art classifiers in combination with novel boosting algorithm. This study shows the use of image classification for clinical diagnosis.

The remainder of this paper is organized as follows: in section 2 we introduce the background for online learning and the algorithms in use. Section 3 provides the reader with technical details of the images acquisition and the data-instances construction. In section 4 the numerical results are presented and analyzed, and section 5 provides discussion and concludes this study.

2   Algorithms

We discuss briefly the basic concepts of *online learning* and provide an overview of the algorithms used. The reader should note that we limit ourselves, at this stage, only to the classical machine learning algorithms, aiming to obtain satisfying experimental results with respect to the given classification task. We prefer the classical algorithms in this study, as they have a full theoretical analysis behind them, lacking in most state-of-the-art algorithms, in particular *neural networks* [9]. This choice would only be valid so long as the selected algorithms deliver and accomplish the given task, and indeed, as will be shown here, they did so and there was no need to use other algorithms.

2.1   Online Supervised Learning

Online supervised learning considers a situation in which instances are given one at a time, in an incremental fashion, and where the learner's task is to learn a hypothesis which classifies the data correctly. Given instances $\{x_i\}_{i=1}^{l}$ in $\mathbb{R}^n$, and their labels set $Y = \{-1, +1\}$, the learning algorithm aims to update its hypothesis $h : \mathbb{R}^n \rightarrow \{\pm 1\}$ in order to minimize the prediction error. The prediction error is simply defined as the total number of mistakes made on the training set. This is the so-called *training phase*. In the realizable case, i.e. where there exists such a hypothesis, the learner will aim to reach a point where there is an upper-bound on the number of mistakes. The training phase is followed by the *testing phase*, where more data is given to the learned hypothesis (ideally *unseen data*), and the accuracy rate is considered.

Statistical learning theory (also known as the Vapnik-Chervonenkis theory) shows explicitly that the class of hypotheses, which is considered by the learning algorithm, must be limited, and in particular adjusted to the size of the training set [10].

2.2   The Perceptron

The Perceptron algorithm (Rosenblatt, 1957) is an online learning algorithm for finding a consistent hypothesis within the class of hyperplanes [11]. It is a classical machine learning algorithm, which can also be considered as an engineering simplification of a simple neural network [9].

Given a sequence $(x^1, y^1), ..., (x^t, y^t), ...$ of instance-label pairs, where instances and labels satisfy $x^t \in \mathbb{R}^n$, $y^t \in \{-1, +1\}$, the Perceptron considers the hypotheses class that corresponds to all possible hyperplane separators:

$$C = \{h(x) = sign\left(w^T \cdot x + b\right) \quad w^t \in \mathbb{R}^n, \ b \in \mathbb{R}\} \tag{1}$$

The Perceptron algorithm maintains a single hyperplane, which it aims to learn, such that the margin between the two sets is maximized. The hyperplane used for prediction at round t is denoted by $\{w^t, b^t\}$, and it is modified only whenever there is a prediction mistake. The learning rule, in case of a prediction mistake, is then defined as:

$$w^{t+1} = w^t + y^t \cdot x^t \tag{2}$$

$$b^{t+1} = b^t - y^t \tag{3}$$

i.e., the hyperplane is adjusted in the direction of the instance which it predicted wrongly. It is guaranteed, by the so-called *Perceptron convergence theorem*, that given linearly-separable data, the Perceptron algorithm will converge into a hyperplane separator within a finite number of steps.

A pseudo-code of the Perceptron is given as Algo. 1.

Initialize: Set $w^1 = 0$, $b^1 = 0$; $w^1 \in \mathbb{R}^n$, $b^1 \in \mathbb{R}$
for $t = 1, 2, ..., l$ **do**
  Get a new instance $x^t \in \mathbb{R}^n$.
  Predict $\hat{y}^t = sign\left(w^t \cdot x^t + b^t\right)$.
  Get a new label y^t.
  if $y^t \neq \hat{y}^t$ **then**
    $w^{t+1} = w^t + y^t \cdot x^t$
    $b^{t+1} = b^t - y^t$
  else
    $w^{t+1} = w^t$
    $b^{t+1} = b^t$
  end
end
Output: $h(x) = sign\left(w^{l+1} \cdot x + b^{l+1}\right)$

Algorithm 1. Perceptron

The *optimal hyperplane* is defined as the one with the maximal margin of separation between the two classes. It should be noted that the problem of maximizing the margin is often considered, even at the cost of allowing some miss-classifications of instances.

2.3   Support Vector Machines and Kernel Functions

Given data which is not linearly-separable, there exists no hyperplane separator hypothesis for the problem, and hence the Perceptron in its original form is useless. However, by using *kernel functions*, the instance (input) space could be mapped onto higher dimensional spaces thereby obtaining non-linear decision boundaries in the original instance space.

The *Support Vector Machines* (SVM) algorithm (Boser, Guyon and Vapnik 1992; see, e.g., [12]) is a *linear* method in a high-dimensional feature space, which is non-linearly interlinked to the instance space, but does not involve any computations in that high-dimensional space. It is a rare combination of two important elements - a statistical learning tool with full theoretical analysis behind it, as well as a strong and robust practical tool for challenging real-world applications. Consider the problem of learning the *optimal hyperplane*, i.e. finding the maximal margin separating hyperplane with margin errors. This can be done by considering a quadratic optimization problem, with a solution

$\boldsymbol{w}^*$ which can be expanded by means of the training instances that lie on the margin:

$$\boldsymbol{w}^* = \sum_i \nu_i \boldsymbol{x}_i \tag{4}$$

Those particular instances carry the critical information concerning the classification problem, and they are called the **support vectors**.

Explicitly, the SVM considers the following optimization problem:

$$\min_{\boldsymbol{w},b,\xi} \quad \frac{1}{2}\boldsymbol{w}^T \cdot \boldsymbol{w} + c\sum_{i=1}^{l} \xi_i \tag{5}$$

subject to

$$y_i\left(\boldsymbol{w}^T \cdot \boldsymbol{x}_i - b\right) \le 1 - \xi_i, \quad \xi_i \ge 0 \tag{6}$$

where $c > 0$ is a penalty weighting factor, and the non-negative variables ξ_i allow data points to be miss-classified.

By deriving the *Lagrangian Dual* of this problem, the obtained *criteria function* involves only the **inner-products of the training instance vectors**. This is the key property which allows the mapping onto non-linear decision surfaces (the dual problem is not necessarily easier than the primal, but it is completely described by the inner-products of the instances).

The common principle of the kernel methods is to construct nonlinear variants of linear algorithms by substituting inner-products by non-linear kernel functions. The function $\phi : \mathbb{R}^n \to \mathbb{F}$ maps the instance vectors onto a higher dimensional space $\mathbb{F}$ (called also *feature space*), and then the SVM aims to find a hyperplane separator with the maximal margin in this space. As noted earlier, this only requires the evaluation of dot-products. Introduce the *kernel function*:

$$k\left(\boldsymbol{x}_i, \boldsymbol{x}_j\right) \equiv \phi(\boldsymbol{x}_i)^T \phi(\boldsymbol{x}_j) \tag{7}$$

In particular, consider the following kernel functions:

- The polynomial kernel:

$$k\left(\boldsymbol{x}_i, \boldsymbol{x}_j\right) = \left(\gamma\left(\boldsymbol{x}_i^T \cdot \boldsymbol{x}_j\right) + r\right)^d \tag{8}$$

- *Radial basis function* (RBF) kernel:

$$k\left(\boldsymbol{x}_i, \boldsymbol{x}_j\right) = \exp\left\{-\frac{1}{2\sigma^2}\|\boldsymbol{x}_i - \boldsymbol{x}_j\|^2\right\} \tag{9}$$

- The *sigmoid* kernel:

$$k\left(\boldsymbol{x}_i, \boldsymbol{x}_j\right) = \tanh\left(\kappa\left(\boldsymbol{x}_i^T \cdot \boldsymbol{x}_j\right) + \Theta\right) \tag{10}$$

Finally, the non-linear decision function will have the form:

$$f(\boldsymbol{x}) = sign\left(\phi\left(\boldsymbol{w}^*\right) \cdot \phi\left(\boldsymbol{x}\right) + b\right) = sign\left(\sum_i \nu_i \cdot k\left(\boldsymbol{x}, \boldsymbol{x}_i\right) + b\right) \tag{11}$$

where the variables ν_i are given as the solutions of the *dual problem*.

3   Applying Online Learning Classification

Visualization of the nuclear lamina was carried out using lamin B-GFP or lamin A-DsRed vectors, after transduction to hMSCs, as described in [4]. Z-stacks images were taken using confocal microscopy, $3D$ image reconstruction was carried out with TeloView [13]. Each $3D$ image was sliced uniformly into 20 X $2D$ images, the orientation of the slicing axis (xz or yz) was chosen with respect to the original orientation of the $3D$ image. For illustration see Fig. 1.

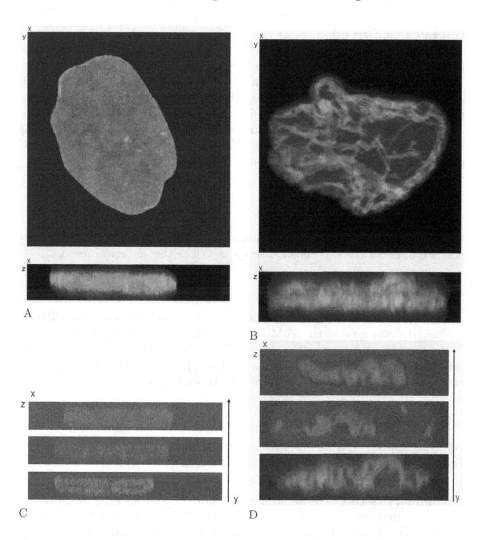

Fig. 1. (A) $3D$ reconstruction of a control cell. (B) $3D$ reconstruction of an apoptotic cell. (C) XZ slicing of a control cell. (D) XZ slicing of an apoptotic cell.

Data Acquisition. In normal cells, the $3D$ imaging revealed a round and flat shape of the lamina. After caspase-8 activation the nuclear lamina shape was significantly deformed. Serial slicing along the xy and xz orientations, taken from an individual nucleus with the DIPimage tool [14], revealed little changes in the spatial organization of the lamina in the normal cell. However, high variations were observed in the serial slicing which was taken from an apoptotic cell along the xz axis. As the nucleus has an ellipse-like shape, for image analysis the xz and the yz axis are indifferent.

Instances Construction. Since the obtained slices are already given as gray-scale matrices (pixel values are normalized within the interval $[0, 1]$), there was no need in image analysis or feature extraction, but only in the rescaling (reducing) the size of the image. Essentially, this means that the given image is introduced directly to the learning algorithm, as a matrix, without any image processing or component analysis. Concerning the final size of the image, there is clearly a trade-off between high resolution and the learning efficiency. Following a preliminary tuning process, the value of $n = 4096$ features per image ($128X32$ grayscale images) turned to be a good compromise.

4   Numerical Results

We have conducted a series of runs using the two algorithms. In this section we describe in detail the experimental setup as well as the numerical results.

4.1   Experimental Setup

The Perceptron algorithm was implemented by us in Matlab 7.0, and the **libsvm** package [15] was used as the SVM implementation (its Matlab interface).

We considered $N_{train} = 2000$ images in the training set, and $N_{test} = 1040$ images in the testing set. The slices were shuffled, and the images were selected randomly. **The testing set contained only *unseen* data.**

4.2   Applying the Perceptron

Applying the Perceptron was straightforward, with respect to parameter settings, and did not require any preliminary tuning.

Preliminary Setup. We would like to report that at an early stage of this study, we considered only *untreated cells* and *cells 6 hours after treatment* not yet considering treated-cells of any stage. For the classification task of those cells the Perceptron algorithm was applied, and obtained satisfying results - test accuracy of 82.2%, far better than we had expected. However, when the additional data was introduced to the algorithm, i.e. the treated-cells of other stages, the classification performance dramatically dropped.

It is possible that the images of early-stage treatment introduced some non-linearity to the classification problem, and hence the Perceptron failed to deliver. For illustration, consider Fig. 2, Fig. 3 and Fig. 4, as instances at different stages of treatment.

Fig. 2. A typical slice of a nucleus, taken from a normal untreated cell

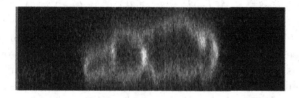

Fig. 3. A typical slice of a nucleus, taken from a cell treated for apoptosis: **4 hours after treatment**

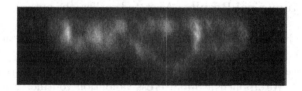

Fig. 4. A typical slice of a nucleus, taken from a cell treated for apoptosis: **6 hours after treatment**

Extended Data. The Perceptron algorithm obtained, after training, a *test accuracy* of 70.38% (732/1040 images were classified correctly). This result led us to the conclusion that the data was not linearly-separable, and a stronger approach was much needed.

4.3   Applying Support Vector Machines

The application of *support vector machines* to the classification problem, with its default settings, yielded poor results (test accuracy of 55% on average, which is only slightly better than using the *random hypothesis*). Thus, the task of tuning the kernel parameters was essential - the penalty weight c (Eq. 5), the profile of the kernel (Eq. 8, 9, 10) and its various appropriate parameters ($\{\gamma, r, d\}$, $\{\sigma\}$ and $\{\kappa, \Theta\}$, respectively).

Evolutionary Optimization. For this task we have chosen Evolution Strategies (ES) [16] as our optimization framework. In particular, we considered the *Covariance Matrix Adaptation* (CMA) Evolution Strategy [17] as the core mechanism. A short description follows. The CMA is a variant of ES that has been successful for treating correlations among object variables. This method tackles the critical element of Evolution Strategies, the adaptation of the mutation parameters. The fundamental property of this method is the exploitation of information obtained from previous successful mutation operations. Given an initial search point x^0, λ offspring are sampled from it by applying the mutation operator. The best search point out of those λ offspring is chosen to become the parent of the next generation. The action of the *mutation operator* for generating the λ samples of search points in generation $g + 1$ is defined as follows:

$$x^{g+1} \sim \mathcal{N}\left(x_k^{(g)}, \sigma^{(g)^2} \mathbf{C}^{(g)}\right), \qquad k = 1, ..., \lambda \qquad (12)$$

where $\mathcal{N}(m, \mathbf{C})$ denotes a normally distributed random vector with mean m and covariance matrix $\mathbf{C}$. The matrix $\mathbf{C}$, the crucial element of this process, is initialized as the *unity matrix* and is learned during the course of evolution, based on cumulative information of successful mutations (the so-called *evolution path*). The global step size, $\sigma^{(g)}$, is based on information from the *principal component analysis* of $\mathbf{C}^{(g)}$ (the so-called *"conjugate" evolution path*). We omit most of the details and refer the reader to Hansen and Ostermeier [17].

Setup and Numerical Results. As stated earlier, the optimization routine is performed on each set of parameters which uniquely defines each kernel, as well as on the penalty weight c. Hence, the dimension of the search space is $\{4, 2, 3\}$ for the Polynomial, RBF and Sigmoid cases, respectively.

The *cross-validation accuracy rate* was chosen as the objective function to be maximized. We applied a 5-fold cross-validation routine. **Each objective function evaluation has the duration of approximately 11 minutes on a Pentium-4 2.6GHz,** and therefore we chose to limit each run to 1000 function evaluations. Given this long evaluation time, we had planned to perform only 5 runs per kernel type, hoping to obtain satisfying results; every run has converged.

Table 1 concentrates the numerical results of the optimization runs - the maximal **cross-validation accuracy values** obtained by the CMA-ES optimization routine for the different kernel types, as well as the **testing-phase classification results** and their best accuracy values respectively of the different kernels. Due to the limited number of runs we do not provide further statistical analysis of the results.

All together, this time-consuming parameters' tuning optimization process obtained a solution with **test accuracy of 97.02% (1009/1040 images were classified correctly)** for the RBF kernel. We consider this numerical result as a highly satisfying one, and as remarkable assessment to the hypothesis that the given images could be classified correctly by a machine. Thus, we have not performed any further optimization.

Table 1. Best test-accuracy values obtained for the different kernels over five runs

Kernel Type	Cross-Validation	Correct Classification	Test-Accuracy Value
Polynomial	86.63%	910/1040	87.50%
RBF	98.90%	1009/1040	97.02%
Sigmoid	52.55%	569/1040	54.71%

Interestingly, the optimization routine found 2 different solutions with high accuracy rates, which are completely different with respect to the penalty term. The observed 2 best solutions, which are associated with the RBF kernel, are given at table 2.

Table 2. Best 2 solutions for the RBF kernel; Note the difference in the penalty terms

Solutions	Optimized σ	Optimized c	Test-Accuracy Value
Best Solution	0.0420	1452.3	97.02%
2nd-best Solution	0.0358	3.3	96.34%

Given this result, we have applied a grid-search of the penalty-term c, with a fixed $\sigma = 0.4$. This search obtained a flat curve, indicating that given such σ, the penalty term can be set arbitrarily in a range of 3 orders of magnitude - in order to yield high 5-fold cross-validation.

5   Discussion and Conclusions

The human hMSCs are pluripotent cells that have the potential to differentiate into various cell types. Therefore, they are attractive candidates for clinical use to treat systemic diseases as well as local tissue and organ defects [18]. Furthermore, they can be used as a vehicle for genes in gene therapy strategies. In clinical practice, however, the rate of success is not satisfying, which is in part due to the status of the transplanted cells. Senescent cells or cells that initiated apoptosis will not succeed after transplantation. Therefore, a quality control of the cells before clinical use may enhance the success rate of treatments with hMSCs. We have previously shown that changes in the nuclear lamina organization are the earliest structural change after activation of caspase-8 in hMSCs [4]. Therefore we aimed to develop a classification system, which is based on the nuclear lamina spatial organization, in order to differentiate between normal and early apoptotic cells. We found that applying a classical machine learning technique, the Support Vector Machines (SVM) method, was sufficient to distinguish between normal and early apoptotic cells. The performance of the Perceptron on the given learning task was limited due to the fact that the data is not linearly separable. The evolutionary-optimized SVM, however, outperformed it and succeeded in classification with highly satisfying results of 97% accuracy. Our choice to use only classical classifiers has been shown to be sufficient for our purposes, as it successfully tackled the given task. Moreover, it provided us with valuable insight concerning the nature of the data, in particular its being not linearly separable.

Our results here further support previous study [8], showing that machine learning techniques can be applied to the task of classifying $3D$ images. We show that machine learning algorithms can also be used to distinguish between different morphologic features of the nuclear lamina, thus classification of cell status can be determined by the learning machine without any human interference or bias. This allows a rapid quality control of hMSC cultures before clinical use.

Outlook. We have several future directions under consideration:

- Testing the classification of images in other cellular processes, where the changes in nuclear lamina spatial organization are less dramatic.
- Applying *unsupervised learning* to this classification problem, aiming to obtain new insights into the biological process.
- Due to the highly dynamic nature of the nuclear lamina, it would be mostly challenging to apply the machine learning to classification of images which are taken at different points of time.

Acknowledgments

Ofer Shir was supported by *FOM*, the Dutch Foundation for *Fundamental Research on Matter*. Vered Raz and Roeland Dirks were supported by the *Cyttron* research program, Bsik grant 03036.

References

1. Gruenbaum, Y., Margalit, A., Goldman, R., Shumaker, D., Wilson, K.: The nuclear lamina comes to an age. Nat Rev Mol Cell Biol **6** (2005) 21–31
2. Kosak, S., Groudine, M.: Gene order and dynamic domains. Science **306** (2004) 644–647
3. Pickersgill, H., Kalverda, B., de Wit, E., Talhout, W., Fornerod, M., van Steensel, B.: Characterization of the drosophila melanogaster genome at the nuclear lamina. Nature Genetics **38** (2006) 1005–1014
4. Raz, V., Carlotti, F., Vermolen, B., van der Poel, E., Sloos, W., Knaän-Shanzer, S., de Vries, A., Hoeben, R., Young, I., Tanke, H., Garini, Y., Dirks, R.: Changes in lamina structure are followed by spatial reorganization of heterochromatic regions in caspase-8-activated human mesenchymal stem cells. J. Cell Sci. **119** (2006) 4247–4256
5. Broers, J., Machiels, B., van Eys, G., Kuijpers, H., Manders, E., van Driel, R., Ramaekers, F.: Dynamics of the nuclear lamina as monitored by gfp-tagged a-type lamins. J. Cell Sci. **112** (1999) 3463–3475
6. Fricker, M., Hollinshead, M., White, N., Vaux, D.: Interphase nuclei of many mammalian cell types contain deep, dynamic, tubular membrane-bound invaginations of the nuclear envelope. J. Cell Biol **136** (1997) 531–544

7. Scaffidi, P., Misteli, T.: Lamin a-dependent nuclear defects in human aging. Science **312** (2006) 1059–1063

8. Zhang, L., Samaras, D., Tomasi, D., Volkow, N., Goldstein, R.: Machine learning for clinical diagnosis from functional magnetic resonance imaging. In: IEEE International Conference Computer Vision and Pattern Recognition. (2005) I:1211–1217

9. Hertz, J., Krogh, A., Palmer, R.G.: Introduction to the Theory of Neural Computation (Santa Fe Institute Studies in the Sciences of Complexity). ASIN (1991)

10. Vapnik, V.N.: The nature of statistical learning theory. Springer-Verlag New York, Inc., New York, NY, USA (1995)

11. Rosenblatt, F.: The perceptron: A perceiving and recognizing automaton. Report 85-460-1, Project PARA, Cornell Aeronautical Laboratory, Ithaca, New York (1957)

12. Boser, B., Guyon, I., Vapnik, V.: A training algorithm for optimal margin classifiers. In: Fifth Annual Workshop on Computational Learing Theory. (1992) 144–152

13. Vermolen, B., Garini, Y., Mai, S., Mougey, V., Fest, T., Chuang, T., Chuang, A., Wark, L., Young, I.: Characterizing the three-dimensional organization of telomeres. Cytometry **67A** (2005) 144–150

14. Hendriks, C.L.L., van Vliet, L.J., Rieger, B., van Ginkel, M., Ligteringen, R.: DIPimage: A Scientific Image Processing Toolbox for MATLAB. (2005) `http://www.ph.tn.tudelft.nl/DIPlib/`.

15. Chang, C.C., Lin, C.J.: LIBSVM: a library for support vector machines. (2001) Software available at `http://www.csie.ntu.edu.tw/~cjlin/libsvm`.

16. Bäck, T.: Evolutionary algorithms in theory and practice. Oxford University Press, New York, NY, USA (1996)

17. Hansen, N., Ostermeier, A.: Completely derandomized self-adaptation in evolution strategies. Evolutionary Computation **9**(2) (2001) 159–195

18. Pittenger, M., Mackay, A., Beck, S., Jaiswal, R., Douglas, R., Mosca, J., Moorman, M., Simonetti, D., Craig, S., Marshak, D.: Multilineage potential of adult human mesenchymal stem cells. Science **284** (1999) 143–147

Reconstructing Linear Gene Regulatory Networks

Jochen Supper, Christian Spieth, and Andreas Zell

Centre for Bioinformatics (ZBIT), University of Tübingen, Germany
jochen.supper@uni-tuebingen.de

Abstract. The ability to measure the transcriptional response after a stimulus has drawn much attention to the underlying gene regulatory networks. Here, we evaluate the application of methods to reconstruct gene regulatory networks by applying them to the SOS response of *E. coli*, the budding yeast cell cycle and *in silico* models. For each network we define an *a priori* validation network, where each interaction is justified by at least one publication. In addition to the existing methods, we propose a SVD based method (NSS). Overall, most reconstruction methods perform well on *in silico* data sets, both in terms of topological reconstruction and predictability. For biological data sets the application of reconstruction methods is suitable to predict the expression of genes, whereas the topological reconstruction is only satisfactory with steady-state measurements. Surprisingly, the performance measured on *in silico* data does not correspond with the performance measured on biological data.

1 Introduction

Measuring the transcriptional response of cellular processes under various conditions provides valuable insight into the global behavior. Such measurements can be analyzed by clustering, thereby providing undirected gene relations. In order to model gene regulatory networks (GRN), directed relations between genes have to be considered. Different approaches, describing how the expression of the genes is controlled, have been proposed. GRNs can be represented by interconnections between genes, each indicating that one gene influences the expression of another gene. The functionality associated with the gene-gene interconnections ranges from simple Boolean interactions to complex differential equations [1–3].

The properties of GRNs have been analyzed on a global scale [4] and recently also on the individual gene scale [5]. The global organization of GRNs seems to follow a power-law distribution, where most genes are controlled by a small number of genes and a few genes are highly connected. At this point it is important to stress that the global structure is a static property of the GRN, and cellular perturbations only activate a subset of this network. Therefore, expression profiles only cover a subset of all possible cellular conditions and thus provide

E. Marchiori, J.H. Moore, and J.C. Rajapakse (Eds.): EvoBIO 2007, LNCS 4447, pp. 270–279, 2007.
© Springer-Verlag Berlin Heidelberg 2007

only partial information about the underlying regulatory program. On a single gene scale quantitative characterization of the promoter activity has been investigated [6]. The common picture from these investigations is, that genes are activated at a certain threshold, then the dose-response curve progresses linearly until it saturates at its maximal fold-change. These investigations allow general questions regarding the proper conceptual and mathematical representation as well as the sufficient amount of detail required for modeling.

The major problem for the reconstruction of GRNs is that it is very hard to validate the hypothesized networks. This makes it difficult to compare methods and yet to judge if certain approaches are helpful at all. In previous publications, GRN models have been validated by co-citation [7], cross-validation [8] or on artificial data [9]. Despite these efforts, no topological validation on biological models, in which each interconnection is *a priori* justified by published work, has been assessed for different reconstruction methods. In this work, we present two biological networks that have been investigated thoroughly, with plenty of known interactions, as well as an *in silico* GRN model. In addition to the topological validation, we predict the gene expression by 5-fold cross-validation.

Several methods have been developed to reverse engineer genetic networks. Since the proper modeling terms of GRNs are not yet known, we avoid specific assumptions by using standard methods and reconstruction algorithms. Therefore, we apply linear regression, greedy search, exhaustive search and methods developed by Weaver *et al.*[10] and Someren *et al.*[11]. The description of these linear models is straightforward, whereas non-linear interactions can be modeled by applying a preprocessing step. Thereby, the gene-gene dependence can be altered, such that the interaction terms are non-linear and multiplicative. To investigate the influence of the preprocessing, we incorporate and evaluate different types of preprocessing.

In addition to comparing different reconstruction algorithms, we propose our own reconstruction method for GRNs, the Null Space Solver (NSS). The core algorithm was previously described by Supper *et al.*[9]. To reconstruct the GRN, we employ a heuristic based on Singular Value Decomposition (SVD). Different groups have previously used SVD to reconstruct GRN [3]. Our approach searches the solution space in an efficient way by exploiting the properties of the SVD. The search steps are probabilistic and provide an ensemble of networks, which we rank according to different criteria. Our first criteria is to choose the solution with the minimal number of interactions. If several such solutions exist, we rank these according to the 1-Norm.

Besides choosing the proper reconstruction method it is critical to obtain applicable data sets. Transcriptional response data is characterized by a large number of measured genes along with a small sampling rate. Consequently, the number of features is high compared to the number of sampling points, rendering the reconstruction of a GRN ambiguous. By restricting ourselves to sub-networks we relax this problem so that the number of measurements is in the range of the number of genes. We also perform a reconstruction on an additional data set containing 800 genes for which this relaxation is not possible.

2   Approach

2.1   Data Sets

We collect data sets from three different sources. The two biological data sets are derived from the cell cycle progression of the budding yeast cell cycle (*Saccharomyces cerevisae*) and the SOS response of *E. coli*. The genes contained in these data sets display only a small subset of the global GRN, however, it is known that these genes play an important role in the respective processes.

The budding yeast cell cycle has been thoroughly investigated over many years. Cho et al.[12] and Spellman et al.[13] contributed to these investigations by publicly providing four transcriptional data sets, measuring the progression of the cell cycle. The sample numbers of these data sets range from 14 to 24 and the samples are taken at equidistant time steps for each series. In our study, we concentrate on the α-factor data set, containing 18 samples and 7 minute time steps. The measurements have been taken on microarrays [14], from which we select two subsets containing 20 and 800 genes, respectively. The gene group containing 800 genes corresponds to the subset chosen by Spellman, meeting his criteria for cell cycle regulation [13]. The second subset is chosen according to the network published by Chen et al.[15].

The SOS response is well understood and regulates cell survival and repair after DNA damage. For the measurement of the SOS response, Collins et al. used the wild-type *E. coli* strain MG1655. Their data set can be obtained from their publication [16]. Unlike Spellman, Collins et al. measured the steady-state concentration of the transcriptional response to a constant perturbation. This perturbation was done for each of the nine genes in the network and hence nine measurements were carried out.

The artificial data set is derived from an *in silico* model. The topology of this model is initialized randomly, while the in-degree of every gene follows a power law distribution with an exponent ranging from 1.1 to 1.8 [4]. The networks are simulated by randomly perturbing the gene expression and subsequently measuring the response of the expression. Three networks with 10, 100 and 1000 genes are generated, and 5, 50 and 500 measurements are taken respectively. To add noise a Gaussian distribution, with a standard deviation of 5 %, is applied to the data sets.

Topological Validation Networks. To evaluate the results of different reconstruction methods, we define a topological network of gene regulation *a priori*. A model for the cell cycle was previously provided by Chen et al.[15] as a set of differential equations, defining the topological structure of the network with 72 edges. In his publication, this model is accompanied by justifying references to literature. From this system of differential equations we extract the topology by assuming an interaction, whenever the differential equation contains one.

A validation network for the SOS response is taken from the publication of Gardner *et al.*[16]. The validation networks can be downloaded as SBML models from our website[1].

Preprocessing. The structure of the network can be largely affected by the preprocessing step. The cell cycle data set is normalized by the average log ratio ($\log_2(ratio)$), which implicitly describes a non-linear relationship between the genes. To assess if such a preprocessing is suitable, we test the reconstruction without preprocessing and with the application of a sigmoidal function. The SOS response data set, on the other hand, is given in relative expression changes, preserving the linear relationship between the genes.

2.2   Reconstruction Methods

We evaluate six GRN reconstruction methods for their applicability to transcriptional data. We start from a gene expression matrix $\mathbf{X} \in \mathbb{R}^{genes \times samples}$, where each row represents a gene and each column represents a sample taken at a specific time step. This means, that an element X_{ij} of $\mathbf{X}$ indicates the expression level of gene i in sample j. This relationship allows to calculate the expression of a gene by summing over all influences, as shown in Eq. 1. For the steady-state data set of the SOS response Eq. 1 has to be interpreted different, while preserving the general structure of the equation. This can be done by assuming that the perturbation is at time point t and that the system returns to steady-state at time point $t + \Delta t$. Thereby, the actual value of the time step Δt can be ignored. By applying the $\log_2(ratio)$ to the cell cycle data set this relationship changes to a non-linear multiplicative relationship between the expression of the genes.

$$g_i(t + \Delta t) = \sum_{j=1}^{N} w_{ij} x_j(t) \tag{1}$$

In this work, we use basic models in combination with standard reconstruction methods. This draws from the fact, that little is known about the proper structure of GRNs and, therefore, the models should be kept general until specific modeling information emerges. Four of our reconstruction methods are basic applications of standard methods, namely exhaustive search, greedy search, linear regression and a random algorithm. The exhaustive search strategy enumerates every possible interconnection and ranks it among all the others, finally returning the best set of interactions as the GRN. The greedy search strategy uses local information to change the current topology, i.e. tries to find the global optimum without enumerating all solutions. Linear regression searches the parameters that cause a linear function to optimally fit the observed data. The random algorithm returns a randomly created network topology; this method is used to compare the other methods to a random guesser. In addition to the standard methods, we also evaluate the performance of methods specifically developed to reconstruct

---

[1] http://www-ra.informatik.uni-tuebingen.de/mitarb/supper/grn/

GRNs, Ref. [9] ,[11] and [10]. The method of Someren *et al.*[11] uses Gaussian elimination followed by free variable manipulation. Weaver *et al.*[10] reconstruct the GRN by applying SVD and afterwards removing the interactions with the lowest weight. Supper *et al.*[9] use SVD with a probabilistic search strategy (NSS) for the null space. Due to the runtime of the algorithms, not all methods can be evaluated on the large data sets.

Validation Methods. To validate the reconstruction methods two different types of validation techniques are applied. The first technique evaluates the topological reconstruction of the GRNs, i.e. how well the reconstructed topology resembles the validation topology. This evaluation is only applied to networks for which such a topology is available. To quantify the quality of the reconstruction, we count the number of true positives TP, false positives FP, false negatives FN and true negatives TN in the reconstructed network and then calculate the Matthews correlation coefficient (MCC, Eq. 2). The MCC combines both sensitivity and specificity into one measure, thereby correcting for unbalanced data sets. The MCC is a measure of correlation between the validation data and the model, where 1 indicates perfect positive correlation and -1 indicates perfect negative correlation. We also calculate the accuracy (ACC) of the reconstructed networks $(ACC = (TP + TN)/(TP + FP + FN + TN))$.

$$\frac{TP \times TN - FP \times FN}{\sqrt{(TP + FN)(TP + FP)(TN + FP)(TN + FN)}} \tag{2}$$

Apart from the qualitative validation of the topology, we are also interested in the predictability of our models. Since the inferred models are quantitative, they can be simulated to predict the expression of genes. To evaluate the predictive power, we use 5-fold cross-validation: We randomly split the data into 5 parts and train our model on 4 of these parts, while leaving the remaining part for testing. Then, the whole procedure is iterated such that each part is left out exactly once. Several measures can be used to evaluate the prediction of gene expression. Here, we distinguish between the two states of increased expression and decreased expression. More precisely, we define the state, s_i, of a gene i at time point t as in Equation 3. This definition cannot be applied to the steady-state SOS response data set, here we similarly define the state, s_{ip}, of a gene i after the perturbation p. The state of a gene is defined as +1 if the gene is upregulated under the respective perturbation, otherwise it is defined as -1.

$$s_i(t) := \begin{cases} +1 \text{ if } g_i(t) \geq g_i(t - \Delta t) \\ -1 \text{ otherwise} \end{cases} \tag{3}$$

Apart from the validation on isolated edges we are interested in indirect connections which are modeled as direct connections. Such a simplification occurs if a transcriptional signal is transduced over one intermediate gene. To calculate the occurrence of direct connections, we implement a graph search algorithm to determine if a FP edge in the reconstructed network can be traversed in the validation network by taking a route over one intermediate gene.

We are fully aware of the fact that the validation network itself will contain incorrect edges and also miss existing edges. Furthermore, we are also aware of the fact that validation can strongly depend on the specific parameter settings of the individual methods. Despite these limitations, we nevertheless think that useful insight can be drawn from this study.

3 Results

3.1 Reconstruction

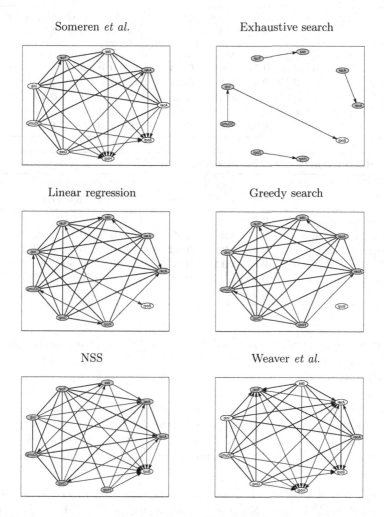

Fig. 1. Topology of the reconstructed *E. coli* SOS response networks networks. Arrows show directed, lines undirected edges and a grey node shows self-regulation.

Table 1. Evaluation of the reconstruction methods. The topology shows the ability of the models to reproduce the validation network. The prediction score is calculated by a 5-fold cross-validation. The results are given as Matthews correlation coefficients (MCC) and accuracy (ACC). **I.** Evaluation of the the SOS response network. **II.** Evaluation of the cell cycle network. **III.** Evaluation of the the *in silico* data sets. The topology column shows how well the the validation network is reproduced. The prediction score column shows the ability of the models to predict the expression of genes.

Reconstruction of the gene regulatory networks

	I. SOS response network				II. Yeast cell cycle network				
	Topology		Prediction		Topology		Prediction		Pred. 800
	MCC	ACC	MCC	ACC	MCC	ACC	MCC	ACC	MCC
Exhaustive search	0.14	0.48	0.56	0.77	0.05	**0.82**	0.35	0.68	-
Linear regression	**0.55**	**0.79**	**0.89**	**0.94**	**0.06**	0.48	0.16	0.62	**0.73**
Greedy search	0.15	0.48	0.87	**0.94**	0.02	**0.82**	0.26	0.63	0.70
NSS	0.40	0.71	0.63	0.77	0.05	0.60	**0.45**	**0.72**	0.72
Weaver *et al.*	0.14	0.59	0.11	0.56	-0.06	0.30	-0.08	0.47	0.50
Someren *et al.*	0.23	0.63	0.38	0.78	0.03	0.51	0.10	0.60	-

III. Reconstruction of the *in silico* networks

	Topology			Prediction		
MCC	10	100	1000	10	100	1000
Exhaustive search	0.82	-	-	0.84	-	-
Linear regression	0.29	0.25	0.03	0.53	0.57	0.35
Greedy search	0.82	0.91	**0.92**	0.75	0.94	**0.94**
NSS	0.75	**0.93**	0.87	0.81	**0.95**	0.37
Weaver *et al.*	0.40	0.18	0.91	0.51	0.43	0.42
Someren *et al.*	**0.93**	-	-	**0.92**	-	-

E. coli SOS Response Network. The results of the reconstruction for the SOS response network are shown in Table 1. **I**. The topology of the network is resembled with an accuracy between 70 % and 80 % by linear regression and the NSS, the remaining methods all outperform the random guesser[2], but not very significantly. The prediction of the gene expression outperforms the topological reconstruction. This is especially apparent for the exhaustive search, where the topological reconstruction is not satisfactory but the predictability is. The reconstructed topologies of the different methods are shown in Figure 2.

Budding Yeast Cell Cycle Network. The topological reconstruction of the 20 gene cell cycle network has a low performance (Table 1 **II**). This is true for every method applied to the data set, with almost no correlation between the reconstructed and the validation network. This also holds for the predictability

---

[2] The random guesser has an MCC value of zero, for topology and predictability.

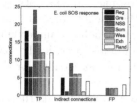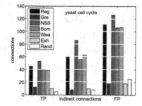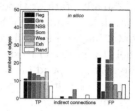

Fig. 2. Analysis of the reconstructed edges for different data sets. The x-label "indirect connections" shows the number of reconstructed edges with an indirect counterpart in the validation network.

of some reconstruction methods. However, the accuracy of the predictability does reach an accuracy of 72 % for the NSS. This is also the case for the 800 gene network, where the linear regression even reaches an accuracy of 73 %.

***In Silico* Networks.** As expected the reconstruction of *in silico* models outperforms the reconstruction of biological networks (Table 1). For the small network with 10 genes, the MCC value of the reconstruction is significantly better than the random reconstruction for all methods. The performance improves for larger data sets with more measurements, whereas the method of Someren *et al.* and Exhausitve search are not listed because the runtime on 100 and 1000 genes is too long. The performance on a network with 1000 genes is in the range of the results obtained on 100 genes.

Indirect Connections. Reconstructed networks frequently contain indirect connections which are reconstructed as direct connections. Figure 2 shows the occurrence of direct connections (TP), indirect connections with one intermediate gene and edges that have no direct or indirect counterpart in the validation network. The reconstructed SOS response networks show a high number of indirect connections, indicating the reconstruction of delayed effects. The random network has far more FP and a low number of indirect connections. The cell cycle network does not show any significant deviation from a random reconstruction, this applies for direct as well as for indirect connections. The *in silico* networks show a different distribution than the biological networks, either the reconstructed connections are TP or they have no connection at all. Thus, the reconstruction of indirect effects does not occur in these networks.

Preprocessing. The application of linear and sigmoidal preprocessing shows no improvement.

4 Discussion and Conclusion

We evaluated six different reconstruction algorithms on transcriptional data sets. Depending on the aim of the reconstruction (topology or prediction) the type of data (steady-state or time-series) and the reconstruction method, the obtained results differ significantly. Surprisingly, this is not only the case for different types of data, but also for different reconstruction methods, although the underlying model (Equation 1) is identical. However, it has to be remarked critically that these results depend on the specific parameter and data format settings for the individual algorithms.

The most significant difference is observed between topological reconstruction and predictability of gene expression. The predictability is acceptable for every data set,e.g. for the SOS response data set a prediction accuracy of 94 % is obtained. For the cell cycle data set, the prediction accuracy is around 70 %. For the *in silico* data sets the discrepancy between topology reconstruction and prediction of gene expression is small, and both measures yield good results. The reason for the successful topological reconstruction might be owed to the benefit of knowing the character of the interaction *a priori*. However, contrary to *in silico* data sets these interactions are not known for biological data sets.

Another apparent difference is observed between the use of time-series data and steady-state data. Steady-state data sets seem to be suitable for network identification. Time-series data, on the other hand, poses many problems hampering reconstruction, e.g. the time dimension has to be accounted for. Another difficulty concerning time-series data is its increased sensitivity to noise, which is caused by the absence of noise compensation through averaging effects.

The performance of the reconstruction methods is quite diverse among the different data sets. We initially assumed, that if a method performs well on *in silico* data sets, it also performs well on biological data sets, this was not the case in our study. It is not clear yet, if this is generally the case, but it poses the question whether validating methods on *in silico* data sets is adequate.

The reconstruction of GRNs can yield satisfactory results as well as poor results in different applications. For instance, the application of GRN models for the prediction of gene expression is satisfactory in most cases. The topological reconstruction on the other hand is far more critical. The cell cycle data set, for instance, does not seem to be suitable for uncovering the topological structure of the network. However, the prediction of gene expression does yield good results.

Overall, the greedy search and NSS method based on steady-state data are suitable for the topological reconstruction. In case of time-series data, we question the applicability of reconstruction methods based solely on transcriptional data; here the incorporation of additional data might improve the reconstruction. For the prediction of gene expression, the choice of the data set seems to be uncritical and basic methods such as linear regression or greedy search provide reasonable results.

Acknowledgement

This work was supported by the National Genome Research Network (NGFN II) of the Federal Ministry of Education and Research in Germany under contract number 0313323.

References

1. Kikuchi, S., Tominaga, D., Arita, M., Takahashi, K., Tomita, M.: Dynamic modeling of genetic networks using genetic algorithm and S-system. Bioinformatics **19**(5) (Mar 2003) 643–50
2. van Someren, E.P., Wessels, L.F., Reinders, M.J.: Linear modeling of genetic networks from experimental data. Proc Int Conf Intell Syst Mol Biol **8** (2000) 355–66
3. Yeung, M.K.S., Tegner, J., Collins, J.J.: Reverse engineering gene networks using singular value decomposition and robust regression. Proc Natl Acad Sci U S A **99**(9) (Apr 2002) 6163–8
4. Bergmann, S., Ihmels, J., Barkai, N.: Similarities and differences in genome-wide expression data of six organisms. PLoS Biol **2**(1) (Jan 2004) E9
5. Pedraza, J.M., van Oudenaarden, A.: Noise propagation in gene networks. Science **307**(5717) (Mar 2005) 1965–1969
6. Kuhlman, T., Z. Zhang, M.H.S., Hwa, T.: Cooperativity, sensitivity, and combinatorial control:quantitative dissection of a bacterial promoter. Submitted to PNAS (2006)
7. van Someren, E.P., Vaes, B.L.T., Steegenga, W.T., Sijbers, A.M., Dechering, K.J., Reinders, M.J.T.: Least Absolute Regression Network Analysis of the murine osteoblast differentiation network. Bioinformatics (Dec 2005)
8. Soinov, L.A., Krestyaninova, M.A., Brazma, A.: Towards reconstruction of gene networks from expression data by supervised learning. Genome Biol **4**(1) (2003) R6
9. Supper, J., Spieth, C., Zell, A.: Reverse engineering non-linear gene regulatory networks based on the bacteriophage lambda ci circuit. In: IEEE Symposium on Computational Intelligence in Bioinformatics and Computational Biology (CIBCB). (2005)
10. Weaver, D.C., Workman, C.T., Stormo, G.D.: Modeling regulatory networks with weight matrices. Pac Symp Biocomput **1** (1999) 112–23
11. van Someren, E., Wessels, L., Reinders, M.: Genetic network models: A comparative study. Proc. of SPIE, Micro-arrays: Optical Technologies and Informatics **4266** (2001)
12. Cho, R.J., Campbell, M.J., Winzeler, E.A., Steinmetz, L., Conway, A., Wodicka, L., Wolfsberg, T.G., Gabrielian, A.E., Landsman, D., Lockhart, D.J., Davis, R.W.: A genome-wide transcriptional analysis of the mitotic cell cycle. Mol Cell **2**(1) (Jul 1998) 65–73
13. Spellman, P.T., Sherlock, G., Zhang, M.Q., Iyer, V.R., Anders, K., Eisen, M.B., Brown, P.O., Botstein, D., Futcher, B.: Comprehensive identification of cell cycle-regulated genes of the yeast Saccharomyces cerevisiae by microarray hybridization. Mol Biol Cell **9**(12) (Dec 1998) 3273–3297
14. Eisen, M.B., Brown, P.O.: DNA arrays for analysis of gene expression. Methods Enzymol **303** (1999) 179–205
15. Chen, K.C., Calzone, L., Csikasz-Nagy, A., Cross, F.R., Novak, B., Tyson, J.J.: Integrative analysis of cell cycle control in budding yeast. Mol Biol Cell **15**(8) (Aug 2004) 3841–3862
16. Gardner, T.S., di Bernardo, D., Lorenz, D., Collins, J.J.: Inferring genetic networks and identifying compound mode of action via expression profiling. Science **301**(5629) (05.05 2003) 102–5

Individual-Based Modeling of Bacterial Foraging with Quorum Sensing in a Time-Varying Environment

W.J. Tang[1], Q.H. Wu[1,*], and J.R. Saunders[2]

[1] Department of Electrical Engineering and Electronics
[2] School of Biological Sciences
The University of Liverpool, Liverpool L69 3GJ, U.K.
qhwu@liv.ac.uk

Abstract. "Quorum sensing" has been described as "the most conse-
quential molecular microbiology story of the last decade" [1][2]. The purpose of this paper is to study the mechanism of quorum sensing, in order
to obtain a deeper understanding of how and when this mechanism works.
Our study focuses on the use of an Individual-based Modeling (IbM)
method to simulate this phenomenon of "cell-to-cell communication" incorporated in bacterial foraging behavior, in both intracellular and population scales. The simulation results show that this IbM approach can
reflect the bacterial behaviors and population evolution in time-varying
environments, and provide plausible answers to the emerging question
regarding to the significance of this phenomenon of bacterial foraging
behaviors.

1   Introduction

Systems biology has been proposed in the past a few years to overcome the existing barriers in biology, as a result, computerized analysis of biological models
is investigated. Individual-based Modeling (IbM) is one of the emerging approaches [3] to this problem. The basic idea of IbM is to simulate the behavior
or dynamics of an individual which interacts with the others synchronously or
asynchronously in a same or different living patterns. IbM is commonly based on
cellular automation models, explaining qualitatively colonial pattern formations
for different nutritional regimes [4]. An essential capability of an IbM enables the
description of all the states, inputs and outputs of an individual and their relationship with that of the other individuals which are living in a same population.
In contrast to the population-based model (PbM), the IbM should possess a more
flexible and robust capability for simulating a complex system where there are a
large number of individuals which have their own behaviors dynamics influenced
by the other individuals and the environment.

It is well recognized that bacterial diseases such as cholera, meningitis, *E.coli*
infection and many others are among the deadliest in the world. However, it has

---

* Correspondence concerning this paper should be made to Professor Q.H. Wu.

E. Marchiori, J.H. Moore, and J.C. Rajapakse (Eds.): EvoBIO 2007, LNCS 4447, pp. 280–290, 2007.
© Springer-Verlag Berlin Heidelberg 2007

been reported recently that bacteria can cause an illness only when there are a sufficient number of them. A cell-to-cell signal is involved in the development of bacterial communities, such as biofilms [5]. This communication between bacteria is often referred as "quorum sensing", which is widespread in nature. It also controls the processes of bacterial behavior, especially those that are usually unproductive when undertaken by an individual bacterium but become effective in a group. For instance, bioluminescence, secretion of virulence factors, sporulation and conjugation are believed to have relationship with quorum sensing. In this sense, bacteria are able to function as multi-cellular organisms with this underlying mechanism.

In the past a few years, quorum sensing has been the research topic addressing a broad audience. Ward *et al.* [6] introduced a general mathematical model of quorum sensing in bacteria. Dockery *et al.* [7] presented a mathematical model of quorum sensing in *P.aeruginosa*, to show how quorum sensing works using a biochemical switch. A number of biologically realistic mechanisms, such as "quorum sensing", have been considered in [8], to study the pattern forming properties of bacteria. A stochastic model to connect intracellular and population scales of the quorum sensing phenomenon has been proposed in [9] to demonstrate that the transition to quorum sensing in an *Agrobacterium* population in liquid medium requires a much higher threshold cell density than in biofilm. Muller *et al.* [10] investigated and analyzed a spatially structured model for a cell population, including a detailed discussion of the regulatory network and its bistable behavior. The modeling approaches done by You [11] and Garcia-Ojalvo [12] also revealed a number of important properties of quorum sensing.

However, all the work discussed above are based on differential equations, which share the same drawbacks, *e.g.*, the parameters of these models have to be determined from the experimental data. Moreover, they are not able to be generalized for analysis of various events of the biological system from which the data were collected. With the aim of systematic approach, it is rare that these models could be able to describe the whole evolution process of biological colonies.

To demonstrate the effectiveness of IbM based models in solving the problems in Systems Biology, we have previously presented a Varying Environment Bacterial Modeling (VEBAM), which is an individual based model of bacterial foraging incorporating mechanisms of chemotactic behaviors [13]. In this paper, we consider a novel VEBAM based approach by which quorum sensing can also utilize biologically realistic mechanisms, including the employment of cell density sensing mechanisms.

This paper is organized as follows: Section 2 introduces the framework and basic components of the previously presented model, VEBAM. Section 3 describes the phenomenon and its underlying mechanism of quorum sensing in detail , then the modeling of quorum sensing based on VEBAM is presented. The bacterial behaviors with and without quorum sensing mechanism are simulated, and the numerical difference between them with a wide variety of intrinsic and extrinsic properties of bacterial foraging are illustrated in Section 4, followed by the conclusion in Section 5.

2   The VEBAM

The principle of VEBAM is described from a macroscopic viewpoint. Such a model represents the foraging patterns of *E.coli* in and between two fundamental elements: the environment and cells. The architecture of this framework is shown in Fig. 1.

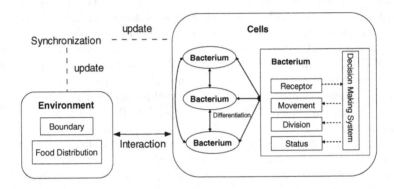

Fig. 1. VEBAM architecture

2.1   The Environment

An environment includes the properties of an artificial surface used for food searching, such as the characteristic of the boundary and the map of food distribution. The food distribution map is initialized by the definition of a specific function.

The environment is defined as

$$Env\langle \sigma, \beta \rangle \tag{1}$$

where σ indicates the nutrient distribution and β represents the characteristic of boundary, indicating whether it is reflective or periodic.

The environment is segmented equally into small niches as a discrete grid, covering a 2-dimensional grid of $M \times N$ units. Thus the Env contains $M \times N$ equally divided units, which can be represented as:

$$Env = \{Env_{ij}^t | i = 1, 2, \cdots, M; j = 1, 2, \cdots, N; t = 1, 2, \cdots, \infty\} \tag{2}$$

where Env_{ij}^t denotes the nutrient or toxin level in the niche of (i, j) and a positive value of Env_{ij}^t indicates nutrient while negative ones represent toxin; i and j are indices of each dimensions along the coordinates; and t denotes the time instant of the step of the foraging process.

2.2   The Cells

Each cell is assumed to have a receptor which detects the value of its surrounding environment. A cell also has a decision making system determining its directions of movements, energy status and activation of division system according to the information obtained by the receptors. These activities last during the cells' lifetime, which in turn, enable them interact with the environment and then modify the food map.

Following the definition of the environment, the cells are defined to consist of a set of bacteria, which is represented as:

$$B = \{B_l^t | l = 1, 2, \cdots, P; t = 1, 2, \cdots, T\} \tag{3}$$

where

$$B_l^t = \langle p_l^t, \varepsilon_l^t, \varphi_l^t \rangle \tag{4}$$

denotes an individual bacterium or cell; P the maximum population in the foraging process and T the final time of this process; B_l^t has its own position p_l^t, $p_l^t = \{i, j | l, t\}$, energy ε_l^t, status φ_l^t, $\varphi_l^t = \{\Phi_I, \Phi_A, \Phi_D\}$, where Φ_I, Φ_A, Φ_D stand for the status of Inactive, Active and Divided, respectively.

3   Modeling of Cell-Cell Communication

3.1   Cell-Cell Communication by Quorum Sensing

Living organisms are always subject to time-varying conditions in the form of environmental changes. That is why they develop the essential ability to sense signal in the environment and adapt their behaviors accordingly. To understand the intracellular molecular machinery that is responsible for the complex collective behavior of multicellular populations is an exigent problem of modern biology.

As a new branch of microbiology, quorum sensing, was discovered by Miller and Bassler [14]. It allows bacteria to activate genetic programs cooperatively, provides an instructive and tractable example illuminating the causal relationships between the molecular organization of gene networks and the complex phenotypes they control. Generally, it is a process that allows bacteria to search for similar cells in their close surroundings using secreted chemical signaling molecules called autoinducers. This is also called "cell-cell communication".

In the quorum sensing process, the population density of bacteria is assessed by detecting the concentration of a particular autoinducer related to the density of a cluster of cells. At a certain stage of the bacterial foraging process, the density would increase. A colony of bacteria would respond to the density by activating specific gene expression programs. This phenomenon, which conducts in an intercellular communication manner, is exhibited commonly in gram-negative bacteria, such as *E. coli*.

The cells, which have their specific gene expression programs activated, could produce autoinducers, *i.e.*, signalling molecules, which initiate the quorum sensing [11]. The "on-off" gene expression switch controls this conjugation phenomenon in response to the increase of autoinducer concentration. Thus, once a certain level of a cluster density is reached, a quorum sensing network of the population is formed and turns to the "on" state. It also amplifies the autoinducer signal of the "quorum" cells. This signal then turns on the expression of the phenotype-specific genes of other cells and boots the production of autoinducers in the entire population. Once it turns to the "on" state, it keeps in this state until the density of the cluster falls below a certain level. The quorum sensing network remains in the "off" state, until a certain concentration of autoinducers is reached.

3.2   Modeling of Quorum Sensing

In the VEBAM, quorum sensing is calculated based on the definition and computation rules of clusters. A set of clusters is an aggregation of population, which is defined as follows:

$$C = \{C_k | k \leq P\} \tag{5}$$

where

$$C_k = \langle D_k, G_k, F_k \rangle \tag{6}$$

denotes the k_{th} cluster in the population, D_k the population density, G_k the position of gravity center, and F_k the diffusion exponent of the autoinducer in cluster C_k.

At each chemotactic step, bacteria move from one position to another, and they tend to move towards the nutrient-rich area. In general, the variation of bacterial population density is affected by four factors: the availability of nutrients, the presence of competitors, the host response and also the quorum sensing network. However, due to lack of the existing knowledge of the competitors and host response behaviors, only the nutrients and the quorum sensing network are considered at this stage. In VEBAM, each cluster, C_k, in the population is classified based on the bacterial population density. The position of gravity center, G_k, are also calculated for each cluster C_k.

At time t, the amount of autoinducers, extracted by cluster C_k with the mechanism of quorum sensing, is calculated as:

$$A_{k,i,j}^t = -\lambda \exp(-(t - t_0)F_k \|G_k - p_l^t\|^2) \tag{7}$$

where t_0 is the time at which the quorum sensing is turned to the "on" state, λ denotes the amplitude of the peak of autoinducer congregation and is to be tuned as appropriate.

The autoinducer could be added to the environment as an attractive factor. Then the environment is modified as

$$Env_{ij}^{t+1} = Env_{ij}^t + A_{k,i,j}^t \tag{8}$$

The energy of each bacterium producing autoinducers also needs recalculation. Let $C_m \subset C$, where C_m is the cluster which produces autoinducers, the energy of the i_{th} individual in C_m is modified as:

$$\varepsilon_i^{t+1} = (1 - \alpha)\varepsilon_i^t \qquad (9)$$

where $i \in [1, 2, \cdots, n_m]$, n_m is the total number of cluster m and α is the coefficient used to control the rate of the energy-autoinducer transformation.

4   Simulation Studies

4.1   Environmental Setting

The VEBAM is set in a multi-modal environment. The nutrient distribution of the toroidal grid environment at $t = 0$ is set by the function $f(x, y)$ as indicated in [13] and this environment is divided into 300×300 grids.

During the whole process of computation, the nutrient in the environment is constantly consumed by bacteria at all grid units where they occupy in each step until the nutrient is lower than a minimal level of consumption.

4.2   Population Evolution with Quorum Sensing

As discussed in section 3, based on the coordinate of the positions, using the method of "density-based spatial clustering" [15], each cluster, C_k, in the population can be classified. If we define an individual as a "point" in the environment, then these points can also be classified as follows.

Above all, Eps, the neighborhood radius of a cluster is defined. Given the parameters: m, the minimal number of points considered as a cluster, x, the positions of all points, and n, the dimension of the positions, the Eps is calculated as:

$$Eps = \left(\prod_{d=1}^{n} (\max(x_d) - \min(x_d)) \times k \times \Gamma(N)) / M \right)^{1/n} \qquad (10)$$

where $\Gamma(a) = \int_0^\infty e^{-t} t^{a-1} dt$, $N = 0.5n + 1$ and $M = m(\pi)^{n/2}$.

Given an arbitrary point p in the environment, let d_s be the distance of p to its s_{th} nearest neighbor, then the d-neighborhood of p contains exactly $s + 1$ points for any p in the environment, and the longest distance is set as s_dist. If and only if $s\_dist \leq Eps$ and $s + 1 \geq m$, the d-neighborhood of p is considered to be a cluster.

For each cluster C_k, all points of C_k fall into two categories, points inside of the cluster (core points) and points on the border of the cluster (border points). Here, the s_dist for a point p with the points belong to C_k is recalculated, all points with the $s\_dist \leq Eps$ will be core points, and the rest points will be border points.

The position of the gravity center g_k for each cluster C_k, is obtained according to the classification of points. Set the position of a core point p_c in C_k as P_{kc}, and $P_{kc} = [P_{kc}^d | d = 1, \cdots, n]$, then the coordinate of g_k in each dimension d is calculated as:

$$g_k^d = \frac{\sum P_{kc}^d}{N} \tag{11}$$

where N is the number of core points in cluster C_k.

4.3   Simulation Results

The interaction between cells and environment: The population evolution of the VEBAM has been simulated and it is illustrated in Fig. 2. The evolution process proceeds 100 steps, featuring the swarming of the cells to the highlighted areas where the nutrient level is high, and Fig. 2 also illustrates that bacteria have the trend to approach the nutrient, avoiding the toxin in the evolution process. If they could not consume enough nutrient in their ambient environment, they will be eliminated gradually.

At the beginning, a population of 100 cells was distributed evenly [16] in the environment. The highlighted parts in Fig. 2(a) indicate the nutrient. The quorum sensing was switched "on" at or before $t = 20$, as shown in Fig. 2(b). From then, the cell steadily produced a small amount of the quorum sensing signaling molecule, autoinducer, that can freely diffuse in and out of the cell. At $t = 40$, three clusters have been identified by the three more highlighted parts in Fig. 2(c). However, the population densities of the top two clusters were low, therefore, most of the autoinducer molecules were washed out and dispersed in the environment by diffusion. That is, at a low cell density, the autoinducer is synthesized at basal levels and diffuses into the surrounding medium, where it is diluted. Much clearer features of quorum sensing are shown in Fig. 2(d), in which the cells in two most highlighted clusters sending autoinducers. When the nutrient is consumed gradually, the quorum sensing is no longer functioned and the active individuals have no effect to each other.

The modification of available nutrient in accordance with the interaction indicated in Fig. 2, without the artificial add-on of autoinducers in the simulation studies, is illustrated in Fig. 3.

Comparison of simulations with and without quorum sensing mechanism: The simulation study of the bacterial foraging behavior without quorum sensing mechanism is undertaken for comparison. Figures 4 and 5 demonstrate the differences of population evolution and their influence to the environment in these two cases, respectively. Three characteristics were featured in Fig. 4: the total population, active population and inactive population of cells. Throughout the process of 100 steps, the total population in the case with quorum sensing is lower than that without it. However, the inactive population is also lower and the peak of active population is much higher than the one without quorum sensing. The simulation results shown in Fig. 4 is consistent with our understanding of biological systems.

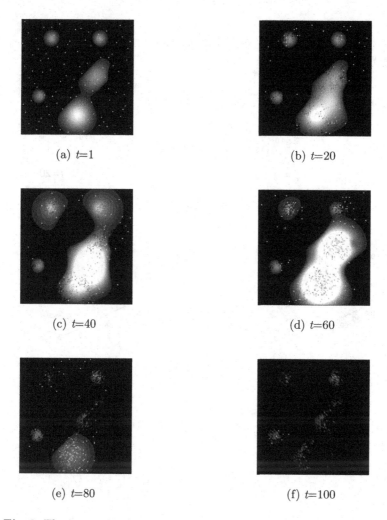

(a) t=1

(b) t=20

(c) t=40

(d) t=60

(e) t=80

(f) t=100

Fig. 2. The interaction between cells and environment in 100 steps

The available nutrient in the environment and that obtained by active and inactive cells are shown in Fig. 5. It was calculated using the following equation:

$$Nutrient = \sum_{i=1}^{300} \sum_{j=1}^{300} f(x_i, y_j)$$

where $x_i, y_j \in [0, 30]$, and $f(x_i, y_j) \geq 0$.

There is no significant difference between the total nutrient lost in the two cases with and without quorum sensing. However, the energy of active cells with

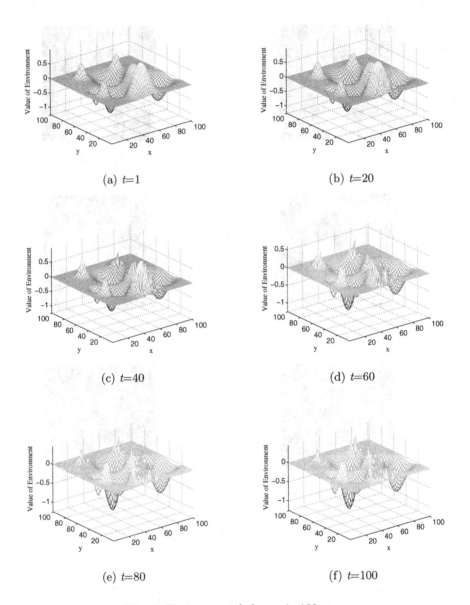

(a) $t=1$

(b) $t=20$

(c) $t=40$

(d) $t=60$

(e) $t=80$

(f) $t=100$

Fig. 3. Environmental change in 100 steps

quorum sensing is much less than the one without, during the peak period of active population as shown in Fig. 4. This result indicates that cells with the quorum sensing mechanism are able to achieve a higher population level with less nutrient consumed than that achieved without quorum sensing.

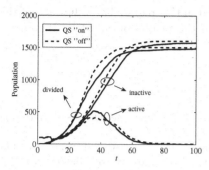

Fig. 4. Population evolution

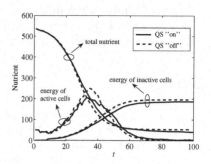

Fig. 5. Energy distribution

4.4 Discussion

The rules for the interaction between the environment and bacteria provide a flexible mechanism for nutrient consumption calculation and bacterial status updating. The bacterial swimming mechanisms are a combination of a rule-based system and stochastic searches. Furthermore, the population modeling includes the rules for initialization, reproduction and termination of bacteria. With these rules which can be specified while used to simulate different biological systems, the VEBAM is a generic and expandable IbM for computation of bacterial foraging patterns, providing a fundamental framework for modeling and simulation studies of a range of bacterial behaviors, and also a scheme which could be constantly updated, such as the add-on of quorum sensing discussed in this paper.

The difference of the situations with and without quorum sensing mechanism is clearly indicated in the simulation. More fundamental results can be obtained if the parameters are chosen from a wider range, however, this is beyond the scope of this paper.

5   Conclusions

The paper has presented the simulation results of quorum sensing using an existing individual-based model, VEBAM, which is for modeling bacterial foraging behaviors in time-varying environments. The details of VEBAM have been introduced. The quorum sensing mechanism has been successfully incorporated into this model and also evaluated in the simulation study which includes modeling a group of bacteria in an evolution process of finite simulation time. A density-based clustering algorithm is adopted to classify the clusters, which assists to identify the timing of quorum sensing switching. The gravity center of each cluster also provides important information for sending and diffusing autoinducer in the environment. The simulation results show that this IbM-based approach is able to provide a plausible methodology to simulate the bacterial foraging with the quorum sensing mechanism investigated in this paper. The validation of this model, as well as the comparison between the results obtained by this model and previous approaches, will be undertaken in the future.

References

1. S. Busby and V. de Lorenzo: Cell regulation putting together pieces of the big puzzle. Current Opinion in Microbiology **4** (2001) 117–118
2. K. Winzer, K. R. Hardie, and P. Williams: Bacterial cell-to-cell communication: sorry, can't talk now - gone to lunch! Current Opinion in Microbiology **5** (2002) 216–222
3. C. Vlachos, R. C. Paton, J. R. Saunders, and Q. H. Wu: A rule-based approach to the modelling of bacterial ecosystems. BioSystems **84** (2005) 49–72
4. E. Ben-Jacob, O. Shochet, A. Tenenbaum, and I. Cohen: Generic modeling of cooperative growth patterns in bacterial colonies. Nature **368** 1994 46–49
5. D. G. Davies, M. R. Parsek, J. P. Pearson, and et al.: The involvement of cell-to-cell signals in the development of a bacterial biofilm. Nature **280** 1998 295–298
6. J. P. Ward, J. R. King, and A. J. Koerber: Mathematical modelling of quorum sensing in bacteria. Journal of Mathematics Applied in Medicine and Biology **18** 2001 263–292
7. J. D. Dockery and J. P. Keener: A mathematical model for quorum sensing in *pseudomonas aeruginosa*. Mathematical Biology 2000 1–22
8. K. J. Painter and T. Hillen: Volume-filling and quorum sensing in models for chemosensitive movement. Canadian Applied Mathematics Quarterly **10** 2002 501–543
9. A. B. Goryachev, D. J. Toh, K. B. Wee, and et al.: Transition to quorum sensing in an agrobacterium population: A stochastic model. PLoS Computational Biology 2005 1–51
10. J. Muller, C. Kuttler, and B. A. Hense: Cell-cell communication by quorum sensing and dimension-reduction. Technical Report Technical University Munich 2005 1–28
11. L. You, R. C. Cox III, R. Weiss, and F. H. Arnold: Programmed population control by cell-cell communication and regulated killing. Nature 2004
12. J. Garcia-Ojalvo, M. B. Elowitz, and S. H. Strogatz: Modeling a synthetic multicellular clock: Repressilators coupled by quorum sensing. Proceedings of the National Academy of Sciences **101** 2004 10 955–10 960
13. W. J. Tang, Q. H. Wu, and J. R. Saunders: A novel model for bacterial foraging in varying environments. Lecture Notes in Computer Science **3980** 2006 556–565
14. M. B. Miller and B. L. Bassler: Quorum sensing in bacteria. Annual Review of Microbiology **55** 2001 165–199
15. M. Ester, H. Kriegel, J. Sander, and X. Xu: A density-based algorithm for discovering clusters in large spatical databases with noise. Proceedings of 2nd International Conference on Knowledge Discovery and Data Mining (KDD-96) 1996
16. Y. J. Cao and Q. H. Wu: Study of initial population in evolutionary programming. Proceedings of the European Control Conference **368** 1997 1–4

Substitution Matrix Optimisation for Peptide Classification

David C. Trudgian and Zheng Rong Yang

School of Engineering, Computer Science and Mathematics,
University of Exeter, Exeter, EX4 4QF, UK
{d.c.trudgian,z.r.yang}@ex.ac.uk

Abstract. The Bio-basis Function Neural Network (BBFNN) is a novel neural architecture for peptide classification that makes use of amino acid mutation matrices and a similarity function to model protein peptide data without encoding. This study presents an Evolutionary Bio-basis network (EBBN), an extension to the BBFNN that uses a self adapting Evolution Strategy to optimise a problem specific substitution matrix for much improved model performance. The EBBN is assessed against BBFNN and multi layer perceptron (MLP) models using three datasets covering cleavage sites, epitope sites, and glycoprotein linkage sites. The method exhibits statistically significant improvements in performance for two of these sets.

1 Introduction

1.1 Background

The study of interactions between proteins within the cell is a major topic of systems biology. Such interactions commonly depend on the successful recognition of suitable functional sites which support them. This recognition process uses the information contained in the 3D conformations and primary structures of the proteins involved. There are many kinds of functional sites which have been examined, including those for protease cleavage, glycosylation, phosphorylation, acetylation and enzymatic catalysis.

Work on the identification of functional sites typically uses sets of fixed length short peptides (short amino acid chains from a protein). These peptides are classified as either functional or non-functional by the trained system. For cases such as protease cleavage sites, the actual functional site will be between two of the residues. In post translational modifications, such as phosphorylation, the functional site is a single residue within the peptide.

A large number of techniques have been applied to functional site recognition. These can be broadly grouped into frequency based, rule based, statistical modelling, and machine learning techniques. Frequency based techniques were the first to be applied using computers and, along with rule based approaches, are simple to interpret. Statistical techniques such as hidden Markov Models (HMMs) and machine learning approaches such as neural networks can attain higher prediction accuracy in many cases.

E. Marchiori, J.H. Moore, and J.C. Rajapakse (Eds.): EvoBIO 2007, LNCS 4447, pp. 291–300, 2007.
© Springer-Verlag Berlin Heidelberg 2007

1.2 Amino Acid Substitution Matrices

The Bio-basis function neural network (BBFNN), defined in [1], makes use of an amino acid substitution matrix to calculate similarity values between input and Bio-basis peptides. These matrices, developed by Dayhoff and others [2], estimate the probability of mutations between amino acids for a given evolutionary distance. For a set of protein sequences, odds values are calculated for the observed probability of mutation from one amino acid into the other, divided by expected mutation rate (the product of the frequencies of these amino acids i and j). These values are presented in logarithmic form such that the calculations for sequence similarity are additive rather than multiplicative:

$$M_{i,j} = \log_2 \left(\frac{q_{i,j}}{p_i p_j} \right)$$

The selection of a substitution matrix for use with a BBFNN is a difficult decision which can influence performance since it alters the transformation from input into feature space. Currently a trial and error approach is adopted in which performance statistics using a variety of standard matrices are compared, the best then being chosen. This is a time consuming process; the current implementation of the BBFNN includes 15 matrices, excluding variation of evolutionary distances.

The PAM, BLOSUM[3], GONNET [4] and other substitution matrices are constructed using all available sequence data, giving rise to probabilities of mutation averaged across all species, protein functions and both functional and non-functional regions. Whilst this generality is a desirable trait for common sequence searching and alignment tasks it is questionable whether it is appropriate for discriminatory use in a BBFNN applied to a specific classification task. Amino acid frequencies and mutations can vary greatly from the average for all sequences. It is reasonable therefore to expect that problem specific matrices could improve BBFNN classification results on the specific datasets by representing the 'closeness' of amino acids in relation to maintaining a required protein primary sequence structure, e.g. for enzyme catalytic activity, or 3-D formation e.g. for protein disorder prediction.

Generating problem specific matrices is certainly possible by limiting the data from which a matrix is produced to the type of peptide we are trying to classify, or transforming the target frequencies of a general matrix [5]. It is reasonable to expect that such a matrix would result in higher Bio-basis scores for functional sites as it would represent the conserved mutations of these sites. PHAT, a transmembrane-specific matrix, was shown to outperform the general matrices in transmembrane specific searches [6].

An alternative to working forward from sequence data to create a problem specific matrix is to start with a random or standard matrix and work backwards from classifier results to optimise the matrix. Research is ongoing into two such methods of optimisation. The first, which will be discussed in this paper, uses an evolutionary computation approach.

2 Methods and Data

In this paper we describe the Evolutionary Bio-basis Network (EBBN), an extension to the BBFNN which retains the benefits of the original architecture, adding an Evolutionary framework for the generation of an optimised problem specific amino acid substitution matrix.

2.1 Network Structure

The EBBN maintains the basic structure of the BBFNN as described in [1]. l input nodes accept input peptides which are l amino acids in length. The input peptides are delivered to each of the N bio-basis neurons in the second layer of the network. Each bio-basis neuron has an accompanying support peptide taken from the training data. At each bio-basis neuron the bio-basis function is applied; the incoming peptide is compared to the support peptide using an amino acid substitution matrix and the resulting value is normalised. The outputs of the bio-basis neurons are passed via multiplicative weighted links to an output neuron, giving an output y . The system can be represented mathematically as below:

$$y = \sum_{n=1}^{N} w_n \Phi(\mathbf{x}, \mathbf{z_n})$$

Where $y \in \Re$ is the output, N is the number of bio-basis neurons, w_n is the weight between the nth bio-basis neuron and the output neuron, $\Phi(\mathbf{x}, \mathbf{z_n})$ is the bio-basis function applied to input peptide $\mathbf{x}$ using support peptide $\mathbf{z_n}$. Optimisation of weights takes place the least squares method with matrix pseudo-inverse [7].

2.2 Evolution Strategy

Evolutionary Algorithms (EAs) have been applied to the optimisation of neural network systems for many years [8]. Various approaches have been used for the optimisation of structural hyperparameters and connection weights across a broad range of network topologies. In this study we optimise both of these aspects with traditional non-evolutionary methods. We choose to apply an EA to the optimisation of the amino acid distance matrix only, so that it can be studied independently of structural changes etc.

The distance matrix is evolved using a $(\mu + \lambda)$-Evolution Strategy with self adaptation [9]. Evolution strategies have been used for a multitude of optimisation problems, including those in bioinformatics such as parameter estimation for biochemical pathways, where they were found to outperform other algorithms [10]. The $(\mu + \lambda)$-ES in particular was chosen for its ability to escape local minima.

A similarity matrix is represented as a vector of 400 real numbers $\mathbf{v}$. We allow asymmetric similarities, as in [5]. This vector, together with a vector of strategy parameters σ, and fitness function value $F(\mathbf{v})$, make up a possible solution $\alpha_k = (\mathbf{v}_k, \sigma_k, F(\mathbf{y}_k))$.

A pool of μ possible solutions are created as initial parents, with random values from a standard Gaussian distribution of $\mathbf{v} \sim N(0, 1)$. Mutation is carried out additively, with a value drawn from a standard Gaussian distribution, multiplied by mutation strength σ_i, added to the previous value. i is the index of the data parameter which is being mutated, thus there are 400 strategy parameters per solution. The mutated data vector is $\hat{\mathbf{v}} = \mathbf{v} + \mathbf{u}$, where $\mathbf{u} = (\sigma_1 N_1(0, 1), \ldots, \sigma_{400} N_{400}(0, 1))$.

The mutation strengths, σ are self adapted during evolution; initialised with random values $\sigma_i \sim N(0, 1)$, they are mutated multiplicatively using the extended log-normal operator described in [9]. In addition, the mutation vector is scaled by a random multiplier to give the new parameter vector $\hat{\sigma}$ as below.

$$\hat{\sigma} = \exp\left(\tau_0 N_0(0, 1)\right) . \left(\sigma_1 \exp\left(\tau N_1(0, 1)\right), \ldots, \sigma_{400} \exp\left(\tau N_{400}(0, 1)\right)\right)$$

τ is the learning parameter controlling the parameter-wise mutation rate, whereas τ_0 controls the mutative multiplier. Here $\tau = 1/\sqrt{40}$ and $\tau_0 = 1/\sqrt{800}$, as suggested by studies summarised in [9].

Marriage of solutions generates offspring by discrete recombination; data values are selected randomly from 2 solutions which have been chosen randomly from the pool of μ parents. This process is repeated to create λ offspring. Crossover of the mutation parameters is (μ/μ) intermediate, i.e. child vectors receive the mean σ_i calculated from the entire parent population.

All solutions are evaluated by training a network on the training set. In this study the fitness function $F(\mathbf{y}_k)$ has been chosen as the area under the ROC curve for the model. The ROC curve is a plot of the fraction of true positive predictions against the fraction of false positive predictions, for varying classification thresholds. The area under the curve has been shown to be non-parametric and is used here as a robust measure of the performance of the model [14]. Selection is elitist, the μ parents selected for the next generation are the highest performing from the entire population of $\mu + \lambda$ solutions.

Stopping is controlled by a validation set. Evolution is stopped when the mean performance of the population on the validation set does not improve for 20 generations, i.e. 20λ candidate solutions have failed to produce a fitness value higher than the population mean. The highest performing solution at the last increase in mean performance on the validation set is selected to give the final evolved matrix.

2.3   Datasets

This paper uses five previously presented datasets to compare performance of the EBBN with BBFNN and MLP models. These cover HIV cleavage sites (HIV), glycoprotein linkage sites (GAL) and T-cell epitope prediction (TCL).

HIV - During the lifecycle of HIV, precursor polyproteins are cleaved by HIV protease. Disruption of cleavage ability causes non-infectious, imperfect virus replication and is therefore a promising target for anti AIDS drugs. Reliable identification of HIV protease cleavage sites is highly important to developing anti-viral drugs for HIV. The dataset presented in [11] consists of 8-residue peptides from HIV protease marked as cleavable or non-cleavable. There are 362 peptides of which 114 are positive, cleavage sites and 248 are negative, non cleavable sites.

GAL - Glycoproteins are an important subset of proteins with a high level of potential pharmacological significance. Carbohydrate groups attached to glyco-proteins affect the solubility and thermal stability and are implicated in impor-tant biological functions such as controlling uptake of glycoproteins into cells. This provides a new possibility for targeting drugs, the design of which requires an understanding of the linkage processes of glycoproteins. Prediction of linkage sites is an important step. Chou et al. [12] presented a dataset of 302 9-residue peptides, of which 190 are linkage sites and 112 non-linkage sites.

TCL - T-cells are a critical part of the immune response to viral infection. Epitopes are sites on viral proteins that are recognised and bound by the T-cell receptor. The TCL dataset consists of 202 10-residue peptides of which 36 are positive T-cell epitope sequences, the remaining 167 are non-epitope sequences. This data was presented in [13].

2.4 Experimental Procedure and Statistics

Experiments are carried out using a 5-fold cross validation approach. The data is split randomly into 5 folds with 4 folds being used as a training set for evolution and network training and the remaining fold being used to assess performance. Training and testing is carried out 5 times with each fold held out for testing on one occasion. The entire procedure is repeated 10 times with means and standard deviations taken for assessment of overall performance.

Five statistics will be used to assess the performance of the system. TNf and TPf are the fractions of true, correct classifications made on the positive and negative test cases respectively. Summary statistics used are the total accuracy across all classes (ACC), the Matthews Correlation Coefficient (MCC) and the area under the ROC curve for the model (AUR).

2.5 Hyperparameter Selection

Population parameters where $\lambda = \mu$, $\lambda = 5\mu$ and $\mu = 10\lambda$ were examined for varying values of μ such that total population sizes in the range $1 \leq \mu + \lambda \leq 300$ were covered.

It was found that the mean AUR value tends to increase as population size is increased. Figure 1 shows the graphs for the $\lambda = 5\mu$ case, displaying continual but diminishing increases of the AUR value as population size is increased. Note that the time taken to complete a cross validation run tends towards a linear relationship with population size, as might be expected since an increase of P population members requires P further fitness evaluations per generation. As population size is increased the number of generations remains roughly constant, the termination condition being met at a similar generation in each case. The mean fitness attained is however higher for larger populations. This is to be expected as the search space is of high dimensionality with the potential for many local optima. A larger population size likely enables the algorithm to avoid becoming trapped in these local optima.

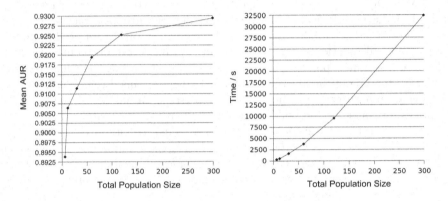

Fig. 1. Effects of Varying Population Size

$\mu = 10$ and $\lambda = 100$ have been chosen as the ES population parameters. This choice represents a trade off between performance and execution time.

The number of bio-basis neurons, N, is set according to prior knowledge and experiment. Although performance generally increases with increasing numbers of bio-basis neurons, the number is limited by the availability of positive examples in the training dataset since we only use positive training examples as bio-basis support peptides, a choice that has been shown to give good performance in previous work with BBFNNs. Peptides chosen for bio-basis neurons are not used in the weight training stage, therefore we must be sure that enough positive peptides remain. Based on these limits we use a rule of thumb that the value of N should be $30 - 50\%$ of the total number of positive peptides in the training set.

Figure 2 summarises an experiment carried out on the GAL dataset to verify this assumption. We clearly see the initial increase in accuracy as the number of basis neurons grows. From around 35 to 60 basis neurons the graph flattens, but with some fluctuations. As we increase the value of N further, performance steadily reduces. Computational cost appears to increase linearly as N is raised. Fluctuations in the execution time graph of Figure 2 are due to the stochastic nature of the evolution strategy and the fact that experiments were, unavoidably, run on machines open to some interactive use.

The number of positive training examples in the HIV and GAL datasets is similar and so 50 bio-basis neurons are used in each case. In the TCL case, there are fewer positive training examples so $N = 20$ is used.

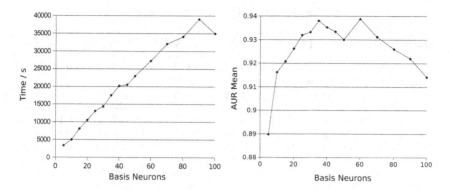

Fig. 2. Effects of Varying Number of Basis Neurons

2.6 BBFNN and MLP Comparisons

A standard BBFNN as first described in [1] is used for comparison purposes. The number of bio-basis neurons, N, is kept in line with that used by the EBBN as it is limited by the same factors.

A multi layer perceptron (MLP) is used for additional comparison. Datasets are pre-processed using a distributed encoding in which every possible amino acid is encoded by a 20 bit binary string, such that these strings are orthonormal [15]. An 8mer peptide is thus converted into a 160 bit binary vector by concatenating the appropriate bit-strings for each amino acid.

The MLP used consists of 3 fully interconnected layers. The number of input neurons is fixed according to the number of bits in the encoded input strings. The hidden layer contains 50 hidden neurons, chosen according to previous experiments with these datasets. There is a single output layer with an associated classification threshold. Weight training is accomplished using a standard feed-forward, back propagation technique.

3 Results and Discussion

Table 1 details the performance of the EBBN, BBFNN and MLP models on the five datasets. Unbracketed figures are means calculated from the cross validation runs. Figures in brackets are the corresponding standard deviations.

In the GAL case the EBBN is strongest in all statistics but TNf. In the HIV case it has the highest values for all statistics. This seems to indicate that for these two datasets the EBBN outperforms the other 2 models. Standard deviations are broadly in line with those of the BBFNN.

Table 2 summarises t-tests carried out on the AUR values to confirm this hypothesis. The AUR statistic is chosen as it is resistant to bias from relative class frequencies and serves as an indicator of the robustness of each classifier. At the 99% significance level the EBBN outperforms the BBFNN on all except the TCL set, and the MLP on all except the HIV set.

Table 1. Classification Performance

Dataset	Method	TNf	TPf	ACC	MCC	AUR
HIV	EBBN	93.17 *(6.01)*	85.89 *(9.62)*	90.72 *(6.49)*	0.79 *(0.14)*	0.95 *(0.05)*
	BBFNN	89.39 *(4.65)*	79.03 *(10.3)*	85.80 *(4.99)*	0.67 *(0.11)*	0.91 *(0.05)*
	MLP	91.16 *(5.38)*	82.57 *(7.86)*	88.27 *(4.86)*	0.72 *(0.10)*	0.94 *(0.04)*
GAL	EBBN	76.69 *(10.1)*	89.58 *(3.56)*	84.77 *(4.32)*	0.67 *(0.10)*	0.92 *(0.04)*
	BBFNN	78.55 *(10.2)*	74.67 *(6.79)*	76.11 *(4.75)*	0.52 *(0.10)*	0.84 *(0.04)*
	MLP	60.73 *(11.4)*	79.36 *(9.68)*	72.40 *(7.33)*	0.41 *(0.15)*	0.79 *(0.07)*
TCL	EBBN	95.11 *(3.91)*	62.62 *(27.4)*	88.40 *(8.48)*	0.60 *(0.27)*	0.91 *(0.10)*
	BBFNN	94.17 *(3.51)*	78.57 *(12.1)*	89.10 *(4.45)*	0.72 *(0.11)*	0.93 *(0.04)*
	MLP	94.42 *(4.68)*	32.50 *(23.8)*	82.90 *(5.98)*	0.31 *(0.24)*	0.86 *(0.07)*

The HIV and GAL datasets are broadly similar in size with 31% and 63% positive cases respectively. For the TCL set there are 18% positive peptides. The set is also much smaller, totalling only 202 peptides. The small size and unbalanced class frequencies represent a difficulty to algorithms that thrive on the availability of training examples. The MLP displays poor performance in the TCL case, likely as a result of the 100 dimensional input and subsequent large number of weights to optimise. The EBBN is also beaten by the BBFNN on the TCL data. There appears to be insufficient data for the evolution of a problem specific matrix. Choosing to use a standard substitution matrix would be a sensible strategy with small datasets.

Table 2. Significance Tests

Dataset	Method			Conclusion	P-value
HIV	EBBN	vs	BBFNN	EBBN outperforms BBFNN	< 0.0001
	EBBN	vs	MLP	Inconclusive at 99% Level	0.0491
	BBFNN	vs	MLP	MLP outperforms BBFNN	0.0018
GAL	EBBN	vs	BBFNN	EBBN outperforms BBFNN	< 0.0001
	EBBN	vs	MLP	EBBN outperforms MLP	< 0.0001
	BBFNN	vs	MLP	BBFNN outperforms MLP	< 0.0001
TCL	EBBN	vs	BBFNN	Inconclusive at 99% Level	0.1283
	EBBN	vs	MLP	EBBN outperforms MLP	0.0016
	BBFNN	vs	MLP	BBFNN outperforms MLP	< 0.0001

In the HIV case, known to be highly separable using linear systems, the EBBN is significantly higher performing than the BBFNN, but versus the MLP the difference is inconclusive. Given that the EBBN can alter the mutation matrix and therefore the nature of the mapping from feature to linear space it is possible that it performs more strongly than the BBFNN as it is able to apply a matrix that is much simpler than the standard matrices. Indeed, on examination of the evolved matrices, we notice that they appear to be simple in structure with few values of significant magnitude.

Of the datasets for which the EBBN is superior it is the GAL dataset which presents the most interesting results. The summary statistics are considerably higher than those for the BBFNN and MLP. The EBBN has an 8.66% higher accuracy compared to the BBFNN, which is in turn considerably stronger than the MLP. This dataset separates the models more than was expected. We hypothesise that the incorporation of mutation knowledge is important to predictors applied to this problem.

One of the main advantages of the BBFNN is its high speed when compared with a standard MLP due to the use of an algebraic method to optimise the weights. The linear classification stage requires a single optimisation step rather than the iterative feed-forward, back propagation technique in the MLP. The EBBN carries out $(\mu + \lambda)g$ of these optimisation steps per run, where g is the number of generations. g is not constant as the ES uses a convergence rule to control stopping.

With the hyperparameters used in this study, the EBBN was found to be in the region of 250 times slower to train than the BBFNN on the GAL dataset; the EBBN taking 12816s per cross validation run vs 48s for the BBFNN. Clearly where training speed is an issue the standard BBFNN is more appropriate. However, for static applications where re-training is carried out rarely, such as prediction servers, the additional computational cost is likely to be acceptable given improved prediction accuracy. Testing costs are identical for both algorithms; a trained EBBN operates identically to a BBFNN.

In section 2.5, we gave an overview of the effect of varying the population size and number of hidden neurons on execution times. The increase in computational cost was linear or sub-linear for increasing population sizes. By reducing the population size it was possible to obtain a mean AUR value of 0.8938 in 314 seconds. Whilst this statistic is below that obtained when using the tuned hyperparameters it is still higher than those of the BBFNN or MLP models (0.84 and 0.79 respectively). It is clear that the population size hyperparameters offer scope to reduce the execution time of the EBBN to an appropriate amount without losing all of the benefits in terms of classification performance.

4   Conclusions

This paper has presented a novel algorithm for peptide classification problems, the Evolutionary Bio-basis Network (EBBN). On two out of three datasets the EBBN shows a statistically significant improvement compared with the BBFNN on which it is based.

Dataset size appears to be the primary limitation to the technique, with improvements having been noted on datasets of > 200 peptides which have broadly balanced class frequencies. Accuracy on the smaller dataset of 200 peptides is not increased, however this dataset is difficult as it also has an imbalance in class frequencies. Execution time is highly dependent on the population hyperparameters for the Evolution Strategy which can be tuned to acheive an appropriate balance between model performance and training cost.

The evolution strategy is clearly a powerful tool that can adapt a large parameter vector. An optimised, problem specific matrix can support higher classification accuracies than standard general matrices when used with the bio-basis kernel.

References

1. R. Thomson, T. C. Hodgman, Z. R. Yang and A. K. Doyle, "Characterising proteolytic cleavage site activity using bio-basis function neural networks", *Bioinformatics*, vol. 19, pp. 1741-1747, 2003.
2. M. O. Dayhoff, R. M. Schwartz and B. C. Orcutt, "A model of evolutionary change in proteins. matrices for detecting distant relationships", in M.O. Dayhoff (ed.), *Atlas of Protein Sequence and Structure*, Nat. Biomed. Res. Found., Washington DC, 5, pp. 345-358, 1978.
3. S. Henikoff, and J. G. Henikoff, "Amino acid substitution matrices from protein blocks", *Proc. Natl. Acad. Sci.*, vol. 89, pp. 10915-10919, 1992.
4. G. H. Gonnet, M. A. Cohen, and S. A. Benner, "Exhaustive matching of the entire protein sequence database", *Science*, vol. 256, pp. 1433-1445, 1992.
5. Y. Yu and J. C. Worron and S. F. Altchul, "The compositional adjustment of amino acid substitution matrices", *PNAS*, vol. 100, pp. 15688-15693, 2003.
6. P. C. Ng and J. G. Henikoff and S. Henikoff, "PHAT: a transmembrane-specific substitution matrix", *Bioinformatics*, vol. 16, pp. 760-766, 2000.
7. R. O. Duda, P. E. Hart and D. G. Stork, "Pattern Classification and Scene Analysis (2nd edition)", Wiley-Interscience, New York, 2002.
8. X. Yao, "Evolving Artificial Neural Networks", *Proc. IEEE*, vol. 87, pp. 1423-1447, 1999.
9. H. Beyer, and H. Schwefel, "Evolution strategies - A comprehensive introduction", *Nat. Comput.*, vol. 1(1), pp. 3 - 52, 2002.
10. C. G. Moles and P. Mendes and J. R. Banga, "Parameter Estimation in Biochemical Pathways: A Comparison of Global Optimization Methods", *Genome Research*, vol. 13, pp. 2467-2474, 2003.
11. Y. Cai and K. Chou, "Artificial neural network model for predicting HIV protease cleavage sites in protein", *Adv. in Eng. Software*, vol. 29(2), pp. 119-128, 1998.
12. K. Chou and C. Zhang and F. J. Kezdy and R. A. Poorman, "A Vector Projection Method for Predicting the Specificity of GalNAc-Transferase", *Proteins*, vol. 21, pp. 118-126, 1995.
13. Y. Zhao, C. Pinilla, D. Valmori, R. Martin and R. Simon, "Application of support vector machines for T-cell epitopes prediction", *Bioinformatics*, vol. 19(15), pp 1978-1984, 2003.
14. J. A. Hanley B. J. McNeil, "The meaning and use of the area under a receiver operating characteristic (ROC) curve", *Radiology*, vol. 143(1), pp. 29-36, 1982.
15. N. Qian, and T. J. Sejnowski, "Predicting the secondary structure of globular proteins using neural network models", *J. Mol. Biol.*, vol. 202, pp. 865-884, 1988.

Author Index

Lecture Notes in Computer Science

For information about Vols. 1–4336

please contact your bookseller or Springer

Vol. 4389: D. Weyns, H.V.D. Parunak, F. Michel (Eds.), Environments for Multi-Agent Systems III. X, 273 pages. 2007. (Sublibrary LNAI).

Vol. 4385: K. Coninx, K. Luyten, K.A. Schneider (Eds.), Task Models and Diagrams for Users Interface Design. XI, 355 pages. 2007.

Vol. 4384: T. Washio, K. Satoh, H. Takeda, A. Inokuchi (Eds.), New Frontiers in Artificial Intelligence. IX, 401 pages. 2007. (Sublibrary LNAI).

Vol. 4383: E. Bin, A. Ziv, S. Ur (Eds.), Hardware and Software, Verification and Testing. XII, 235 pages. 2007.

Vol. 4381: J. Akiyama, W.Y.C. Chen, M. Kano, X. Li, Q. Yu (Eds.), Discrete Geometry, Combinatorics and Graph Theory. XI, 289 pages. 2007.

Vol. 4380: S. Spaccapietra, P. Atzeni, F. Fages, M.-S. Hacid, M. Kifer, J. Mylopoulos, B. Pernici, P. Shvaiko, J. Trujillo, I. Zaihrayeu (Eds.), Journal on Data Semantics VIII. XV, 219 pages. 2007.

Vol. 4378: I. Virbitskaite, A. Voronkov (Eds.), Perspectives of Systems Informatics. XIV, 496 pages. 2007.

Vol. 4377: M. Abe (Ed.), Topics in Cryptology – CT-RSA 2007. XI, 403 pages. 2006.

Vol. 4376: E. Frachtenberg, U. Schwiegelshohn (Eds.), Job Scheduling Strategies for Parallel Processing. VII, 257 pages. 2007.

Vol. 4374: J.F. Peters, A. Skowron, I. Düntsch, J. Grzymała-Busse, E. Orłowska, L. Polkowski (Eds.), Transactions on Rough Sets VI, Part I. XII, 499 pages. 2007.

Vol. 4373: K. Langendoen, T. Voigt (Eds.), Wireless Sensor Networks. XIII, 358 pages. 2007.

Vol. 4372: M. Kaufmann, D. Wagner (Eds.), Graph Drawing. XIV, 454 pages. 2007.

Vol. 4371: K. Inoue, K. Satoh, F. Toni (Eds.), Computational Logic in Multi-Agent Systems. X, 315 pages. 2007. (Sublibrary LNAI).

Vol. 4370: P.P Lévy, B. Le Grand, F. Poulet, M. Soto, L. Darago, L. Toubiana, J.-F. Vibert (Eds.), Pixelization Paradigm. XV, 279 pages. 2007.

Vol. 4369: M. Umeda, A. Wolf, O. Bartenstein, U. Geske, D. Seipel, O. Takata (Eds.), Declarative Programming for Knowledge Management. X, 229 pages. 2006. (Sublibrary LNAI).

Vol. 4368: T. Erlebach, C. Kaklamanis (Eds.), Approximation and Online Algorithms. X, 345 pages. 2007.

Vol. 4367: K. De Bosschere, D. Kaeli, P. Stenström, D. Whalley, T. Ungerer (Eds.), High Performance Embedded Architectures and Compilers. XI, 307 pages. 2007.

Vol. 4366: K. Tuyls, R. Westra, Y. Saeys, A. Nowé (Eds.), Knowledge Discovery and Emergent Complexity in Bioinformatics. IX, 183 pages. 2007. (Sublibrary LNBI).

Vol. 4364: T. Kühne (Ed.), Models in Software Engineering. XI, 332 pages. 2007.

Vol. 4362: J. van Leeuwen, G.F. Italiano, W. van der Hoek, C. Meinel, H. Sack, F. Plášil (Eds.), SOFSEM 2007: Theory and Practice of Computer Science. XXI, 937 pages. 2007.

Vol. 4361: H.J. Hoogeboom, G. Păun, G. Rozenberg, A. Salomaa (Eds.), Membrane Computing. IX, 555 pages. 2006.

Vol. 4360: W. Dubitzky, A. Schuster, P.M.A. Sloot, M. Schroeder, M. Romberg (Eds.), Distributed, High-Performance and Grid Computing in Computational Biology. X, 192 pages. 2007. (Sublibrary LNBI).

Vol. 4358: R. Vidal, A. Heyden, Y. Ma (Eds.), Dynamical Vision. IX, 329 pages. 2007.

Vol. 4357: L. Buttyán, V. Gligor, D. Westhoff (Eds.), Security and Privacy in Ad-Hoc and Sensor Networks. X, 193 pages. 2006.

Vol. 4355: J. Julliand, O. Kouchnarenko (Eds.), B 2007: Formal Specification and Development in B. XIII, 293 pages. 2006.

Vol. 4354: M. Hanus (Ed.), Practical Aspects of Declarative Languages. X, 335 pages. 2006.

Vol. 4353: T. Schwentick, D. Suciu (Eds.), Database Theory – ICDT 2007. XI, 419 pages. 2006.

Vol. 4352: T.-J. Cham, J. Cai, C. Dorai, D. Rajan, T.-S. Chua, L.-T. Chia (Eds.), Advances in Multimedia Modeling, Part II. XVIII, 743 pages. 2006.

Vol. 4351: T.-J. Cham, J. Cai, C. Dorai, D. Rajan, T.-S. Chua, L.-T. Chia (Eds.), Advances in Multimedia Modeling, Part I. XIX, 797 pages. 2006.

Vol. 4349: B. Cook, A. Podelski (Eds.), Verification, Model Checking, and Abstract Interpretation. XI, 395 pages. 2007.

Vol. 4348: S.T. Taft, R.A. Duff, R.L. Brukardt, E. Ploedereder, P. Leroy (Eds.), Ada 2005 Reference Manual. XXII, 765 pages. 2006.

Vol. 4347: J. Lopez (Ed.), Critical Information Infrastructures Security. X, 286 pages. 2006.

Vol. 4346: L. Brim, B. Haverkort, M. Leucker, J. van de Pol (Eds.), Formal Methods: Applications and Technology. X, 363 pages. 2007.

Vol. 4345: N. Maglaveras, I. Chouvarda, V. Koutkias, R. Brause (Eds.), Biological and Medical Data Analysis. XIII, 496 pages. 2006. (Sublibrary LNBI).

Vol. 4344: V. Gruhn, F. Oquendo (Eds.), Software Architecture. X, 245 pages. 2006.

Vol. 4342: H. de Swart, E. Orłowska, G. Schmidt, M. Roubens (Eds.), Theory and Applications of Relational Structures as Knowledge Instruments II. X, 373 pages. 2006. (Sublibrary LNAI).

Vol. 4341: P.Q. Nguyen (Ed.), Progress in Cryptology - VIETCRYPT 2006. XI, 385 pages. 2006.

Vol. 4340: R. Prodan, T. Fahringer, Grid Computing. XXIII, 317 pages. 2007.

Vol. 4339: E. Ayguadé, G. Baumgartner, J. Ramanujam, P. Sadayappan (Eds.), Languages and Compilers for Parallel Computing. XI, 476 pages. 2006.

Vol. 4338: P. Kalra, S. Peleg (Eds.), Computer Vision, Graphics and Image Processing. XV, 965 pages. 2006.

Vol. 4337: S. Arun-Kumar, N. Garg (Eds.), FSTTCS 2006: Foundations of Software Technology and Theoretical Computer Science. XIII, 430 pages. 2006.